# Nonlinear Mechanics for Composite Heterogeneous Structures

# Nonlinear Mechanics for Composite Heterogeneous Structures

Georgios A. Drosopoulos

Georgios E. Stavroulakis

CRC Press is an imprint of the
Taylor & Francis Group, an **informa** business

First published 2022
by Routledge
4 Park Square, Milton Park, Abingdon, Oxon OX14 4RN

and by Routledge
605 Third Avenue, New York, NY 10158

*Routledge is an imprint of the Taylor & Francis Group, an informa business*



*British Library Cataloguing-in-Publication Data*
A catalogue record for this book is available from the British Library

**Library of Congress Cataloging-in-Publication Data**

Names: Drosopoulos, Georgios A., author. | Stavroulakis, G. E. (Georgios E.), author.
Title: Nonlinear mechanics for composite heterogeneous structures / Georgios A. Drosopoulos, Georgios E. Stavroulakis.
Description: Abingdon, Oxon ; New York, NY : Routledge, 2022. | Includes bibliographical references and index.
Identifiers: LCCN 2021054373 (print) | LCCN 2021054374 (ebook) | ISBN 9780367861551 (hardback) | ISBN 9781032257358 (paperback) | ISBN 9781003017240 (ebook)
Subjects: LCSH: Inhomogeneous materials. | Composite materials. | Nonlinear mechanics.
Classification: LCC TA418.9.I53 D76 2022 (print) | LCC TA418.9.I53 (ebook) | DDC 620.1/1--dc23/eng/20211227
LC record available at https://lccn.loc.gov/2021054373
LC ebook record available at https://lccn.loc.gov/2021054374

ISBN: 978-0-367-86155-1 (hbk)
ISBN: 978-1-032-25735-8 (pbk)
ISBN: 978-1-003-01724-0 (ebk)

DOI: 10.1201/9781003017240

Typeset in CMR
by KnowledgeWorks Global Ltd.

Access the Support Material: www.routledge.com/9780367861551

*To my wife Elena*
*To my children Aristeidis and Christina*
*To my father Aristeidis*
*G. D.*

*To my wife Maria*
*To my children Eleftheria, Vassiliki, and Theodoros*
*To my mother Eleftheria*
*G. S.*

# Contents

# *Foreword*

The examination of materials and structures shows that homogeneity is often a simplification of reality. Most materials depict at micromechanical level a heterogenous structure which governs properties and structural responses. Thus, great attention has to be paid to the characterization of the mechanical behaviour of heterogeneous materials when real structures are designed. Generally, material and structural behaviour is non-linear including large deformations, inelastic features, like damage, and failure of parts and constructions.

In civil engineering, quasi-brittle materials like concrete or masonry have highly heterogeneous and disordered microstructures. Material interfaces between aggregates and cement paste are responsible for strength and toughness. Interfaces between bricks and mortar or steel reinforcement and concrete influence structural behaviour. In masonry structures, cracks often propagate along the interfaces between bricks and mortar and are relevant for the static and dynamic response.

These material and structural challenges require sophisticated models and due to the non-linearity also numerical simulation tools that are efficient, robust and reliable. The book at hand contains necessary theoretical background, mathematical tools and discretization technologies to tackle the aforementioned engineering problems. It is well written and covers the broad range of continuum and contact mechanics including failure analysis. Multi-scale approaches are discussed on the basis of specific homogenization strategies. The authors introduce as discretization technique the finite element method for linear and non-linear applications including special formulations like the extended finite element method for fracture analysis. Finally, the emerging field of data driven approaches is discussed and applied to problems in the area of structural mechanics.

In conclusion, the book contains a collection of state-of-the-art descriptions of modern engineering approaches in the field of numerical analysis of heterogeneous structures. Based on a solid theoretical background, the book provides a clear strategy for the derivation of numerical methods for students and researchers. Additionally, the authors designed computer codes for the study of heterogeneous structures which are freely available and thus especially of value for students.

Peter Wriggers
Hannover, Germany
January, 2022

# Preface

Composites are found in every modern structure. The advent of additive manufacturing or 3D printing with various materials and the possibility to create graded composites and microstructures have pushed the limit of applications and require new analysis and design tools. Prediction of the mechanical behaviour, especially beyond linearity, becomes a challenging task. Established theories, like large displacements, plasticity, damage and contact mechanics, can still be used.

In several cases, plenty of experimental measurements are available while a clear application of a classical theoretical framework is questionable. A kind of interpolation from the available experimental data seems to be in right place. In other cases, the need for quick development of new materials and applications with reduced cost does now allow the execution of many experiments.

Therefore, fusion of classical computational mechanics and machine learning tools, the artificial neural networks being the most known of these, constitutes a modern tool of choice. Alternatively, new approaches utilize directly the available experimental data, mainly for the missing material constitutive law, within the non-linear incremental-iterative solution scheme.

From another point of view, classical micromechanical models and the prediction of the overall behaviour have been exploited mainly within the framework of linear homogenization. For more complex, non-linear behaviour, and under the presence of non-linear interfaces, new approaches are required.

The multi-scale or concurrent, $FE^2$ computational homogenization tool has been proposed and proven to be an efficient technique for the study of composites. Unfortunately, it requires significant computational effort and is not directly available within commercial finite element codes.

Towards the exploitation of the non-linear response, localization of damage needs to be captured by computational mechanics descriptions. Advanced concepts are required in this case to properly address the ultimate behaviour of composite materials.

Intention of the authors is to provide holistically, the whole range of computational descriptions emphasizing in the non-linear response, using a simple but efficient writing style. This provides the opportunity even to the non-experienced reader, to realize key concepts of non-linear mechanics, involving treating both simple but also sophisticated problems. The steps of the finite element method are also presented, starting from linear analysis and proceeding towards non-linear descriptions. In this framework, the reader will be able to effectively apply the provided theory to commercial finite element software, or to develop structural analysis programming codes.

For these reasons, the book can be used as a textbook for a graduate course on non-linear computational methods for composite materials, modern multi-scale techniques and data-driven schemes. Final year undergraduate, MSc and PhD students as well as professional engineers and scientists in Civil, Mechanical and Aerospace Engineering, who work with advanced non-linear problems, will be benefited from the book.

The outlined thoughts describe the need and the contents of the present book. Extensions to multi-physics and time-dependent problems should be straightforward and are not considered, due to space limitations.

After a short introduction into continuum mechanics, in chapter 1, linear and non-linear finite element analysis are presented in chapter 2. Continuum failure laws, cohesive zone models and contact mechanics are outlined in chapters 3 and 4, respectively. The Extended Finite Element Method that allows for the consideration of discontinuous fields, is outlined in chapter 5 and completes the presentation of classical computational mechanics tools. Homogenization theory for composites is outlined in chapter 6, and multi-scale analysis in chapter 7. Data-driven tools are described in chapter 8.

A set of Matlab codes that demonstrate the usage of the presented techniques in two-dimensional problems are provided through the accompanying publisher's webpage of the book. The computer programs, including key Matlab functions and scripts, are described analytically in Appendix A. This offers the opportunity to the interested reader, to apply sophisticated concepts of non-linear mechanics, including contact, multi-scale ($FE^2$) analysis, cohesive laws, the XFEM method and machine learning, including artificial neural networks. Extension of the codes is also possible.

The book is based on the research and teaching activity of the authors at various places during the last years. The authors will appreciate any feedback from the readers of this book.

Georgios A. Drosopoulos
Georgios E. Stavroulakis

# Acknowledgements

The authors would like to thank stimulating environment and discussions with students and colleagues at the Technical University of Crete, Chania, Greece, the University of Central Lancashire, Preston, United Kingdom, the Technical University of Carolo Wilhelmina, Braunschweig, Germany, the Leibniz University of Hannover, Germany, the University of Ioannina, Greece and the University of KwaZulu-Natal (UKZN), in Durban, South Africa. Thanks, are especially addressed to our mentors, Prof. Dr.-Ing. Habil. Peter Wriggers, Leibniz University of Hannover, Professor Christos Massalas, University of Ioannina, Prof.Dr.h.c. Heinz Antes, TU Braunschweig, Professor Charalampos Baniotopoulos, University of Birmingham and Aristotle University and Professor Sarp Adali, UKZN. Special thanks to Prof. Dr. Stéphane Bordas, University of Luxembourg and Cardiff University for kindly providing an initial version of the XFEM code. Our research activity during the last years has been supported by the Greek Ministry of Research and Technology, the Alexander von Humboldt Foundation, the Stavros Niarchos Foundation through the Greek Diaspora Fulbright visiting scholarship, UKZN, and local research and development funds.

# *List of Figures*

# *List of Tables*

# Symbol Description

## Linear algebra

| | |
|---|---|
| $\|\|.\|\|$ | Euclidian norm of a vector. |
| $\mathbf{u} \cdot \mathbf{v}$ | Inner product between two vectors $\mathbf{u}$ and $\mathbf{v}$. |
| $\mathbf{u} \times \mathbf{v}$ | Cross product between two vectors $\mathbf{u}$ and $\mathbf{v}$. |
| $\mathbf{u} \otimes \mathbf{v}$ | Dyadic product between two vectors $\mathbf{u}$ and $\mathbf{v}$. |
| $\square^T$ | Transpose of a matrix. |
| $\mathbf{A} : \mathbf{B}$ | Double dot product between two tensors $\mathbf{A}$ and $\mathbf{B}$. |
| $\nabla \cdot \boldsymbol{\sigma}$ | Divergence operator for the second-order tensor $\boldsymbol{\sigma}$. |
| $\mathbf{e}$ | Unit vector along cartesian axis. |
| $\mathbf{I}$ | Identity matrix. |
| $tr(\square)$ | Trace of matrix. |
| $\delta_{ij}$ | Kronecker-delta. |

## Continuum mechanics

| | |
|---|---|
| $\mathbf{T}$ | Transformation matrix. |
| $\mathbf{F}$ | Deformation gradient tensor. |
| $J$ | Jacobian defined as the determinant of the deformation gradient tensor. |
| $\mathbf{E}$ | Green strain tensor. |
| $\epsilon^*$ | Euler strain tensor. |
| $\boldsymbol{\varepsilon}$ | Strain tensor within small displacement analysis. |
| $\mathbf{n}$ | Unit normal vector. |
| $\mathbf{t}$ | Traction or stress vector. |
| $\boldsymbol{\sigma}$ | Cauchy stress tensor. |
| $\mathbf{P}$ | First Piola-Kirchhoff stress tensor. |
| $\mathbf{S}$ | Second Piola-Kirchhoff stress tensor. |
| $I_1, I_2, I_3$ | Invariants of a tensor (stress or strain). |
| $J_1, J_2, J_3$ | Invariants of a deviatoric tensor (stress or strain). |
| $p$ | Hydrostatic or volumetric pressure. |
| $\mathbf{s}$ | Deviatoric stress tensor. |
| $\mathbf{e}$ | Deviatoric strain tensor. |
| $\mathbf{C}$ | Elasticity tensor. |
| $\mathbf{S}^*$ | Compliance tensor. |
| $E$ | Young's modulus. |
| $G$ | Shear modulus. |
| $\nu$ | Poisson's ratio. |
| $C^\square$ | Degree of continuity. |

## Finite element analysis

| | |
|---|---|
| $\dot{\square}$ | First derivative of a variable with respect to time. |
| $\ddot{\square}$ | Second derivative of a variable with respect to time. |
| $\mathbf{u}$ | Displacement field. |
| $\mathbf{u}_e$ | Element's nodal displacement vector. |
| $\mathbf{N}$ | Matrix of shape functions. |
| $\mathbf{B}$ | Strain-displacement matrix. |
| $\mathbf{F}_{ext}$ | External force vector of a structure. |
| $\mathbf{F}_{int}$ | Internal force vector of a structure. |
| $\mathbf{U}$ | Nodal displacement vector of a structure. |
| $\mathbf{M}$ | Mass matrix of a structure. |
| $\mathbf{K}$ | Global stiffness matrix of a structure. |
| $\mathbf{k}$ | Element stiffness matrix. |
| $\xi, \eta, \zeta$ | Natural coordinates. |
| $\mathbf{J}$ | Jacobian matrix. |
| $\Delta\square$ | Incremental value of quantity. |
| $\delta\square$ | Variation of quantity. |

| | |
|---|---|
| $\mathbf{K}_T$ | Tangent stiffness matrix of a structure. |
| $\mathbf{G}$ | Residual vector of a structure. |
| $\mathbf{D}$ | Instantaneous or consistent (tangent) stiffness tensor. |

**Non-linear continuum laws**

| | |
|---|---|
| $\boldsymbol{\xi}$ | Strain-like internal variables for plasticity. |
| $\boldsymbol{\chi}$ | Stress-like internal variables for plasticity. |
| $f$ | Yield criterion for plasticity. |
| $\square^e$ | Elastic part of quantity. |
| $\square^p$ | Plastic part of quantity. |
| $c$ | Cohesion. |
| $\phi$ | Angle of internal friction. |
| $\tau$ | Shear stress. |
| $\mathbf{n}$ | Flow vector. |
| $\mathbf{h}$ | Hardening modulus. |
| $\omega, \boldsymbol{\omega}, \boldsymbol{\Omega}$ | Internal damage variables as a scalar, second-order and fourth-order tensor. |
| $\mathbf{D}^s$ | Secant elasticity tensor. |
| $\hat{\boldsymbol{\sigma}}$ | Effective stress tensor. |
| $f$ | Damage loading function for damage mechanics. |
| $\tilde{\varepsilon}, \tilde{\sigma}$ | Equivalent strain and stress. |
| $\kappa$ | History dependent variable. |
| $\dot{\boldsymbol{\sigma}}$ | Stress rate. |
| $\dot{\boldsymbol{\varepsilon}}$ | Strain rate. |
| $\|\|\mathbf{u}\|\|$ | Separation or displacement jump vector. |
| $\mathbf{k}_{coh_e}$ | Tangent stiffness matrix for cohesive element. |
| $\mathbf{f}_{coh_e}$ | Internal force vector for cohesive element. |
| $\bar{\varepsilon}(\mathbf{x})$ | Non-local equivalent strain. |

**Contact mechanics**

| | |
|---|---|
| $u_N$ | Normal displacement to the contact interface. |
| $S_N$ | Normal traction to the contact interface. |
| $u_T$ | Tangential displacement to the contact interface. |
| $S_T$ | Tangential traction to the contact interface. |
| $\tau_T$ | Critical shear traction required to initiate sliding in the contact interface. |
| $\mu$ | Friction coefficient. |
| $g$ | Initial distance (gap) in the contact interface. |
| $\boldsymbol{\lambda}$ | Vector of Lagrange multipliers. |

**Fracture mechanics**

| | |
|---|---|
| $K_I$ | Stress intensity factor for mode I failure type. |
| $K_{II}$ | Stress intensity factor for mode II failure type. |
| $\mathcal{G}$ | Energy release rate. |
| $J$ | J-integral. |
| $I^{(1,2)}$ | Interaction integral for the present state (1) and the auxiliary state (2). |
| $\phi(\mathbf{x})$ | Signed distance function. |
| $\|\phi(\mathbf{x})\|$ | Ramp function. |
| $H(\mathbf{x})$ | Heaviside function. |
| $\psi(r, \theta)$ | Crack tip enrichment function for strong discontinuities. |

**Homogenization**

| | |
|---|---|
| $\epsilon$ | Ratio between the microscopic and macroscopic characteristic dimensions. |
| $\mathbf{x}$ | Coordinates of the macro domain. |

| | |
|---|---|
| $\mathbf{y}$ | Coordinates of the micro domain. |
| $< \boldsymbol{\varepsilon} >_{V_m}$ | Average or effective strain tensor, over the microstructural volume $V_m$. |
| $< \boldsymbol{\sigma} >_{V_m}$ | Average or effective stress tensor, over the microstructural volume $V_m$. |

**Multi-scale analysis**

| | |
|---|---|
| $\boldsymbol{\sigma}^M$ | Effective macroscopic stress tensor. |
| $\boldsymbol{\varepsilon}^M$ | Effective macroscopic strain tensor. |
| $\boldsymbol{\sigma}^m$ | Microscopic stress tensor. |
| $\boldsymbol{\varepsilon}^m$ | Microscopic strain tensor. |
| $\mathbf{D}^M$ | Consistent tangent stiffness tensor at the macroscopic level. |
| $\|\|\mathbf{u}^M\|\|$ | Displacement jump at the macroscopic level. |
| $\boldsymbol{t}^M$ | Average traction at the macroscopic level. |

**Data-driven analysis**

| | |
|---|---|
| $w_{ij}$ | Weights used in neural networks. |
| $b$ | Bias used in neural networks. |
| $\sigma$ | Activation function used in neural networks. |
| $E$ | Least squares error function. |

# Chapter 1

# Introduction

## 1.1 Introduction to composite materials

Composite materials are widely used in various applications, including among others the civil, mechanical and aerospace engineering field. The increasing demand for modern as well as traditional composite materials poses the need for the detailed and accurate representation of their mechanical response. Several methods, both numerical and experimental, have been developed during the last years, to succeed in this goal. Emphasis is given among others, in the ultimate failure behaviour of the materials under different loading conditions. Goal of this book is to offer an insight on different computational mechanics approaches which are widely adopted to represent the non-linear response of composite materials and structures. Both the state-of-the-art and cutting-edge research concepts will be highlighted.

The term *composite* indicates that different constituent materials are used to integrate the overall structural system. There are various types of composite materials, applied to different structural systems. Masonry and concrete are among the most known, with several applications in civil engineering. Masonry structures consist of bricks and mortar joints. Concrete is a mixture of cement, water, aggregates and additives. A concrete specimen and an image of its microstructure are depicted in figure 1.1.

Another category of heterogeneous materials is defined by the *composite laminates* and *nanocomposites*, which are widely known as *fibre reinforced polymers* (FRP). The main concept which summarizes the structural behaviour of these materials, is the usage of reinforcing elements (fibres) with excellent mechanical properties, embedded in a polymer matrix, as shown in figure 1.2. FRP composites are used in several applications, among others within the field of mechanical, aerospace, automotive and biomechanical engineering.

Several computational mechanics methodologies have been proposed for the evaluation of the structural behaviour of heterogeneous composite materials. They target in the accurate representation of the interaction between the constituent materials as well as in the determination of the ultimate strength. Goal of this book is to provide a holistic view of those methodologies, enabling post-graduate students and researchers to apply them. Emphasis is given in their numerical implementation, in the context of static, linear and

DOI: 10.1201/9781003017240-1

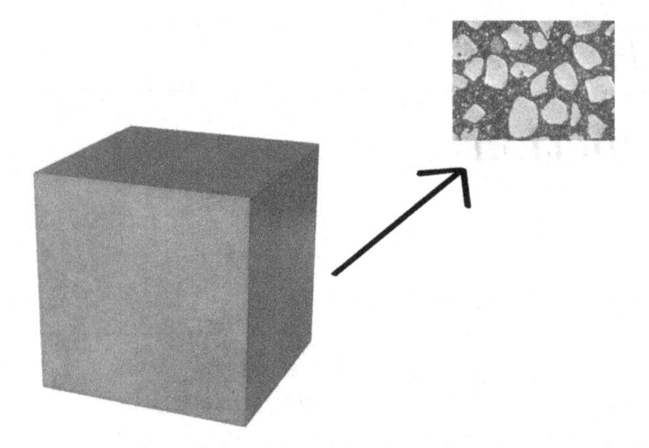

**FIGURE 1.1**: A concrete specimen and its microstructure.

**FIGURE 1.2**: Carbon fibre reinforced composite.

non-linear, mechanical behaviour. Extensions of those methodologies to multi-physics techniques can also be found in literature. Below is presented a brief description of the methods which are discussed analytically in the chapters of the book.

Among the different approaches which have been developed in the last years, someone may recognize two major categories. The first, relies on the phenomenological representation of the structural response, by adopting proper constitutive descriptions taken from continuum mechanics and used to determine the failure of the system. Within this approach, the response in the macroscopic (or coarse) scale of the structure is depicted. For the calibration of the parameters of the adopted constitutive laws, experimental data may be used.

The second category, which is also used to evaluate the ultimate response of composite structures, considers the interaction between the constituent materials, taking place in a microscopic (or fine) scale, in order to derive the constitutive law of the system. For the representation of the behaviour in the microscopic scale, the internal boundaries of the constituent materials can be taken into account and principles of continuum mechanics may also be adopted. This mechanical characterization of the composite material in the fine scale, will allow for the detailed derivation of the constitutive response which is then adopted in the macroscopic, structural scale. Since an interaction between the length scales takes place, the term *multi-scale* is widely used for this class of methods. Both approaches present advantages and disadvantages which are discussed in the book and the proper selection depends on the problem under investigation.

From another point of view, someone may recognize different computational mechanics techniques, depending on the mechanical response of the composite structure which is studied. Some of these techniques have been included in this book, and for this reason are briefly mentioned here.

A core concept of representing damage in composite materials is the usage of non-linear continuum laws. In particular, when the plastic strains which appear during unloading of a structure are negligible, then laws of *continuum damage mechanics* can be adopted to provide a description of the failure response. On the contrary, when non-negligible plastic strains arise, then *plasticity* models should be considered.

In addition, the need for simulating discontinuities such as cracks which are developed and propagate within heterogeneous structures may arise. To cover this class of problems, methodologies adopting some enhancement of the traditional continuum mechanics solutions to provide the response of the discontinuities, are discussed in the book. Thus, the method of *cohesive zone models*, as well as the *Extended Finite Element Method* (XFEM) are analytically presented.

Another computational approach which is discussed in the book, relies on the exploitation of principles of non-smooth mechanics, for the simulation of the interaction between different bodies. Unilateral contact and friction laws can be used in this case to provide the non-linear response of the interfaces, identified between the constituents of composite materials.

Methods which consider the different length scales for the evaluation of the mechanical response, have also been included in the book. Emphasis is given to *homogenization* and *multi-scale* schemes, which are developed to provide the behaviour of composite materials. The core concept of these approaches, is that the average (effective) material properties are derived, by conducting an investigation of a representative microstructure, on the microscopic, fine length scale. All the constitutive laws mentioned above, like non-smooth mechanics, plasticity and damage mechanics, may be considered in this scale. Then, the effective response is introduced in the macroscopic, coarse scale and used to derive the mechanical behaviour of the equivalent homogeneous structure, which is now less complex.

It should be noted that the classical structural theories applied to composite laminates, which consist of stacking layers with different material or fibre orientation, have not been included in the book. These theories constitute a big, but independent class of problems and include among others, the *equivalent single layer theories*, like the classical laminated plate theory and the shear deformation laminated plate theory, as well as the *three-dimensional elasticity theories*, such as the layerwise theory [180]. These are initially formulated to solve linear problems but they can be extended to solve even non-linear ones [160], [197]. Within these categories, the study of nanomaterials, including *carbon nano-tubes* and *graphene* reinforced composites, can also be depicted [98], [182], [208], [234].

The book also aims in highlighting cutting-edge research concepts. The main steps of *multi-scale computational homogenization*, also known as $FE^2$, are presented. The concept of *localization of damage* is discussed, in the context of continuum damage laws, as well as within the described multi-scale schemes. Descriptions of advanced numerical methods developed in a *data-driven* framework are also included in the book. The usage of modern numerical tools taken from *machine learning*, such as *artificial neural networks*, is highlighted. These state-of-the-art techniques support and extend the traditional multi-scale formulations, by introducing large databases directly in the numerical simulation.

In the appendix of the book, analytical descriptions of *Matlab* codes are provided, for applications on composite structures. These codes can be considered as cases studies, since they present specific examples of the theory which is found in the main body of the book. The reader will gain hands-on experience by solving two-dimensional problems with the provided complete Matlab codes, which can be adapted and extended in various ways.

Finally, the content of each chapter of the book is briefly presented below. In chapter 1, an introduction to continuum mechanics principles is presented. Elements of linear algebra are initially discussed, emphasizing in vectors and matrices. It follows an introduction to tensors and related operations. The deformation is then defined, including descriptions for the deformation gradient tensor, as well as for the Green and Euler strain tensors. The stress is also defined and the corresponding, Cauchy, first and second Piola-Kirchhoff stress tensors are presented. The elasticity tensor and generalized Hooke's law are eventually discussed in this chapter.

In chapter 2, the finite element method, which is the main numerical tool adopted in the book for the implementation of non-linear mechanics, is presented. The concept of equilibrium using the strong and the weak form, including the principle of virtual work, is initially discussed. Then, the main steps of the finite element method are provided. The chapter concludes with core concepts of non-linear finite element analysis, including the Newton-Raphson incremental-iterative process, as well as the formulation of the finite element method for geometrical and material non-linearities.

In chapter 3, continuum laws which can be used to predict damage in composite structures are presented. The core concept of plasticity is initially provided in the chapter. It follows a detailed representation of continuum damage mechanics within numerical analysis. Localization of damage is discussed including how computational mechanics approaches can be enhanced to depict localization phenomena. The principles and main steps of the cohesive zone model, its implementation within non-linear finite element analysis, as well as the concept of the non-local damage model, are also provided in the chapter.

Chapter 4 presents the principles of contact mechanics. The definition of non-smooth functions are discussed and linear, as well as non-linear complementarity problems are presented. The concepts of unilateral contact and friction are also included in the chapter. The theoretical framework of non-smooth mechanics, including descriptions for variational inequalities are eventually added.

Chapter 5 presents the formulation of the Extended Finite Element Method (XFEM). First, an introduction to linear fracture mechanics is provided and concepts such as the stress intensity factor and the interaction integral are discussed. Then, the formulation of the XFEM method within linear fracture mechanics is added, including descriptions for weak versus strong discontinuities, discretization of the governing equations, crack propagation and crack tip enrichment. The chapter concludes with the presentation of the cohesive XFEM approach.

In chapter 6, the concept of homogenization, as an alternative approach to simulate composite materials, is provided. After an introduction to homogenization, the asymptotic expansion homogenization scheme is developed, including the formulation within finite element analysis. A brief introduction to numerical homogenization is then presented.

In chapter 7, the formulation of multi-scale computational homogenization is provided. First, the steps of first-order homogenization are presented, including details of using averaging relations and implementing periodic boundary conditions. It follows an introduction to the concept of localization of damage, as it appears within multi-scale analysis. An approach for multi-scale analysis considering localization phenomena is eventually presented in the chapter.

In chapter 8, detailed descriptions of data-driven schemes, which can be used to provide the response of composite materials, are added. The chapter focuses on three approaches of applying data-driven concepts. The first, relies on the usage of interpolation schemes, used to introduce databases providing the constitutive non-linear response in the structural analysis code. The second, presents core concepts of machine learning elements, emphasizing in artificial neural networks, as tools which are used to provide the data-driven aspect to the simulation. The third is a recently developed data-driven approach, which relies on using a distance minimization scheme, between every value of a database, representing the constitutive description, and a corresponding value which satisfies equilibrium and compatibility.

The book is completed with an appendix providing applications, based on the presented theory, as well as descriptions for relevant Matlab codes which have been developed by the authors. In particular, four examples are analytically discussed in the appendix. The first example refers to a multi-scale computational homogenization scheme, developed to capture the response of fibre reinforced composites. The second example provides the framework of a data-driven scheme which adopts the distance minimization concept. The third example highlights another data-driven approach which introduces machine learning principles, in the form of neural networks. The fourth example refers to a multi-scale scheme presented in the main body of the book, providing the response of composite structures when localization phenomena arise.

---

## 1.2 Principles of continuum mechanics

In this section, the necessary mathematical framework is provided, with elements of linear algebra. Then, principles of continuum mechanics are presented.

### 1.2.1 Elements from linear algebra: Vectors and matrices

#### 1.2.1.1 Vectors

In the framework of numerical methods which are adopted to represent the mechanical behaviour of composite materials, elements taken from linear algebra, such as vectors and matrices are widely used. Therefore, a short introduction to these elements is presented in the following lines. A vector is a one-dimensional array of scalar elements, given here by the bold face letter $\mathbf{u}$. A vector is represented by an arrow, the length of which indicates the magnitude of the vector, usually given by $||\mathbf{u}||$ (or $|\mathbf{u}|$, $u$). Quite often, a vector is expressed in terms of corresponding unit basis vectors, usually given by the letter $\mathbf{e}$. Unit vectors have a length equal to 1. According to this description, the vector $\mathbf{u}$ is defined in the three-dimensional space by equations (1.1) and (1.2). The length of vector $\mathbf{u}$, which is provided in equation (1.3) is also the Euclidian norm of the vector. The unit vector along $\mathbf{e}$ is given by equation (1.4).

$$\mathbf{u} = u_1\mathbf{e_1} + u_2\mathbf{e_2} + u_3\mathbf{e_3} \tag{1.1}$$

$$\mathbf{u} = \begin{bmatrix} u_1 \\ u_2 \\ u_3 \end{bmatrix} \tag{1.2}$$

$$u = ||\mathbf{u}|| = \sqrt{u_1^2 + u_2^2 + u_3^2} = \sqrt{\mathbf{u^T u}} \tag{1.3}$$

$$\mathbf{e} = \frac{\mathbf{u}}{u} \tag{1.4}$$

Next, some basic calculations with vectors are presented. If $\mathbf{u}$, $\mathbf{v}$ and $\mathbf{w}$ are any vectors, then addition of vectors is defined in equation (1.5).

$$\mathbf{w} = \mathbf{u} + \mathbf{v} \tag{1.5}$$

The multiplication of vector $\mathbf{v}$ by the scalar $\lambda$ is defined by equation (1.6).

$$\mathbf{u} = \lambda\mathbf{v} \tag{1.6}$$

The scalar or inner product between two vectors is a scalar quantity, which is given by equation (1.7). According to this definition, the scalar product of two vectors is given by the multiplication between the magnitude of the first vector and the magnitude of the projection of the second vector into the direction of the first. Another convenient way for reproducing the scalar product between two vectors is given in equation (1.8).

$$\mathbf{u} \cdot \mathbf{v} = uvcos(\mathbf{u}, \mathbf{v}) = uvcos(\theta),\ 0 \leqslant \theta \leqslant \pi \tag{1.7}$$

$$\mathbf{u} \cdot \mathbf{v} = \mathbf{u}^T \mathbf{v} \ = \sum_{i=1}^{n} u_i v_i \tag{1.8}$$

Both the addition and the scalar product between two vectors possess the commutative property, as shown in equations (1.9) and (1.10).

$$\mathbf{u} + \mathbf{v} = \mathbf{v} + \mathbf{u} \tag{1.9}$$

$$\mathbf{u} \cdot \mathbf{v} = \mathbf{v} \cdot \mathbf{u} \tag{1.10}$$

Next, the cross product (or outer or vector product) between two vectors, **u** and **v**, is defined as a third vector **w**, according to equation (1.11). The cross product **w** is a vector orthogonal to **u** and **v** and its direction is found by the right-hand rule.

$$\mathbf{w} = \mathbf{u} \times \mathbf{v} = uvsin(\mathbf{u}, \mathbf{v})\mathbf{e} = uvsin(\theta)\mathbf{e} \tag{1.11}$$

The components of the cross product **w** are provided in equality (1.12).

$$\mathbf{w} = \begin{bmatrix} u_2 v_3 - u_3 v_2 \\ u_3 v_1 - u_1 v_3 \\ u_1 v_2 - u_2 v_1 \end{bmatrix} \tag{1.12}$$

It is noted that as shown in equation (1.13), the cross product does not possess the commutative property.

$$\mathbf{u} \times \mathbf{v} = -\mathbf{v} \times \mathbf{u} \tag{1.13}$$

Often in mechanics, it is necessary to assign a direction to a plane. Since the cross product **w** of two vectors **u** and **v** is a vector perpendicular to the plane, which is defined by the two vectors, it can be used to represent the direction of this plane and thus, the orientation of the plane in space. For this case, the unit normal vector **n** can be used to represent the orientation of the plane, as given by equation (1.14).

$$\mathbf{n} = \frac{\mathbf{u} \times \mathbf{v}}{\|\mathbf{u} \times \mathbf{v}\|} \tag{1.14}$$

Another useful operation in mechanics is given by the gradient of a scalar-valued function $\phi(x, y, z)$, which is presented in equation (1.15). When

$\phi(x,y,z)$ is a function of the components of the vector **u**, then, the gradient **v** is obtained by differentiation of $\phi(x,y,z)$ with respect to **u**, as given in equation (1.16). This gradient **v** is an orthogonal vector to the surface, which is defined by $\phi(x,y,z) = constant$.

$$\mathbf{grad}\phi(x,y,z) = \nabla\phi(x,y,z) = \begin{bmatrix} \frac{\partial\phi}{\partial x} \\ \frac{\partial\phi}{\partial y} \\ \frac{\partial\phi}{\partial z} \end{bmatrix} \tag{1.15}$$

$$\mathbf{v} = \frac{\partial\phi}{\partial\mathbf{u}} = \begin{bmatrix} \frac{\partial\phi}{\partial u_1} \\ \frac{\partial\phi}{\partial u_2} \\ \frac{\partial\phi}{\partial u_3} \end{bmatrix} \tag{1.16}$$

#### 1.2.1.2 Matrices

Matrices are defined as two-dimensional arrays of scalars. A matrix of $m$ rows and $n$ columns has dimensions $m \times n$, as given in (1.17).

$$\mathbf{A} = \begin{bmatrix} A_{11} & A_{12} & \dots & A_{1n} \\ A_{21} & A_{22} & \dots & A_{2n} \\ \vdots & & \ddots & \\ A_{m1} & A_{m2} & \dots & A_{mn} \end{bmatrix} \tag{1.17}$$

An element of matrix $A$ in the $i_{th}$ row and $j_{th}$ column is denoted by $A_{ij}$. A single column matrix with dimensions $m \times 1$ is equivalent to a vector. A single row $1 \times n$ matrix is equivalent to a row vector.

Next, some basic types of matrices are identified. A square matrix has the same number of rows and columns. The elements of a square matrix with the same index of row and column, thus, elements $A_{ij}$ with $i = j$ are called diagonal elements. A matrix which has zeros for all the off-diagonal elements is called diagonal matrix. A diagonal matrix which has ones for all the diagonal elements, is called identity matrix, denoted by **I**.

Some basic calculations with matrices are now presented. Addition takes place for matrices with the same dimensions and is provided by the sum of the corresponding elements of each matrix, as denoted by equation (1.18). It is noted that addition of matrices possesses the commutative property, thus $\mathbf{A} + \mathbf{B} = \mathbf{B} + \mathbf{A}$.

$$\mathbf{C}_{ij} = \mathbf{A}_{ij} + \mathbf{B}_{ij}, \ for\ all\ i,\ j \tag{1.18}$$

Multiplication between a scalar $\lambda$ (constant number) and a matrix **A** is equal to a new matrix **B** of dimensions equal to **A**, where its elements are obtained by multiplying the scalar $\lambda$ by all elements of **A**.

Multiplication between a matrix $\mathbf{A}$ of dimensions $m \times n$ and a vector $\mathbf{b}$ of dimension $n$ is equal to a vector $\mathbf{c}$ of dimension $m$, as given in equation (1.19). Each of the $m$ elements of vector $\mathbf{c}$ is calculated as provided in equation (1.20).

$$\mathbf{c} = \mathbf{A}\mathbf{b} \tag{1.19}$$

$$\mathbf{c}_i = \sum_{j=1}^{n} A_{ij} b_j \tag{1.20}$$

Similar to the product between a matrix and a vector is defined the product between two matrices. A matrix $\mathbf{A}$ of dimensions $m \times n$ can be multiplied by a matrix $\mathbf{B}$ of dimensions $n \times o$ and the result is a matrix $\mathbf{C}$ of dimensions $m \times o$ as shown in equations (1.21) and (1.22).

$$\mathbf{C} = \mathbf{A}\mathbf{B} \tag{1.21}$$

$$\mathbf{C}_{ij} = \sum_{e=1}^{n} A_{ie} B_{ej} \tag{1.22}$$

Next, the transpose operation for matrices is defined as $\mathbf{B} = \mathbf{A}^T$ with the components of matrices $\mathbf{B}$ and $\mathbf{A}$ provided by $B_{ij} = A_{ji}$. In addition, the transpose of the product of two matrices is given in equation (1.23).

$$(\mathbf{A}\mathbf{B})^T = \mathbf{B}^T \mathbf{A}^T \tag{1.23}$$

A square matrix has the same number of rows and columns and is often used in numerical analysis. For a square matrix $\mathbf{A}$, an inverse matrix $\mathbf{B} = \mathbf{A}^{-1}$ is defined, according to (1.24).

$$\mathbf{A}\mathbf{B} = \mathbf{A}\mathbf{A}^{-1} = \mathbf{I} \tag{1.24}$$

The inverse operation is needed when a system of equations expressing for instance equilibrium of a structural system is solved, within finite element analysis. For this case, the equation (1.25) is solved in respect to the unknown vector $\mathbf{x}$.

$$\mathbf{A}\mathbf{x} = \mathbf{b} \Rightarrow \mathbf{x} = \mathbf{A}^{-1}\mathbf{b} \tag{1.25}$$

Another scalar quantity which is related to square matrices is the determinant. For a matrix matrix $\mathbf{A} = [A_{ij}]$ of dimensions $n \times n$, the determinant $det(\mathbf{A})$ or $|\mathbf{A}|$ is defined in equation (1.26).

$$det(\mathbf{A}) = |\mathbf{A}| = \sum_{j=1}^{n} (-1)^{i+j} A_{ij} |\mathbf{A}_{ij}| \tag{1.26}$$

In the above equation, $|\mathbf{A}_{ij}|$ is the determinant of the $(n-1) \times (n-1)$ matrix that remains after deleting the $i_{th}$ row and the $j_{th}$ column of matrix $\mathbf{A}$.

The inverse of a matrix does not always exist in classical sense and a matrix is called singular if it has no inverse. In addition to this definition, a matrix is singular if and only if its determinant is zero. A real-symmetric $n \times n$ matrix $\mathbf{A}$ is positive definite when inequality (1.27) applies for every non-zero vector $\mathbf{x}$ of dimensions $n \times 1$ [142].

$$\mathbf{x}^T \mathbf{A} \mathbf{x} > 0 \tag{1.27}$$

When for a real-symmetric $n \times n$ matrix $\mathbf{A}$ inequality (1.28) applies, the matrix is called positive semidefinite. If for matrix $\mathbf{A}$ relation (1.28) is an equality, then matrix $\mathbf{A}$ is singular. When inequality (1.29) applies, matrix $\mathbf{A}$ is negative definite. A matrix which is not positive definite and not negative definite is called indefinite. This means that for matrix $\mathbf{A}$, there may be found two vectors $\mathbf{x}_1$ and $\mathbf{x}_2$ for which both inequalities (1.27) and (1.29) are satisfied.

$$\mathbf{x}^T \mathbf{A} \mathbf{x} \geq 0 \tag{1.28}$$

$$\mathbf{x}^T \mathbf{A} \mathbf{x} < 0 \tag{1.29}$$

The property of matrices being positive definite, singular, negative definite and indefinite, significantly influences the numerical response of composite heterogeneous structures within finite element analysis. When the stiffness matrix of a structural model is positive definite, a unique and accurate solution is obtained from the finite element solution. This is the case for instance, when a linear elasticity problem is solved. However, for advanced non-linear problems, the tangent stiffness matrix may become singular or indefinite. For this case, no unique solution may be obtained or the convergence to a solution may not be achieved. Material instability and localization phenomena may result in these conditions as it will be discussed later.

### 1.2.2 Tensors

A tensor is a matrix with physical meaning, with its components being dependent on a given coordinate system. A tensor $\mathbf{A}$ is of order $n$ when each of its components $A_{ijk...m}$ has $n$ indices and obeys certain transformation rules from an initial to a new coordinate system.

A tensor of order $n$ which is defined in a two-dimensional space has $2^n$ components while a tensor in a three-dimensional space has $3^n$ components. Thus, a scalar can be considered as a tensor of zero order with one component (e.g. temperature). Provided that it obeys transformation rules, a vector in two-dimensional space is a first-order tensor with $2^1 = 2$ components, while a vector in three-dimensional space is a first-order tensor with $3^1 = 3$ components. Similar to this definition, a second-order tensor in the two-dimensional space has $2^2 = 4$ components while a second-order tensor in the three-dimensional space has $3^2 = 9$ components.

An example which is often presented in literature to introduce first-order tensors is that of a force applied to a surface. To define the force someone needs to describe not only its magnitude but also its direction. In a given cartesian coordinate system defined in a two-dimensional space, a force vector **f** will have an $x$ and a $y$ component $f_x$ and $f_y$ as given in equation (1.30). In a rotated coordinate system shown in figure 1.3, the components of the force vector **f** along the rotated axes $\hat{x}$ and $\hat{y}$ are $\hat{f}_x$ and $\hat{f}_y$, equation (1.30). The relation between the components of the vector **f** in the rotated and in the initial coordinate system is obtained by basic trigonometric calculations and is given in equation (1.31). This relation defines the transformation rule for the first-order tensor **f**. The matrix **T** is called transformation matrix and it satisfies the condition presented in equation (1.32). Matrices obeying (1.32) are called orthogonal matrices.

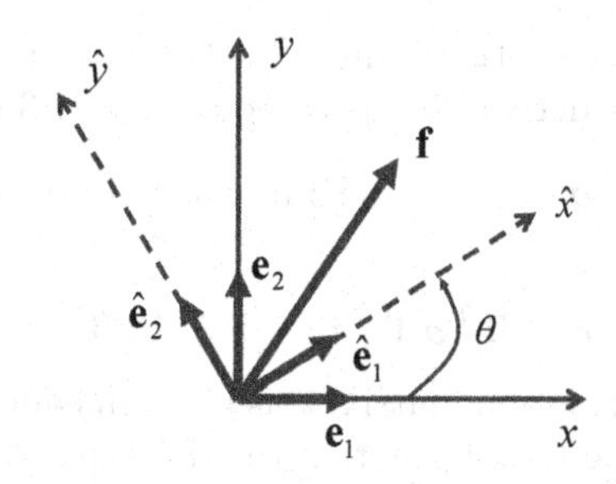


**FIGURE 1.3**: Initial and rotated coordinate systems.

$$\mathbf{f} = f_x\mathbf{e}_1 + f_y\mathbf{e}_2 = \hat{f}_x\hat{\mathbf{e}_1} + \hat{f}_y\hat{\mathbf{e}_2} \tag{1.30}$$

$$\begin{bmatrix} \hat{f}_x \\ \hat{f}_y \end{bmatrix} = \begin{bmatrix} cos(\theta) & sin(\theta) \\ -sin(\theta) & cos(\theta) \end{bmatrix} \begin{bmatrix} f_x \\ f_y \end{bmatrix} \Rightarrow \hat{\mathbf{f}} = \mathbf{T}\mathbf{f} \tag{1.31}$$

$$\mathbf{T}^{-1} = \mathbf{T}^T \tag{1.32}$$

An example of a second-order tensor is the stress tensor, which can be defined in the two and three-dimensional space. Stress is defined as the force per unit area and depends on the orientation of the plane where the force is acting, as well as on the orientation of the force. According to strength of materials conventions, a stress quantity, for instance $\sigma_{xy}$, is applied on a plane perpendicular to axis $x$ and is oriented towards axis $y$, for an infinitesimal cube representing a material point on a cartesian coordinate system. Therefore, for the specification of a second-order tensor such as the stress tensor, two vectors should be defined: One perpendicular to the plane on which the force is acting, to define the orientation of that plane and another to define the direction and the magnitude of the force per unit area. The stress tensor will relate these two vectors, according to the definitions which are provided below.

First, the relation between the second-order tensor $\boldsymbol{\sigma}$ and two vectors (or first-order tensors) **t** and **n** is given in equation (1.33). As it will further be explained in the following sections, $\mathbf{t}$ represents the stress vector (or traction vector) on a surface and $\mathbf{n}$ is the unit normal vector to that surface. It will also be given that the stress tensor $\boldsymbol{\sigma}$ is symmetric, indicating that $\boldsymbol{\sigma}^T = \boldsymbol{\sigma}$.

$$\mathbf{t} = \boldsymbol{\sigma}^T \mathbf{n} = \boldsymbol{\sigma} \mathbf{n} \tag{1.33}$$

In the rotated coordinate system $\hat{x}$, $\hat{y}$, relation (1.33) turns into (1.34). In addition, transformation relation (1.31) applies for both $\hat{\mathbf{t}}$, $\mathbf{t}$ and $\hat{\mathbf{n}}$, $\mathbf{n}$ as given in equation (1.35).

$$\hat{\mathbf{t}} = \hat{\boldsymbol{\sigma}} \hat{\mathbf{n}} \tag{1.34}$$

$$\hat{\mathbf{t}} = \mathbf{T}\mathbf{t}, \hat{\mathbf{n}} = \mathbf{T}\mathbf{n} \tag{1.35}$$

Equations (1.35) are substituted into (1.34), resulting in equation (1.36). Then, relation (1.37) is derived by using equations (1.36) and (1.33).

$$\hat{\mathbf{t}} = \hat{\boldsymbol{\sigma}} \hat{\mathbf{n}} \Rightarrow \mathbf{T}\mathbf{t} = \hat{\boldsymbol{\sigma}} \mathbf{T}\mathbf{n} \Rightarrow \mathbf{t} = \mathbf{T}^T \hat{\boldsymbol{\sigma}} \mathbf{T}\mathbf{n} \tag{1.36}$$

$$\boldsymbol{\sigma} = \mathbf{T}^T \hat{\boldsymbol{\sigma}} \mathbf{T} \quad \text{or} \quad \hat{\boldsymbol{\sigma}} = \mathbf{T} \boldsymbol{\sigma} \mathbf{T}^T \tag{1.37}$$

Substitution of transformation matrix $\mathbf{T}$ as given in equation (1.31) into equation (1.37) results in the transformation rule for the second-order tensor $\boldsymbol{\sigma}$ in two-dimensional space.

Another, more general definition of the transformation rules for first and second-order tensors can be formulated, when Einstein summation convention is used. Following Einstein summation, the coordinates of a rotated cartesian system in respect to the original system, are given by equation (1.38):

$$\hat{x}_i = \alpha_{ij} x_j \tag{1.38}$$

where summation takes place in terms of the repeated index $j$ on the right-hand part of this equation. In the three-dimensional space, equation (1.38) is extended for $i = 1$ to $i = 3$, according to equations (1.39).

$$\begin{aligned} \hat{x}_1 &= \alpha_{11} x_1 + \alpha_{12} x_2 + \alpha_{13} x_3 \\ \hat{x}_2 &= \alpha_{21} x_1 + \alpha_{22} x_2 + \alpha_{23} x_3 \\ \hat{x}_3 &= \alpha_{31} x_1 + \alpha_{32} x_2 + \alpha_{33} x_3 \end{aligned} \tag{1.39}$$

After proper manipulation of the transformation relations between the rotated and the initial coordinates, the following relation is formulated:

$$\hat{x}_i = (\hat{\mathbf{e}}_i \cdot \mathbf{e}_j) x_j \tag{1.40}$$

From equations (1.38) and (1.40), it is derived that the components $\alpha_{ij}$ are given by the inner product of the unit vectors $\hat{\mathbf{e}}_i$, $\mathbf{e}_j$, along the rotated and the initial coordinate system:

$$\begin{aligned} \alpha_{ij} =& \hat{\mathbf{e}}_i \cdot \mathbf{e}_j = \hat{e}_i \, e_j \, cos(\hat{\mathbf{e}}_i, \mathbf{e}_j) = cos(\hat{\mathbf{e}}_i, \mathbf{e}_j) \Rightarrow \\ \alpha_{ij} =& cos(\hat{\mathbf{e}}_i, \mathbf{e}_j) \end{aligned} \tag{1.41}$$

According to (1.41), the components $\alpha_{ij}$ are given by the *cosine* of the angle between vectors $\hat{\mathbf{e}}_i$, $\mathbf{e}_j$ and are called direction cosines.

Equation (1.39) can be rewritten in a matrix notation as:

$$\hat{\mathbf{x}} = \mathbf{T}\mathbf{x} \tag{1.42}$$

where $\mathbf{T}$ is the $3 \times 3$ transformation matrix for a three-dimensional cartesian system, which includes the direction cosines $\alpha_{ij}$. For a two-dimensional space, matrix $\mathbf{T}$ reduces to $2 \times 2$, with its components given in equation (1.31).

An alternative definition for transformation rules of tensors can now be provided, according to table 1.1. The relation given in this table for second-order tensors is equivalent to equation (1.37), which expresses the transformation relation for the stress tensor using matrix notation.

**TABLE 1.1:** Transformation rules for zero, first and second-order tensors

| Order | Transformation rule |
|---|---|
| 0 | $c(\hat{x})=c(x)$ |
| 1 | $\hat{x}_i = \alpha_{ij} x_j$ |
| 2 | $\hat{A}_{ij}=\alpha_{ik}\alpha_{jl}A_{kl}$ |

### 1.2.3 Additional operations and notations for tensors

Three main notations are widely used, to depict operations between tensors. The first is the matrix notation, where tensors are represented by matrices and rules for matrix operations are applied. The stress or traction vector $\mathbf{t}$ in equation (1.33) is given in terms of multiplication of the matrix $\boldsymbol{\sigma}^T$ or $\boldsymbol{\sigma}$ and the unit normal vector $\mathbf{n}$.

The second is the direct tensor notation, where the dot product between a second-order tensor and a vector, produces another vector. Using direct tensor notation, the traction vector can be provided using the dot product between the transpose of the stress tensor and the unit normal vector, as follows:

$$\mathbf{t} = \boldsymbol{\sigma}^T \cdot \mathbf{n} = \mathbf{n} \cdot (\boldsymbol{\sigma}^T)^T = \mathbf{n} \cdot \boldsymbol{\sigma} \tag{1.43}$$

The third notation implies that Einstein summation convention is adopted, according to the following equation:

$$a_{ij} = b_{ik} c_{kj} \tag{1.44}$$

where summation with respect to the repeated index $k$, which appears twice in the right-hand part of the equation, is considered. In the two-dimensional space, the indices $i$, $j$ vary from 1 to 2, while in the three-dimensional space, from 1 to 3. The repeated index is often called "dummy" index.

Using Einstein summation convention, the relation between the traction vector, the stress tensor and the unit normal vector is expressed as follows:

$$t_j = n_i \sigma_{ij} \tag{1.45}$$

The multiplication product between two matrices, which is given in equation (1.22), can also be expressed in direct tensor notation, using the dot product between tensors as given in equations (1.46) and (1.47):

$$\mathbf{C} = \mathbf{A} \cdot \mathbf{B} \tag{1.46}$$

$$(\mathbf{A} \cdot \mathbf{B})^T = \mathbf{B}^T \cdot \mathbf{A}^T, \quad (\mathbf{A} \cdot \mathbf{B} \cdot \mathbf{C})^T = \mathbf{C}^T \cdot \mathbf{B}^T \cdot \mathbf{A}^T \tag{1.47}$$

In Einstein notation, it is given as:

$$C_{ij} = A_{ie} B_{ej} \tag{1.48}$$

The double dot product between two tensors is also defined, as a scalar:

$$c = \mathbf{A} : \mathbf{B} \tag{1.49}$$

In Einstein notation, the double dot product is given as:

$$c = A_{ie} B_{ei} \tag{1.50}$$

In addition to operations between vectors which are presented in section 1.2.1.1, the dyadic or outer product or tensor product between two vectors **a** and **b** is defined here, as the second-order tensor expressed using vector multiplication:

$$\mathbf{C} = \mathbf{a}\mathbf{b}^T \tag{1.51}$$

The dyadic product between two vectors is also expressed as follows:

$$\mathbf{C} = \mathbf{a} \otimes \mathbf{b} \tag{1.52}$$

where each component of the second-order tensor $\mathbf{C}$ is given by $C_{ij} = a_i b_j$.

Another significant operation for vectors and tensors is expressed by the *Divergence* or *Gauss' theorem*, which states that a volume integral can be transformed into a surface integral, according to the following equations:

$$\int_V div\mathbf{v} dV = \int_V \nabla \cdot \mathbf{v} dV = \int_S \mathbf{n} \cdot \mathbf{v} dS \tag{1.53}$$

$$\int_V div\boldsymbol{\sigma} dV = \int_V \nabla \cdot \boldsymbol{\sigma} dV = \int_S \mathbf{n} \cdot \boldsymbol{\sigma} dS \tag{1.54}$$

Equation (1.53) provides the transformation for the divergence operation of the vector (or first-order tensor) $\mathbf{v}$ from volume integral $V$ to surface integral

$S$, with outward unit normal $\mathbf{n}$ to surface $S$. Equation (1.54) describes the corresponding transformation for the second-order tensor $\boldsymbol{\sigma}$.

The divergence operator $div$ is defined below for the vector $\mathbf{v}$ and the second-order tensor $\boldsymbol{\sigma}$.

$$div\mathbf{v} = \nabla \cdot \mathbf{v} = v_{i,i} = \frac{\partial v_i}{\partial x_i} = \frac{\partial v_1}{\partial x_1} + \frac{\partial v_2}{\partial x_2} + \frac{\partial v_3}{\partial x_3} \quad (1.55)$$

$$div\boldsymbol{\sigma} = \nabla \cdot \boldsymbol{\sigma} = \sigma_{ji,j} = \frac{\partial \sigma_{ji}}{\partial x_j} = \begin{bmatrix} \frac{\partial \sigma_{11}}{\partial x_1} + \frac{\partial \sigma_{21}}{\partial x_2} + \frac{\partial \sigma_{31}}{\partial x_3} \\ \frac{\partial \sigma_{12}}{\partial x_1} + \frac{\partial \sigma_{22}}{\partial x_2} + \frac{\partial \sigma_{32}}{\partial x_3} \\ \frac{\partial \sigma_{13}}{\partial x_1} + \frac{\partial \sigma_{23}}{\partial x_2} + \frac{\partial \sigma_{33}}{\partial x_3} \end{bmatrix} \quad (1.56)$$

According to equation (1.55), the divergence operator of a vector is a scalar. Equation (1.56) indicates that the divergence operator for a second-order tensor is a vector.

### 1.2.4 Deformation

A significant parameter of continuum mechanics is related to the geometric changes of structural elements. These geometric changes are generally expressed by the term *deformation.* Figure 1.4 shows the initial, undeformed and the final, deformed position of a body, in a three-dimensional cartesian system. The first is usually called undeformed and the second, deformed *configuration.*

A time equal to zero has been assigned to the undeformed configuration and a time $t$ to the deformed one. Time plays no role in static problems, however, it is still used to assign an order in a sequence of events. For nonlinear analysis of composite materials, the concept of time can be used to denote the incremental loading process using load steps.

According to figure 1.4, the position of a material point on a body in the undeformed and in the deformed configuration, can be expressed in the same coordinate system, by the initial coordinates $\mathbf{X}$ of the undeformed configuration and by the final coordinates $\mathbf{x}$ of the deformed configuration, respectively. The initial coordinates are also called *material coordinates*, while the final coordinates are called *spatial coordinates.*

#### 1.2.4.1 Lagrangian and Eulerian description of motion

Two approaches are used to define the deformation of a body: a) the material or Lagrangian description and b) the spatial or Eulerian description. According to the Lagrangian description, the position of a body refers to a reference configuration, such as the undeformed configuration. This means that the current coordinates are expressed in respect to the reference coordinates:

$$\mathbf{x} = \mathbf{x}(\mathbf{X}, t) \quad (1.57)$$

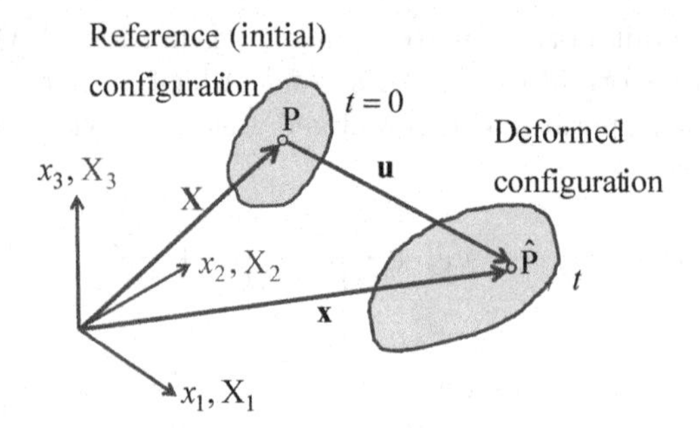


**FIGURE 1.4**: Deformation of a body in cartesian coordinates.

The Lagrangian description implies that the position of a material point is recorded on different configurations and times. Therefore, emphasis is given on different positions of a material point in the continuum.

The Eulerian description states that the position of a body refers to a current configuration:

$$\mathbf{X} = \mathbf{X}(\mathbf{x}, t) \tag{1.58}$$

The Eulerian description focuses on a spatial position and not on a material point, since different material points may occupy the current coordinates $\mathbf{x}$ at different times. This description is preferred for fluid mechanics applications.

On the solid body which is shown in figure 1.5, the position of two material points $P$ and $Q$ is examined in an initial and in a current configuration. The displacement for a material point on the body is defined as follows:

$$\mathbf{u} = \mathbf{x} - \mathbf{X} \tag{1.59}$$

The displacement within the Lagrangian and Eulerian descriptions is given by equations (1.60) and (1.61), respectively:

$$\mathbf{u}(\mathbf{X}, t) = \mathbf{x}(\mathbf{X}, t) - \mathbf{X} \tag{1.60}$$

$$\mathbf{u}(\mathbf{x}, t) = \mathbf{x} - \mathbf{X}(\mathbf{x}, t) \tag{1.61}$$

### 1.2.4.2 Deformation gradient tensor

Next, the deformation gradient tensor $\mathbf{F}$ is defined, as the tensor which relates the deformed position of a line $d\mathbf{x}$ (figure 1.5) with the undeformed position $d\mathbf{X}$ of the same line on a body, according to relation (1.62).

$$d\mathbf{x} = \mathbf{F} \cdot d\mathbf{X} = d\mathbf{X} \cdot \mathbf{F}^T \tag{1.62}$$

The deformation gradient tensor is then provided by equation (1.63):

$$\mathbf{F} = \left( \frac{\partial \mathbf{x}}{\partial \mathbf{X}} \right)^T = (\nabla_0 \mathbf{x})^T \tag{1.63}$$

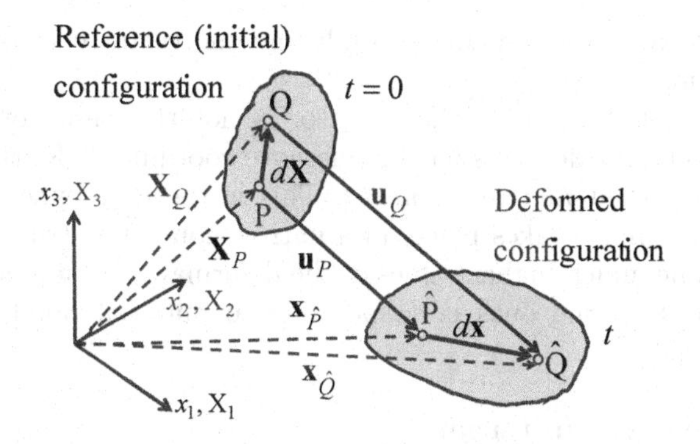


**FIGURE 1.5**: Deformation of two neighboring points on a body.

where $\nabla_0$ is the gradient operator with respect to reference coordinates $\mathbf{X}$. The deformation gradient tensor can also be provided by index notation as well as in the form of a matrix as follows:

$$F_{ij} = \left(\frac{\partial x_i}{\partial X_j}\right) \tag{1.64}$$

$$\mathbf{F} = \begin{bmatrix} \frac{\partial x_1}{\partial X_1} & \frac{\partial x_1}{\partial X_2} & \frac{\partial x_1}{\partial X_3} \\ \frac{\partial x_2}{\partial X_1} & \frac{\partial x_2}{\partial X_2} & \frac{\partial x_2}{\partial X_3} \\ \frac{\partial x_3}{\partial X_1} & \frac{\partial x_3}{\partial X_2} & \frac{\partial x_3}{\partial X_3} \end{bmatrix} \tag{1.65}$$

Next the *Jacobian* is defined as the determinant of the deformation gradient tensor:

$$J = det(\mathbf{F}) = \left|\frac{\partial x_i}{\partial X_j}\right| \tag{1.66}$$

The mapping between spatial and material coordinates given in equation (1.62) exists when $\mathbf{F}$ is a non-singular tensor. Hence, $J \neq 0$ and the inverse of the deformation gradient tensor exists.

The deformation gradient tensor $\mathbf{F}$ can also be expressed in terms of the displacement vector $\mathbf{u}$:

$$\mathbf{F} = (\nabla_0 \mathbf{x})^T = [\nabla_0(\mathbf{u} + \mathbf{X})]^T = (\nabla_0 \mathbf{u} + \mathbf{I})^T = \left(\frac{\partial \mathbf{u}}{\partial \mathbf{X}} + \mathbf{I}\right)^T \tag{1.67}$$

where $\mathbf{x} = \mathbf{u} + \mathbf{X}$ from relation (1.59) has been substituted in the above equation.

When all material points on a body have the same deformation gradient $\mathbf{F}$, indicating that $\mathbf{F}$ is independent of the reference coordinates $\mathbf{X}$, the underlying deformation of the body is called *homogeneous.* In addition to this definition, a mapping between spatial and material coordinates corresponds to homogeneous deformation gradient, if and only if:

$$\mathbf{x} = \mathbf{F} \cdot \mathbf{X} + \mathbf{c} \tag{1.68}$$

where the deformation gradient tensor $\mathbf{F}$ and the vector $\mathbf{c}$, depicting a rigid body translation, are constants.

When the deformation gradient tensor is not the same for all material points on a body but depends on the reference coordinates $\mathbf{X}$, the underlying deformation is called *nonhomogeneous.* This is the case, when for instance localization of damage takes place, in a narrow zone of a structural element. Within this zone, much higher values of the deformation gradient components arise, in respect to the ones corresponding to material points outside the localization zone.

#### 1.2.4.3 Green strain tensor

Next, the study focuses on two material points $P$, $Q$ of the reference configuration, occupying positions $\hat{P}$ and $\hat{Q}$ in the deformed configuration as shown in figure 1.5. The distance $dS$ of the two points in the reference configuration and $ds$ in the deformed configuration, can be expressed by the reference coordinates as follows:

$$(dS)^2 = d\mathbf{X} \cdot d\mathbf{X} \tag{1.69}$$

$$(ds)^2 = d\mathbf{x} \cdot d\mathbf{x} = d\mathbf{X} \cdot (\mathbf{F}^T \cdot \mathbf{F}) \cdot d\mathbf{X} \tag{1.70}$$

where $d\mathbf{x}$ in the right-hand part of (1.70) has been substituted by equation (1.62). The change in the distance can then be defined by the following relation using the reference coordinates and equations (1.69), (1.70):

$$(ds)^2 - (dS)^2 = 2\, d\mathbf{X} \cdot \mathbf{E} \cdot d\mathbf{X} \tag{1.71}$$

where $\mathbf{E}$ is called *Green or Lagrangian strain tensor* given by:

$$\mathbf{E} = \frac{1}{2}(\mathbf{F}^T \cdot \mathbf{F} - \mathbf{I}) \tag{1.72}$$

The Green strain tensor is a symmetric second-order tensor and can be expressed in respect to the displacement vector, when relation (1.67) is substituted in (1.72):

$$\begin{aligned} \mathbf{E} &= \frac{1}{2}\left[(\nabla_0 \mathbf{u} + \mathbf{I}) \cdot (\nabla_0 \mathbf{u} + \mathbf{I})^T - \mathbf{I}\right] \\ &= \frac{1}{2}\left[\nabla_0 \mathbf{u} + (\nabla_0 \mathbf{u})^T + (\nabla_0 \mathbf{u}) \cdot (\nabla_0 \mathbf{u})^T\right] \end{aligned} \tag{1.73}$$

The components of the Green strain tensor in a cartesian coordinate system can be written in index notation as follows:

$$E_{ij} = \frac{1}{2}\left(\frac{\partial u_i}{\partial X_j} + \frac{\partial u_j}{\partial X_i} + \frac{\partial u_k}{\partial X_i}\frac{\partial u_k}{\partial X_j}\right) \tag{1.74}$$

As it is shown in equations (1.73) and (1.74), the Green strain tensor is expressed in relation to the reference coordinates $X$ indicating that it refers to the undeformed configuration. By expanding the terms of (1.74), all the components of the Green strain tensor are obtained:

$$
\begin{aligned}
E_{11} &= \frac{\partial u_1}{\partial X_1} + \frac{1}{2}\left[\left(\frac{\partial u_1}{\partial X_1}\right)^2 + \left(\frac{\partial u_2}{\partial X_1}\right)^2 + \left(\frac{\partial u_3}{\partial X_1}\right)^2\right] \\
E_{22} &= \frac{\partial u_2}{\partial X_2} + \frac{1}{2}\left[\left(\frac{\partial u_1}{\partial X_2}\right)^2 + \left(\frac{\partial u_2}{\partial X_2}\right)^2 + \left(\frac{\partial u_3}{\partial X_2}\right)^2\right] \\
E_{33} &= \frac{\partial u_3}{\partial X_3} + \frac{1}{2}\left[\left(\frac{\partial u_1}{\partial X_3}\right)^2 + \left(\frac{\partial u_2}{\partial X_3}\right)^2 + \left(\frac{\partial u_3}{\partial X_3}\right)^2\right] \\
E_{12} &= \frac{1}{2}\left(\frac{\partial u_1}{\partial X_2} + \frac{\partial u_2}{\partial X_1} + \frac{\partial u_1}{\partial X_1}\frac{\partial u_1}{\partial X_2} + \frac{\partial u_2}{\partial X_1}\frac{\partial u_2}{\partial X_2} + \frac{\partial u_3}{\partial X_1}\frac{\partial u_3}{\partial X_2}\right) \\
E_{13} &= \frac{1}{2}\left(\frac{\partial u_1}{\partial X_3} + \frac{\partial u_3}{\partial X_1} + \frac{\partial u_1}{\partial X_1}\frac{\partial u_1}{\partial X_3} + \frac{\partial u_2}{\partial X_1}\frac{\partial u_2}{\partial X_3} + \frac{\partial u_3}{\partial X_1}\frac{\partial u_3}{\partial X_3}\right) \\
E_{23} &= \frac{1}{2}\left(\frac{\partial u_2}{\partial X_3} + \frac{\partial u_3}{\partial X_2} + \frac{\partial u_1}{\partial X_2}\frac{\partial u_1}{\partial X_3} + \frac{\partial u_2}{\partial X_2}\frac{\partial u_2}{\partial X_3} + \frac{\partial u_3}{\partial X_2}\frac{\partial u_3}{\partial X_3}\right)
\end{aligned}
\tag{1.75}
$$

The components $E_{ii}$ are called normal strains and the components $E_{ij}$ are called shear strains.

The above relations can be significantly simplified if the assumption of small displacement gradients can be adopted for a structural problem. This can be a valid assumption for several types of materials and structural systems, such as masonry. For this case, infinitesimal strains arise and no distinction is made between the material coordinates $\mathbf{X}$ and the spatial coordinates $\mathbf{x}$, or between different configurations.

For small displacement gradients, the non-linear terms $\frac{\partial u_k}{\partial X_i}\frac{\partial u_k}{\partial X_j}$ of the Green strain tensor shown in equation (1.74) can be neglected. For this type of analysis, which is usually called *small displacement analysis*, the following simplified formulation can then be considered for the Green strain tensor:

$$
\boldsymbol{\varepsilon} = \frac{1}{2}\left[\nabla\mathbf{u} + (\nabla\mathbf{u})^T\right] = \frac{1}{2}\left(\frac{\partial u_i}{\partial X_j} + \frac{\partial u_j}{\partial X_i}\right) \tag{1.76}
$$

By expanding the right-hand part of equation (1.76), the following components arise for the Green strain tensor under small displacements assumption:

$$\begin{aligned}
\varepsilon_{11} &= \frac{\partial u_1}{\partial X_1}, \quad \varepsilon_{22} = \frac{\partial u_2}{\partial X_2}, \quad \varepsilon_{33} = \frac{\partial u_3}{\partial X_3} \\
\varepsilon_{12} &= \frac{1}{2}\left(\frac{\partial u_1}{\partial X_2} + \frac{\partial u_2}{\partial X_1}\right) \\
\varepsilon_{13} &= \frac{1}{2}\left(\frac{\partial u_1}{\partial X_3} + \frac{\partial u_3}{\partial X_1}\right) \\
\varepsilon_{23} &= \frac{1}{2}\left(\frac{\partial u_2}{\partial X_3} + \frac{\partial u_3}{\partial X_2}\right)
\end{aligned} \tag{1.77}$$

The physical interpretation of Green strain tensor can be stated, by observing equations (1.77). The normal strain components $\varepsilon_{ii}$ express the change in the length of a line element parallel to $\mathbf{X}_i$ axis in the undeformed state, over the initial length.

The shear strain components can also be expressed as *engineering shear strains*: $\gamma_{ij} = 2\varepsilon_{ij}$. Then, the physical interpretation for shear strains is formulated, such that the shear strain components $\gamma_{ij}$ define the change in the angle between line elements which were perpendicular in the undeformed state.

#### 1.2.4.4 Euler strain tensor

The distance $dS$ of two points from figure 1.5 in the reference configuration and $ds$ in the deformed configuration, can be expressed by the current coordinates as follows:

$$(dS)^2 = d\mathbf{X} \cdot d\mathbf{X} = d\mathbf{x} \cdot (\mathbf{F}^{-T} \cdot \mathbf{F}^{-1}) \cdot d\mathbf{x} \tag{1.78}$$

$$(ds)^2 = d\mathbf{x} \cdot d\mathbf{x} \tag{1.79}$$

where $d\mathbf{X}$ in the right-hand part of (1.78) has been substituted by equation (1.62). The change in the distance can then be defined by the following relation using the current coordinates and equations (1.78), (1.79):

$$(ds)^2 - (dS)^2 = 2\ d\mathbf{x} \cdot \boldsymbol{\epsilon}^* \cdot d\mathbf{x} \tag{1.80}$$

where $\boldsymbol{\epsilon}^*$ is called *Euler or Almansi-Hamel strain tensor* given by:

$$\boldsymbol{\epsilon}^* = \frac{1}{2}(\mathbf{I} - \mathbf{F}^{-T} \cdot \mathbf{F}^{-1}) \tag{1.81}$$

Then, equation (1.62) is solved in respect to d$\mathbf{X}$ using the inverse of $\mathbf{F}$:

$$d\mathbf{x} = \mathbf{F} \cdot d\mathbf{X} = d\mathbf{X} \cdot \mathbf{F}^T \Rightarrow d\mathbf{X} = \mathbf{F}^{-1} \cdot d\mathbf{x} = d\mathbf{x} \cdot \mathbf{F}^{-T} \tag{1.82}$$

where:

$$\mathbf{F}^{-T} = \frac{\partial \mathbf{X}}{\partial \mathbf{x}} = \nabla \mathbf{X} \tag{1.83}$$

with $\nabla$ being the gradient operator with respect to current coordinates $\mathbf{x}$.

The Euler strain tensor can be expressed in respect to the displacement vector, by substituting relations (1.59) and (1.83), in (1.81):

$$\begin{aligned} \boldsymbol{\epsilon}^* &= \frac{1}{2}\left[(\mathbf{I} - (\mathbf{I} - \nabla\mathbf{u}) \cdot (\mathbf{I} - \nabla\mathbf{u})^T\right] \\ &= \frac{1}{2}\left[\nabla\mathbf{u} + (\nabla\mathbf{u})^T - (\nabla\mathbf{u}) \cdot (\nabla\mathbf{u})^T\right] \end{aligned} \tag{1.84}$$

The components of the Euler strain tensor in a cartesian coordinate system can be written in index notation as follows:

$$\boldsymbol{\epsilon}^*{}_{ij} = \frac{1}{2}\left(\frac{\partial u_i}{\partial x_j} + \frac{\partial u_j}{\partial x_i} + \frac{\partial u_k}{\partial x_i}\frac{\partial u_k}{\partial x_j}\right) \tag{1.85}$$

It is clearly seen that Euler strain tensor is determined in respect to current coordinates, contrary to Green strain tensor, which is determined in respect to initial coordinates.

It is also noted that when small displacement gradients can be considered within small displacement analysis, the non-linear terms $\frac{\partial u_k}{\partial x_i}\frac{\partial u_k}{\partial x_j}$ of the Euler strain tensor can be neglected. For this case, the expression for the Euler strain tensor coincides the expression for the Green strain tensor and relations (1.77) can be adopted.

#### 1.2.4.5 Principal strains

For every point in three-dimensional space, there will be three strain vectors defined by the corresponding unit normal vectors $\mathbf{n}$, for which the maximum value of the normal strain component, with zero values of shear components, arise. These maximum normal strain values and the corresponding unit normal vectors can be found if the following eigenvalue problem is solved:

$$(\boldsymbol{E} - \lambda\mathbf{I})\mathbf{n} = 0 \Rightarrow \begin{bmatrix} E_{11} - \lambda & E_{12} & E_{13} \\ E_{21} & E_{22} - \lambda & E_{23} \\ E_{31} & E_{32} & E_{33} - \lambda \end{bmatrix} \begin{bmatrix} n_1 \\ n_2 \\ n_3 \end{bmatrix} = 0 \tag{1.86}$$

Unknown quantities of this eigenvalue problem are the eigenvalues $\lambda$ representing the maximum normal strains and the eigenvectors $\mathbf{n}$. The maximum normal strains are called *principal strains* and the corresponding planes are called *principal planes.*

When $\mathbf{n} \neq 0$, a non-trivial solution to the eigenvalue problem exists, if and only if the determinant of $(\boldsymbol{E} - \lambda\mathbf{I})$ is equal to zero:

$$det(\boldsymbol{E} - \lambda\mathbf{I}) = 0 \tag{1.87}$$

Equation (1.87) yields 3 values for $\lambda$, each representing a principal strain. For every principal strain, one eigenvector is calculated from equation (1.86).

### 1.2.5 Stress

When a force $\mathbf{f}$ is applied to a surface $\Delta A$ of a body, then the stress vector or traction vector $\mathbf{t}$ on that surface, is defined according to equation (1.88).

$$\mathbf{t} = \lim_{\Delta A \to 0} \frac{\Delta \mathbf{f}}{\Delta A} = \frac{d\mathbf{f}}{dA} \tag{1.88}$$

Since $\Delta A \to 0$, the traction vector indicates the stress intensity for a distinct point of the structure. To provide an overall description of the stress state at a material point, an infinitesimal cube is considered in a three-dimensional cartesian coordinate system, and stresses acting on each plane of the cube are defined according to figure 1.6. The traction vector on each plane of the cube, is resolved into three components along the three axes of the cartesian system, according to equation (1.89).

$$\mathbf{t}_i = \sigma_{ij}\mathbf{e}_j \tag{1.89}$$

In relation (1.89), vectors $\mathbf{e}$ represent the unit vectors along each axis. Then, the second-order stress tensor is defined from all the stress components shown in figure 1.6, according to equation (1.90).

$$\boldsymbol{\sigma} = \begin{bmatrix} \sigma_{11} & \sigma_{12} & \sigma_{13} \\ \sigma_{21} & \sigma_{22} & \sigma_{23} \\ \sigma_{31} & \sigma_{32} & \sigma_{33} \end{bmatrix} \tag{1.90}$$

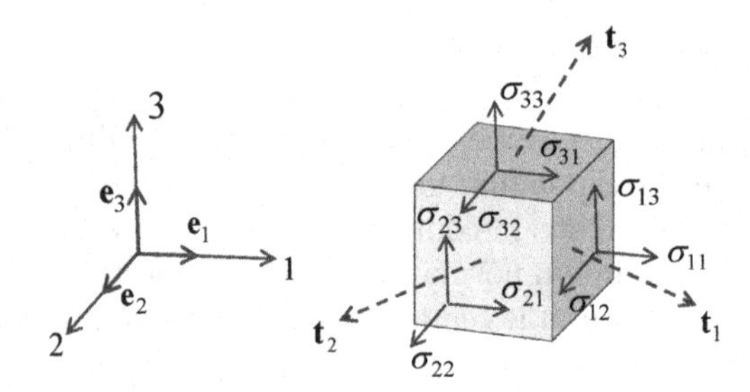


**FIGURE 1.6**: Stress components on infinitesimal cube.

As shown in figure 1.6, one of the three stress components acting on a plane is normal to that plane, while the other two components lie on the plane. The normal stress vectors are generally called normal stresses while the other two stress vectors which lie on the plane are called shear stresses.

There is a specific notation, which is adopted for the indices used on each stress component. For the normal stress vectors, the same indices $ii$ are used. For the shear stress components, two different indices $ij$ are assigned. For a normal or shear stress component, the first index shows the axis of the coordinate system, which is perpendicular to the plane of the cube where the stress is acting. Thus, the first index refers the unit normal $\mathbf{e}_i$ which denotes the plane on which the stresses on that face acts. The second index indicates the direction of the stress in the given cartesian system. For example, stress $\sigma_{33}$ is the normal stress on a plane, perpendicular to axis 3 of figure 1.6 oriented towards axis 3, while stress $\sigma_{31}$ is the shear stress on the same plane, acting towards axis 1.

The following definitions are formulated for the sign of each stress component. A normal stress is positive when it creates tension to the plane it acts and negative when it creates compression. An alternative definition for the sign of normal stress can be adopted, by considering that a normal stress is positive when is acting on a plane with a normal vector $\mathbf{e}_i$ that points in the positive direction of the coordinate system and is oriented towards the positive axis. A normal stress is also positive, when is acting on a plane with a normal vector pointing the negative direction of the coordinate system and is oriented towards the negative axis. Thus, when the direction of the normal vector on the plane and the orientation of the stress are both positive or negative, then the stress is positive. When one of the two is positive and the other is negative, the stress is negative.

The same definition is used to define the sign for a shear stress, indicating that a shear stress is positive when the direction of the normal vector corresponding to the plane and the direction of the stress are both positive or negative. When one of the two is positive and the other is negative, the shear stress is negative.

#### 1.2.5.1 Cauchy stress tensor

When the stress tensor on a point is known, then the stress or traction vector on an arbitrary but known direction on the same point can be derived. Figure 1.7 shows a tetrahedron representing a stress state, according to which three traction (or stress) vectors $\mathbf{t}_i$ are acting on three planes defined by corresponding unit normals $\mathbf{e}_i$ and another traction vector $\mathbf{t}_n$ is acting on a fourth plane, which is defined by its unit normal vector $\mathbf{n}$.

Equilibrium of all the forces acting on the tetrahedron leads to the following equation:

$$\mathbf{t}_n \Delta A_n - \mathbf{t}_1 \Delta A_1 - \mathbf{t}_2 \Delta A_2 - \mathbf{t}_3 \Delta A_3 = 0 \tag{1.91}$$

noticing that $\Delta A_n$ is the area of the plane where the traction vector $\mathbf{t}_n$ is acting and $\Delta A_1$, $\Delta A_2$, $\Delta A_3$ the corresponding areas of the three planes, where the traction vectors $\mathbf{t}_1$, $\mathbf{t}_2$ and $\mathbf{t}_3$ are acting. Each of the areas is further resolved according to:

$$\begin{aligned} \Delta A_1 &= (\mathbf{n} \cdot \mathbf{e}_1) \Delta A_n \\ \Delta A_2 &= (\mathbf{n} \cdot \mathbf{e}_2) \Delta A_n \\ \Delta A_3 &= (\mathbf{n} \cdot \mathbf{e}_3) \Delta A_n \end{aligned} \tag{1.92}$$

By substituting equations (1.92) into (1.91), it is derived that:

$$\mathbf{t}_n - \mathbf{t}_1 (\mathbf{n} \cdot \mathbf{e}_1) - \mathbf{t}_2 (\mathbf{n} \cdot \mathbf{e}_2) - \mathbf{t}_3 (\mathbf{n} \cdot \mathbf{e}_3) = 0 \tag{1.93}$$

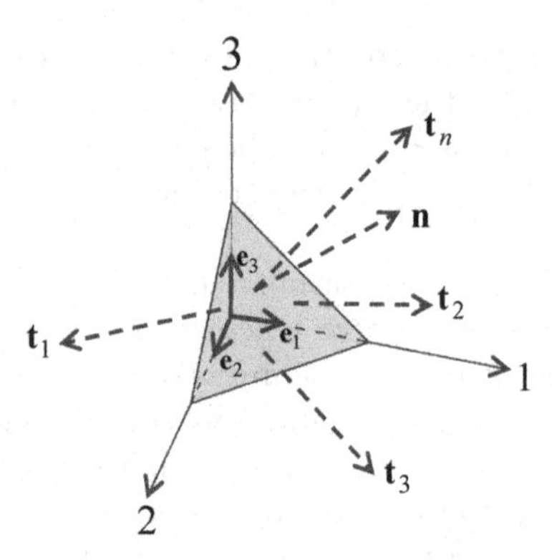


**FIGURE 1.7**: Tetrahedron representing a stress state in cartesian coordinates.

Vectors $\mathbf{n}$ and $\mathbf{e}_i$ are given by:

$$\mathbf{n} = \begin{bmatrix} n_1 \\ n_2 \\ n_3 \end{bmatrix} \tag{1.94}$$

$$\mathbf{e}_1 = \begin{bmatrix} 1 \\ 0 \\ 0 \end{bmatrix} \quad \mathbf{e}_2 = \begin{bmatrix} 0 \\ 1 \\ 0 \end{bmatrix} \quad \mathbf{e}_3 = \begin{bmatrix} 0 \\ 0 \\ 1 \end{bmatrix} \tag{1.95}$$

By substituting vectors $\mathbf{n}$ and $\mathbf{e}_i$ from equations (1.94) and (1.95) into equation (1.93), the following relation is obtained.

$$\mathbf{t}_n - \mathbf{t}_1 n_1 - \mathbf{t}_2 n_2 - \mathbf{t}_3 n_3 = 0 \Rightarrow \mathbf{t}_n = \mathbf{t}_1 n_1 + \mathbf{t}_2 n_2 + \mathbf{t}_3 n_3 \tag{1.96}$$

Equation (1.96) can be rewritten as:

$$\mathbf{t}_n = \begin{bmatrix} t_1^1 \\ t_2^1 \\ t_3^1 \end{bmatrix} n_1 + \begin{bmatrix} t_1^2 \\ t_2^2 \\ t_3^2 \end{bmatrix} n_2 + \begin{bmatrix} t_1^3 \\ t_2^3 \\ t_3^3 \end{bmatrix} n_3 = \begin{bmatrix} t_1^1 & t_1^2 & t_1^3 \\ t_2^1 & t_2^2 & t_2^3 \\ t_3^1 & t_3^2 & t_3^3 \end{bmatrix} \begin{bmatrix} n_1 \\ n_2 \\ n_3 \end{bmatrix} \tag{1.97}$$

Finally, by using equation (1.89), it is derived that the matrix of the traction vectors in the right-hand part of equation (1.97) is the transpose of the stress tensor given in equation (1.90). Thus, the traction vector $\mathbf{t}_n$ acting in an arbitrary direction defined by a unit normal $\mathbf{n}$, is given by the following equation.

$$\mathbf{t}_n = \begin{bmatrix} \sigma_{11} & \sigma_{21} & \sigma_{31} \\ \sigma_{12} & \sigma_{22} & \sigma_{32} \\ \sigma_{13} & \sigma_{23} & \sigma_{33} \end{bmatrix} \begin{bmatrix} n_1 \\ n_2 \\ n_3 \end{bmatrix} \Rightarrow \mathbf{t}_n = \boldsymbol{\sigma}^T \mathbf{n} \tag{1.98}$$

Equation (1.98) is known as *Cauchy theorem* and the second-order stress tensor $\boldsymbol{\sigma}$ as *Cauchy stress tensor*. Cauchy tensor refers to the force in the current configuration per deformed area. For a problem with small displacement gradients, Cauchy tensor can be used in numerical analysis, since the current, unknown configuration coincides with the initial, known configuration. However, when large displacement gradients are involved, the initial and the current configuration are different and Cauchy tensor may not be found, since it refers to the unknown current configuration. To overcome this issue, alternative stress measures have been introduced, referring to the initial, known configuration. These stress measures are presented in the next sections.

By applying moment equilibrium or balance of moment of momentum, it is proved that not all the stress components of the stress tensor are independent. Indeed, the following relations are derived:

$$\sigma_{12} = \sigma_{21}, \quad \sigma_{13} = \sigma_{31}, \quad \sigma_{23} = \sigma_{32} \tag{1.99}$$

which states that out of nine components, the number of independent components in the stress tensor is six. Thus, the stress tensor is equal to its transpose:

$$\boldsymbol{\sigma} = \boldsymbol{\sigma}^T \tag{1.100}$$

indicating that Cauchy tensor is symmetric and the *Cauchy theorem* can be formulated as follows:

$$\mathbf{t}_n = \boldsymbol{\sigma}\mathbf{n} \quad or \quad \mathbf{t}_n = \boldsymbol{\sigma} \cdot \mathbf{n} \tag{1.101}$$

where tensor notation is used in the right part.

#### 1.2.5.2 First Piola-Kirchhoff stress tensor

It is assumed that a force vector $d\mathbf{f}$ is applied to the deformed surface $d\alpha$ of the current coordinate system with unit normal $\mathbf{n}$ and corresponding traction (or stress) vector $\mathbf{t}$. In addition, it is suggested that there is a surface $dA$ with unit normal $\mathbf{N}$ and a traction vector $\mathbf{T}$ of the body in the reference coordinate system, such that it corresponds to the same force vector $d\mathbf{f}$. The following equation is then used to relate the reference and the current traction vectors:

$$d\mathbf{f} = \mathbf{T}dA = \mathbf{t}d\alpha \tag{1.102}$$

Similar to the definition of the Cauchy stress tensor given in (1.101), the first Piola-Kirchhoff stress tensor $\mathbf{P}$ is defined as:

$$\mathbf{T} = \mathbf{P} \cdot \mathbf{N} \tag{1.103}$$

Using (1.102) and (1.103), the relation between the current force vector $\mathbf{df}$ and the first Piola-Kirchhoff stress tensor $\mathbf{P}$ is obtained:

$$d\mathbf{f} = \mathbf{P} \cdot \mathbf{N}dA \tag{1.104}$$

According to (1.104), the first Piola-Kirchhoff stress tensor $\mathbf{P}$ relates the force vector df which is applied on the current configuration and the undeformed area $dA$ on the reference configuration.

Then, the force vector df given in (1.104) and the traction vector $\mathbf{t}$ given in (1.101) are substituted in equation (1.102) and the following relation appears:

$$\mathbf{P} \cdot \mathbf{N}dA = \boldsymbol{\sigma} \cdot \mathbf{n}d\alpha \Rightarrow \mathbf{P} \cdot d\mathbf{A} = \boldsymbol{\sigma} \cdot d\boldsymbol{\alpha} \tag{1.105}$$

where

$$\mathbf{N}dA = d\mathbf{A}, \quad \mathbf{n}d\alpha = d\boldsymbol{\alpha} \tag{1.106}$$

With $\boldsymbol{\sigma}$ is denoted the Cauchy stress tensor defined in equation (1.101). Vectors $d\boldsymbol{\alpha}$ and $d\mathbf{A}$ can be related by the following equation [181]:

$$d\boldsymbol{\alpha} = det(\mathbf{F})\mathbf{F}^{-T} \cdot d\mathbf{A} \tag{1.107}$$

where $\mathbf{F}$ is the deformation gradient tensor defined in equations (1.63), (1.65) and $det(\mathbf{F})$ represents its determinant.

By substituting (1.107) into (1.105), the relation which connects the first Piola-Kirchhoff stress tensor $\mathbf{P}$ and the Cauchy stress tensor $\boldsymbol{\sigma}$ is derived:

$$\mathbf{P} \cdot d\mathbf{A} = \boldsymbol{\sigma} \cdot det(\mathbf{F})\mathbf{F}^{-T} \cdot d\mathbf{A} \Rightarrow \mathbf{P} = det(\mathbf{F})\boldsymbol{\sigma} \cdot \mathbf{F}^{-T} \tag{1.108}$$

According to (1.108), the first Piola-Kirchhoff stress tensor $\mathbf{P}$ is not symmetric, despite the fact that the Cauchy stress tensor $\boldsymbol{\sigma}$ is symmetric.

#### 1.2.5.3 Second Piola-Kirchhoff stress tensor

In the framework of non-linear finite element analysis involving large displacement gradients, arises the need of using stress measures that are expressed in the undeformed configuration, since this configuration is known. The second Piola-Kirchhoff stress tensor which is defined in this section, is widely used for these types of problems.

First, a force vector $d\mathcal{F}$ in the undeformed area is introduced such that it corresponds to the force vector df in the deformed area. Similar to equation (1.62), which relates the deformed coordinates $d\mathbf{x}$ and the undeformed coordinates $d\mathbf{X}$ using the deformation gradient tensor $\mathbf{F}$, the force vector $d\mathbf{f}$ in the deformed area and the force vector $d\mathcal{F}$ in the undeformed area, can also be related using the deformation gradient tensor as follows:

$$d\mathbf{f} = \mathbf{F} \cdot d\mathcal{F} \Rightarrow d\mathcal{F} = \mathbf{F}^{-1} \cdot d\mathbf{f} \tag{1.109}$$

Using (1.104) and (1.106), the force vector df can also be expressed as:

$$d\mathbf{f} = \mathbf{P} \cdot \mathbf{N}dA = \mathbf{P} \cdot d\mathbf{A} \tag{1.110}$$

with $\mathbf{P}$ being the first Piola-Kirchhoff stress tensor. By substituting (1.110) into (1.109), the second Piola-Kirchhoff stress tensor $\mathbf{S}$ is defined:

$$d\mathcal{F} = \mathbf{F}^{-1} \cdot \mathbf{P} \cdot d\mathbf{A} \quad or \quad d\mathcal{F} = \mathbf{S} \cdot d\mathbf{A} \tag{1.111}$$

where

$$\mathbf{S} = \mathbf{F}^{-1} \cdot \mathbf{P} \tag{1.112}$$

According to (1.111), the second Piola-Kirchhoff stress tensor relates the force vector $d\mathcal{F}$ of the initial configuration, which corresponds to the force vector df on the current configuration and the undeformed area $d\mathbf{A}$. Thus, it is defined from quantities which appear in the reference configuration and this makes it suitable for problems involving large displacement gradients.

By substituting the first Piola-Kirchhoff stress tensor $\mathbf{P}$ given in (1.108) into (1.112), it is obtained that:

$$\mathbf{S} = det(\mathbf{F})\mathbf{F}^{-1} \cdot \boldsymbol{\sigma} \cdot \mathbf{F}^{-T} \tag{1.113}$$

According to (1.113), the second Piola-Kirchhoff stress tensor $\mathbf{S}$ is symmetric.

#### 1.2.5.4 Principal stresses

Some stress tensors with special properties can now be defined. As it can be concluded by observing figure 1.7, the stress or traction vector $\mathbf{t}_n$ will receive the maximum value for the normal stress component, when the other two stress components which lie on the plane vanish. This is equivalent to a state that the stress vector acts in parallel, in regard to the unit normal $\mathbf{n}$, thus $\mathbf{t}_n = \lambda\mathbf{n}$, where $\lambda$ is a scalar value representing the maximum normal stress to a plane with unit normal $\mathbf{n}$.

For every point in three-dimensional space, there will be three stress vectors $\mathbf{t}_n$, defined by corresponding unit normal vectors $\mathbf{n}$, for which the maximum value of the normal stress component, with zero values of shear components arise. These maximum normal stress values and the corresponding unit normal vectors can be found if the following equation, is solved:

$$\mathbf{t}_n = \boldsymbol{\sigma}\mathbf{n} = \lambda\mathbf{n} \tag{1.114}$$

where Cauchy theorem has been used to express the stress vector $\mathbf{t}_n$ in respect to the second-order stress tensor $\boldsymbol{\sigma}$. Equation (1.114) can be rewritten in the following form:

$$(\boldsymbol{\sigma} - \lambda\mathbf{I})\mathbf{n} = 0 \Rightarrow \begin{bmatrix} \sigma_{11} - \lambda & \sigma_{12} & \sigma_{13} \\ \sigma_{21} & \sigma_{22} - \lambda & \sigma_{23} \\ \sigma_{31} & \sigma_{32} & \sigma_{33} - \lambda \end{bmatrix} \begin{bmatrix} n_1 \\ n_2 \\ n_3 \end{bmatrix} = 0 \tag{1.115}$$

Equation (1.115) defines an eigenvalue problem, with unknown quantities the eigenvalues $\lambda$ and the eigenvectors $\mathbf{n}$. For each eigenvalue, which represents the maximum normal stress in a plane with zero shear stresses, the corresponding eigenvector denotes the unit normal to that plane.

These maximum normal stresses are called *principal stresses* and the corresponding planes are called *principal planes.* Principal stresses and principal planes are often needed when failure is studied for composite heterogeneous structures. The reason for this, is that some damage is expected in points that stresses become maximum, for instance in a crack tip.

When $\mathbf{n} \neq 0$, a non-trivial solution to the eigenvalue problem exists, if and only if the determinant of $(\boldsymbol{\sigma} - \lambda\mathbf{I})$ is equal to zero:

$$det(\boldsymbol{\sigma} - \lambda\mathbf{I}) = 0 \tag{1.116}$$

Equation (1.116) results in a cubic equation for $\lambda$, the solution of which yields three values for $\lambda$, each representing a principal stress. Thus, the solution for the eigenvalue problem of the second-order stress tensor given in (1.115), results in the principal stress tensor given in equation (1.117). For every principal stress, one eigenvector is calculated from equation (1.115). It is noted that since the system of equations given in (1.115) is linearly dependent, one additional equation is needed. This is obtained by the condition that each unit normal vector has a length equal to 1.

$$\boldsymbol{\sigma}_p = \begin{bmatrix} \lambda_1 & 0 & 0 \\ 0 & \lambda_2 & 0 \\ 0 & 0 & \lambda_3 \end{bmatrix} \tag{1.117}$$

#### 1.2.5.5 Stress invariants and deviatoric stress

For every stress tensor there are three quantities which remain invariant in respect to the chosen coordinate system. These quantities are functions of the stress tensor components and can be determined by reformulating equation (1.116) as follows:

$$\lambda^3 - I_1\lambda^2 + I_2\lambda - I_3 = 0 \tag{1.118}$$

where $I_1$, $I_2$ and $I_3$ take the following values in relation to the stress tensor components:

$$\begin{aligned} I_1 =& \sigma_{11} + \sigma_{22} + \sigma_{33} \\ I_2 =& \sigma_{11}\sigma_{22} + \sigma_{22}\sigma_{33} + \sigma_{33}\sigma_{11} - \sigma_{12}^2 - \sigma_{23}^2 - \sigma_{31}^2 \\ I_3 =& \sigma_{11}\sigma_{22}\sigma_{33} + 2\sigma_{12}\sigma_{23}\sigma_{31} - \sigma_{11}\sigma_{23}^2 - \sigma_{22}\sigma_{31}^2 - \sigma_{33}\sigma_{12}^2 \end{aligned} \tag{1.119}$$

or using summation notation:

$$\begin{aligned} I_1 =& \sigma_{ii} = tr(\boldsymbol{\sigma}) \\ I_2 =& \frac{1}{2}(\sigma_{ii}\sigma_{jj} - \sigma_{ij}\sigma_{ij}) \\ I_3 =& |\sigma_{ij}| = det(\boldsymbol{\sigma}) \end{aligned} \tag{1.120}$$

In equations (1.119) and (1.120), the values of $I_1$, $I_2$ and $I_3$ remain invariant when a coordinate transformation is considered. Stress invariants are often used in constitutive material laws describing damage for composite structures.

It is noted that the concept of invariants of the stress tensor as well as the concept for principal stresses and principal directions, apply for every second-order tensor.

When a stress tensor takes the form of equation (1.121), is called hydrostatic stress tensor.

$$\boldsymbol{\sigma}_H = p \begin{bmatrix} 1 & 0 & 0 \\ 0 & 1 & 0 \\ 0 & 0 & 1 \end{bmatrix} \tag{1.121}$$

If it is assumed that $p = \sigma_{ii}/3 = tr(\boldsymbol{\sigma})/3$ is the *hydrostatic* or *volumetric pressure*, then the deviatoric stress tensor $\mathbf{s}$ is defined as follows:

$$\mathbf{s} = \boldsymbol{\sigma} - p\mathbf{I} = \begin{bmatrix} \sigma_{11} - p & \sigma_{12} & \sigma_{13} \\ \sigma_{21} & \sigma_{22} - p & \sigma_{23} \\ \sigma_{31} & \sigma_{32} & \sigma_{33} - p \end{bmatrix} \tag{1.122}$$

### 1.2.6 Constitutive description of the material behaviour

#### 1.2.6.1 Generalized Hooke's law and elasticity tensor

The constitutive description of a structural system, aims to provide a mathematical model representing the material behaviour, in terms of stress versus strain relations. These mathematical models are validated using experimental output and are included within finite element analysis computer codes to express the stress-strain response of the used material(s).

Within small displacements assumption, the relation between the stress and the strain tensor is given by the *generalized Hooke's law* equation:

$$\sigma_{ij} = C_{ijkl}\varepsilon_{kl} \tag{1.123}$$

where the components $C_{ijkl}$ are called elastic stiffness coefficients and they represent the fourth-order tensor $\mathbf{C}$, which is called *elasticity tensor.* The elasticity tensor has $3^4 = 81$ components.

However, when symmetry of the stress and strain tensors is considered, indicating that each of these tensors have six instead of nine components due to the symmetry condition $\sigma_{ij} = \sigma_{ji}$ and $\varepsilon_{ij} = \varepsilon_{ji}$, the following equalities apply to the elasticity tensor components and their number reduces to 36:

$$C_{ijkl} = C_{jikl}, \quad C_{ijkl} = C_{jilk} \tag{1.124}$$

The number of independent components reduces further to 21, when according to equations (1.125) symmetry of the elasticity tensor is considered:

$$C_{ijkl} = C_{klij} \tag{1.125}$$

The elasticity tensor with 21 components represents the most general description of the material behaviour, called *anisotropic*. The elasticity tensor which corresponds to this behaviour, is given by the following matrix:

$$\boldsymbol{C} = \begin{bmatrix} C_{1111} & C_{1122} & C_{1133} & C_{1123} & C_{1113} & C_{1112} \\ & C_{2222} & C_{2233} & C_{2223} & C_{2213} & C_{2212} \\ & & C_{3333} & C_{3323} & C_{3313} & C_{3312} \\ & \text{Sym.} & & C_{2323} & C_{2313} & C_{2312} \\ & & & & C_{1313} & C_{1312} \\ & & & & & C_{1212} \end{bmatrix} \tag{1.126}$$

To further elaborate the stress versus strain relation given in (1.123), the stress and strain tensors can be expressed in a vector format, using six independent components due to the symmetry of these tensors:

$$\begin{aligned} \boldsymbol{\sigma} &= \begin{bmatrix} \sigma_{11} & \sigma_{22} & \sigma_{33} & \sigma_{23} & \sigma_{13} & \sigma_{12} \end{bmatrix}^T \\ \boldsymbol{\varepsilon} &= \begin{bmatrix} \varepsilon_{11} & \varepsilon_{22} & \varepsilon_{33} & 2\varepsilon_{23} & 2\varepsilon_{13} & 2\varepsilon_{12} \end{bmatrix}^T \end{aligned} \tag{1.127}$$

In addition, an alternative notation for the indices of the stress, strain and elasticity tensors is adopted. According to this notation, one index is used to substitute two indices of the initial formulation:

$$11 \rightarrow 1, \quad 22 \rightarrow 2, \quad 33 \rightarrow 3, \quad 23 \rightarrow 4, \quad 13 \rightarrow 5, \quad 12 \rightarrow 6 \tag{1.128}$$

$$\begin{aligned} &\sigma_1 = \sigma_{11}, \quad \sigma_2 = \sigma_{22}, \quad \sigma_3 = \sigma_{33}, \quad \sigma_4 = \sigma_{23}, \quad \sigma_5 = \sigma_{13}, \quad \sigma_6 = \sigma_{12} \\ &\varepsilon_1 = \varepsilon_{11}, \quad \varepsilon_2 = \varepsilon_{22}, \quad \varepsilon_3 = \varepsilon_{33}, \quad \varepsilon_4 = 2\varepsilon_{23}, \quad \varepsilon_5 = 2\varepsilon_{13}, \quad \varepsilon_6 = 2\varepsilon_{12} \end{aligned} \tag{1.129}$$

This notation is called *engineering* or *Voigt* notation. Following Voigt notation, equation (1.123) can be rewritten in index as well as in matrix notation:

$$\sigma_i = C_{ij}\varepsilon_j \tag{1.130}$$

$$\begin{bmatrix} \sigma_1 \\ \sigma_2 \\ \sigma_3 \\ \sigma_4 \\ \sigma_5 \\ \sigma_6 \end{bmatrix} = \begin{bmatrix} C_{11} & C_{12} & C_{13} & C_{14} & C_{15} & C_{16} \\ & C_{22} & C_{23} & C_{24} & C_{25} & C_{26} \\ & & C_{33} & C_{34} & C_{35} & C_{36} \\ & \text{Sym.} & & C_{44} & C_{45} & C_{46} \\ & & & & C_{55} & C_{56} \\ & & & & & C_{66} \end{bmatrix} \begin{bmatrix} \varepsilon_1 \\ \varepsilon_2 \\ \varepsilon_3 \\ \varepsilon_4 \\ \varepsilon_5 \\ \varepsilon_6 \end{bmatrix} \tag{1.131}$$

Relations (1.130) and (1.131) can be inverted resulting in a new equation which relates the strain tensor components to stress tensor components as follows:

$$\varepsilon_i = S^*_{ij}\sigma_j \tag{1.132}$$

The tensor $\mathbf{S}^*$ in equation (1.132) is called *compliance* tensor and is equal to the inverse of the elasticity tensor, $\mathbf{S}^* = \mathbf{C}^{-1}$.

### 1.2.6.2 Material symmetry

When the elastic tensor components remain the same in two different coordinate systems, one of which is considered as the initial and the other is obtained by mirroring one or more axes of the initial coordinate system in respect to certain planes, it is stated that the material presents symmetry with respect to these planes.

The transformation relations between a new coordinate system $\hat{x}_i$ and an initial $x_i$ are given below, for the second-order stress and strain tensors as well as for the fourth-order elasticity tensor:

$$\begin{aligned} \hat{\sigma}_{ij} &= \alpha_{ip}\alpha_{jq}\sigma_{pq} \\ \hat{\varepsilon}_{ij} &= \alpha_{ip}\alpha_{jq}\varepsilon_{pq} \\ \hat{C}_{ijkl} &= \alpha_{ip}\alpha_{jq}\alpha_{kr}\alpha_{ls}C_{pqrs} \end{aligned} \tag{1.133}$$

where terms $\alpha_{ij}$ are the direction cosines. The simplest case of symmetry arises when only one plane of symmetry between the initial and the new coordinate system is considered. This is the case, when axes $x_1$ and $x_2$ of the initial coordinate system remain the same in the new coordinate system, thus $\hat{x}_1 = x_1$, $\hat{x}_2 = x_2$ while for the third axis $\hat{x}_3 = -x_3$, indicating that the new axis $\hat{x}_3$ is the mirror of the initial $x_3$ axis in respect to a plane parallel to $x_1$-$x_2$. Materials whose elastic tensor components remain the same in two coordinate systems presenting one plane of symmetry, are called *monoclinic materials.*

The transformation matrix given in (1.134) is obtained for the two coordinate systems.

$$\mathbf{T} = \begin{bmatrix} 1 & 0 & 0 \\ 0 & 1 & 0 \\ 0 & 0 & -1 \end{bmatrix} \tag{1.134}$$

The transformation of the stress and strain components using the first two equations in (1.133) and the transformation matrix $\mathbf{T}$, results in the following relations:

$$\begin{aligned} &\hat{\sigma}_{23} = -\sigma_{23}, \quad \hat{\sigma}_{31} = -\sigma_{31} \\ &\hat{\sigma}_{11} = \sigma_{11}, \quad \hat{\sigma}_{22} = \sigma_{22}, \quad \hat{\sigma}_{33} = \sigma_{33}, \quad \hat{\sigma}_{12} = \sigma_{12} \\ &\hat{\varepsilon}_{23} = -\varepsilon_{23}, \quad \hat{\varepsilon}_{31} = -\varepsilon_{31} \\ &\hat{\varepsilon}_{11} = \varepsilon_{11}, \quad \hat{\varepsilon}_{22} = \varepsilon_{22}, \quad \hat{\varepsilon}_{33} = \varepsilon_{33} \quad \hat{\varepsilon}_{12} = \varepsilon_{12} \end{aligned} \tag{1.135}$$

If Hooke's law (1.123) is implemented in the initial and in the new coordinate system, taking into account equalities (1.135) and the fact that the same elasticity tensor components appear in the two systems, then the following relations are obtained for the elasticity tensor components:

$$
\begin{aligned}
\hat{C}_{1113} &= C_{1113} = -C_{1113}, \quad \hat{C}_{1123} = C_{1123} = -C_{1123} \\
\hat{C}_{1213} &= C_{1213} = -C_{1213}, \quad \hat{C}_{2213} = C_{2213} = -C_{2213} \\
\hat{C}_{2223} &= C_{2223} = -C_{2223}, \quad \hat{C}_{2312} = C_{2312} = -C_{2312} \\
\hat{C}_{3323} &= C_{3323} = -C_{3323}, \quad \hat{C}_{3313} = C_{3313} = -C_{3313}
\end{aligned}
\tag{1.136}
$$

The above equations are only valid when these terms are equal to zero, indicating that in Voigt notation the following components become zero.

$$
C_{15} = 0,\ C_{14} = 0,\ C_{65} = 0,\ C_{25} = 0,\ C_{24} = 0,\ C_{46} = 0,\ C_{34} = 0,\ C_{35} = 0 \tag{1.137}
$$

Therefore, the number of independent components of the elasticity tensor reduces for this case to $21 - 8 = 13$ and the corresponding elasticity tensor is obtained by substituting these zeros in (1.126).

When there are three orthogonal planes of symmetry between the initial and the new coordinate system, for which the components of the elasticity tensor remain the same, the material is called *orthotropic*. For this case, the relations between the new and the initial coordinates are $\hat{x}_1 = -x_1$, $\hat{x}_2 = -x_2$ and $\hat{x}_3 = -x_3$ and the transformation matrices are given by:

$$
\mathbf{T}_1 = \begin{bmatrix} 1 & 0 & 0 \\ 0 & 1 & 0 \\ 0 & 0 & -1 \end{bmatrix} \quad \mathbf{T}_2 = \begin{bmatrix} -1 & 0 & 0 \\ 0 & 1 & 0 \\ 0 & 0 & 1 \end{bmatrix} \quad \mathbf{T}_3 = \begin{bmatrix} 1 & 0 & 0 \\ 0 & -1 & 0 \\ 0 & 0 & 1 \end{bmatrix} \tag{1.138}
$$

If the procedure presented above for the case of one symmetry plane is repeated for the case of three symmetry planes, by using subsequently the transformation equations for stress and strain tensors and the transformation matrices of (1.138), some additional equalities given in (1.139) arise, resulting in four more zero values for the corresponding elasticity tensor components.

$$
\begin{aligned}
\hat{C}_{1112} &= C_{1112} = -C_{1112}, \quad \hat{C}_{2212} = C_{2212} = -C_{2212} \\
\hat{C}_{3312} &= C_{3312} = -C_{3312}, \quad \hat{C}_{2313} = C_{2313} = -C_{2313}
\end{aligned}
\tag{1.139}
$$

These zero values of the elasticity tensor components, are given below in Voigt notation:

$$
C_{16} = 0,\ C_{26} = 0,\ C_{36} = 0,\ C_{45} = 0 \tag{1.140}
$$

Thus, the total number of independent elasticity tensor components reduces to $13 - 4 = 9$ for an orthotropic material and Hooke's law is expressed as given in (1.141), using Voigt notation.

$$
\begin{bmatrix} \sigma_1 \\ \sigma_2 \\ \sigma_3 \\ \sigma_4 \\ \sigma_5 \\ \sigma_6 \end{bmatrix} = \begin{bmatrix} C_{11} & C_{12} & C_{13} & 0 & 0 & 0 \\ & C_{22} & C_{23} & 0 & 0 & 0 \\ & & C_{33} & 0 & 0 & 0 \\ & & & C_{44} & 0 & 0 \\ & \text{Sym.} & & & C_{55} & 0 \\ & & & & & C_{66} \end{bmatrix} \begin{bmatrix} \varepsilon_1 \\ \varepsilon_2 \\ \varepsilon_3 \\ \varepsilon_4 \\ \varepsilon_5 \\ \varepsilon_6 \end{bmatrix} \tag{1.141}
$$

The inverse of the elasticity tensor given in (1.141) can also be used, to produce the compliance matrix $\mathbf{S}^*$:

$$\begin{bmatrix} \varepsilon_1 \\ \varepsilon_2 \\ \varepsilon_3 \\ \varepsilon_4 \\ \varepsilon_5 \\ \varepsilon_6 \end{bmatrix} = \begin{bmatrix} S^*_{11} & S^*_{12} & S^*_{13} & 0 & 0 & 0 \\ & S^*_{22} & S^*_{23} & 0 & 0 & 0 \\ & & S^*_{33} & 0 & 0 & 0 \\ & \text{Sym.} & & S^*_{44} & 0 & 0 \\ & & & & S^*_{55} & 0 \\ & & & & & S^*_{66} \end{bmatrix} \begin{bmatrix} \sigma_1 \\ \sigma_2 \\ \sigma_3 \\ \sigma_4 \\ \sigma_5 \\ \sigma_6 \end{bmatrix} \tag{1.142}$$

The components of the elasticity and compliance tensors for an orthotropic material can alternatively be expressed in terms of Young's modulus, Poisson's ratio and Shear modulus, which are called *engineering constants.* Young's modulus $E$ is defined as the material constant which relates the normal stress and strain, using Hooke's law for a material under uniaxial behaviour:

$$\sigma_{ii} = E\varepsilon_{ii} \tag{1.143}$$

The corresponding law for shear stress and strain defines the Shear modulus $G_{ij}$:

$$\sigma_{ij} = G_{ij}\gamma_{ij} = G_{ij}2\varepsilon_{ij} \tag{1.144}$$

Poisson's ratio $\nu_{ij}$ is defined as the ratio of the transverse normal strain in respect to the axial normal strain, created by a load applied along the axial direction:

$$\nu_{ij} = \frac{-\varepsilon_{jj}}{\varepsilon_{ii}} \tag{1.145}$$

where $i$ indicates the axial direction of the load and $j$ the direction of the transverse normal strain.

Using the principle of superposition, the following relations between engineering constants are built for an orthotropic material:

$$\begin{aligned} \varepsilon_{11} &= \frac{\sigma_{11}}{E_1} - \nu_{21}\frac{\sigma_{22}}{E_2} - \nu_{31}\frac{\sigma_{33}}{E_3} \\ \varepsilon_{22} &= -\nu_{12}\frac{\sigma_{11}}{E_1} + \frac{\sigma_{22}}{E_2} - \nu_{32}\frac{\sigma_{33}}{E_3} \\ \varepsilon_{33} &= -\nu_{13}\frac{\sigma_{11}}{E_1} - \nu_{23}\frac{\sigma_{22}}{E_2} + \frac{\sigma_{33}}{E_3} \\ 2\varepsilon_{12} &= \frac{\sigma_{12}}{G_{12}}, \quad 2\varepsilon_{13} = \frac{\sigma_{13}}{G_{13}}, \quad 2\varepsilon_{23} = \frac{\sigma_{23}}{G_{23}} \end{aligned} \tag{1.146}$$

These equations can then be presented in a matrix form, definining the compliance tensor $\mathbf{S}^*$ for an orthotropic material, in terms of engineering constants:

$$\begin{bmatrix} \varepsilon_1 \\ \varepsilon_2 \\ \varepsilon_3 \\ \varepsilon_4 \\ \varepsilon_5 \\ \varepsilon_6 \end{bmatrix} = \begin{bmatrix} \frac{1}{E_1} & -\frac{\nu_{21}}{E_2} & -\frac{\nu_{31}}{E_3} & 0 & 0 & 0 \\ & \frac{1}{E_2} & -\frac{\nu_{32}}{E_3} & 0 & 0 & 0 \\ & & \frac{1}{E_3} & 0 & 0 & 0 \\ & & & \frac{1}{G_{23}} & 0 & 0 \\ & \text{Sym.} & & & \frac{1}{G_{13}} & 0 \\ & & & & & \frac{1}{G_{12}} \end{bmatrix} \begin{bmatrix} \sigma_1 \\ \sigma_2 \\ \sigma_3 \\ \sigma_4 \\ \sigma_5 \\ \sigma_6 \end{bmatrix} \tag{1.147}$$

This is a convenient way to build the compliance tensor and its inverse (elasticity tensor), since the engineering constants used in equation (1.147) have a direct physical interpretation.

#### 1.2.6.3 Isotropic materials

Materials whose properties are independent of the material direction are called *isotropic*. For these materials, only three engineering constants appear, according to the following relations:

$$\begin{aligned} E_1 &= E_2 = E_3 = E \\ G_{12} &= G_{13} = G_{23} = G \\ \nu_{12} &= \nu_{13} = \nu_{23} = \nu \end{aligned} \tag{1.148}$$

In addition, equation (1.149) can be used for isotropic materials to provide one of the Shear modulus, Young's modulus and Poisson's ratio constants, when the other two are known:

$$G = \frac{E}{2(1+\nu)} \tag{1.149}$$

The following equations between engineering constants are then built for an isotropic material, where relation (1.149) is used to substitute the Shear modulus G:

$$\begin{aligned} \varepsilon_{11} &= \frac{\sigma_{11}}{E} - \nu\frac{\sigma_{22}}{E} - \nu\frac{\sigma_{33}}{E} \\ \varepsilon_{22} &= -\nu\frac{\sigma_{11}}{E} + \frac{\sigma_{22}}{E} - \nu\frac{\sigma_{33}}{E} \\ \varepsilon_{33} &= -\nu\frac{\sigma_{11}}{E} - \nu\frac{\sigma_{22}}{E} + \frac{\sigma_{33}}{E} \\ 2\varepsilon_{23} &= 2(1+\nu)\frac{\sigma_{23}}{E}, \quad 2\varepsilon_{13} = 2(1+\nu)\frac{\sigma_{13}}{E}, \quad 2\varepsilon_{12} = 2(1+\nu)\frac{\sigma_{12}}{E} \end{aligned} \tag{1.150}$$

Equations (1.150) are expressed in matrix form using Voigt notation, as follows:

$$\begin{bmatrix} \varepsilon_1 \\ \varepsilon_2 \\ \varepsilon_3 \\ \varepsilon_4 \\ \varepsilon_5 \\ \varepsilon_6 \end{bmatrix} = \frac{1}{E} \begin{bmatrix} 1 & -\nu & -\nu & 0 & 0 & 0 \\ & 1 & -\nu & 0 & 0 & 0 \\ & & 1 & 0 & 0 & 0 \\ & & & 2(1+\nu) & 0 & 0 \\ & \text{Sym.} & & & 2(1+\nu) & 0 \\ & & & & & 2(1+\nu) \end{bmatrix} \begin{bmatrix} \sigma_1 \\ \sigma_2 \\ \sigma_3 \\ \sigma_4 \\ \sigma_5 \\ \sigma_6 \end{bmatrix} \tag{1.151}$$

By inversing the compliance tensor given in (1.151) for an isotropic material, the elasticity tensor is obtained:

$$\mathbf{C} = \frac{E}{(1+\nu)(1-2\nu)} \begin{bmatrix} 1-\nu & \nu & \nu & 0 & 0 & 0 \\ & 1-\nu & \nu & 0 & 0 & 0 \\ & & 1-\nu & 0 & 0 & 0 \\ & & & \frac{1-2\nu}{2} & 0 & 0 \\ & \text{Sym.} & & & \frac{1-2\nu}{2} & 0 \\ & & & & & \frac{1-2\nu}{2} \end{bmatrix} \tag{1.152}$$

The linear constitutive relations can be extended and used in *rate-dependent* problems, by assuming linear relations between rates of stresses and strains.

# Chapter 2

# *Linear and non-linear finite element analysis*

## 2.1 Introduction

Despite the usefulness of linear theories, a great variety of structural applications in civil, mechanical or aerospace engineering are characterized by a non-linear response. Quite often this type of response leads to failure of the structural system. A simplified description of the non-linear response is given here for a structure subjected to a gradually increased loading. In small values of loading, the relation between the load and the corresponding displacement, which is developed due to this load remains linear. However, after the load reaches a critical level, the structure ceases to follow this linear behaviour, indicating that for a relatively small increase of the load, a significant increase of the displacement arises. The load versus displacement diagram becomes non-linear in this case and the capacity of the structure to resist external actions decreases.

The response of a structure may become non-linear, when specific conditions characterize the underlying problem. Thus, the following categories of non-linear problems in mechanics are identified:

- *Material non-linearity*: This type of non-linearity is mainly expressed by non-linear stress versus strain laws representing the response of structural materials. For example, the stress-strain behaviour of concrete or steel is non-linear.

- *Geometrical non-linearity*: This type of non-linearity arises when large displacements and rotations need to be considered in structural systems. The strains however, remain small. This type of non-linearity may appear in steel or shell structures.

- *Finite deformations*: This type of non-linear analysis takes place, when both large displacements and large strains arise. Such problems appear in metal forming or tyre mechanics.

- *Boundary non-linearity*: This type of non-linearity is attributed to non-linear boundary conditions, which appear when problems involving unilateral contact and friction arise. These conditions may appear in the interface

DOI: 10.1201/9781003017240-2

between dry masonry units or between matrix and inclusions in a composite material.

The finite element method is widely used, for the solution of linear and non-linear, structural analysis problems. In the following sections, an introduction to the method is initially presented. Then, the mathematical formulation of non-linear finite element analysis is provided, for the case of geometrical and material non-linearities.

## 2.2 Equilibrium

### 2.2.1 Strong form

The principle of conservation of linear momentum, which is also known as Newton's second law of motion, states that the sum of forces applied externally to a body is equal to the time rate of change of the linear momentum:

$$\mathbf{F} = \frac{d(m\mathbf{v})}{dt} = m\mathbf{a} \tag{2.1}$$

where vector $\mathbf{F}$ represents the sum of all the external forces, $\mathbf{v}$ the velocity and $\mathbf{a}$ the acceleration of a body with mass $m$.

If $\mathbf{f}$ is the vector of the body forces per unit mass, $\mathbf{t}$ is the vector of the surface forces per unit area, also known as traction or stress vector and $\rho$ is the density of the body, then the body forces applied to an infinitesimal volume $dV$ are given by $m\mathbf{f} = \rho\mathbf{f}dV$ and the surface forces applied to an infinitesimal surface $dS$ by $\mathbf{t}dS$.

The total external force applied to the body can be found by integrating the given body and surface forces over the volume and surface respectively, as follows:

$$\mathbf{F} = \int_V \rho\mathbf{f}dV + \int_S \mathbf{t}dS \tag{2.2}$$

The stress vector $\mathbf{t}$ is then substituted by equation (1.43) which expresses the Cauchy theorem:

$$\mathbf{F} = \int_V \rho\mathbf{f}dV + \int_S \mathbf{n}\cdot\boldsymbol{\sigma}dS \tag{2.3}$$

with $\boldsymbol{\sigma}$ being the stress tensor and $\mathbf{n}$ the unit normal vector to the boundary surface of the body.

Considering equation (2.3), the principle of conservation of linear momentum is expressed as follows:

$$\int_V \rho\mathbf{f}dV + \int_S \mathbf{n}\cdot\boldsymbol{\sigma}dS = \frac{d}{dt}\int_V \rho\mathbf{v}dV \tag{2.4}$$

The velocity $\mathbf{v}$ is generally a function of time and position variables $t$ and $x$, respectively. Thus, the derivative of the velocity with respect to time is provided by:

$$\begin{aligned} \frac{d(\rho\mathbf{v})}{dt} &= \rho(\frac{\partial \mathbf{v}}{\partial t} + \mathbf{v}\cdot\nabla\mathbf{v}) \quad or \\ &= \rho(\frac{\partial v_i}{\partial t} + v_j\frac{\partial v_i}{\partial x_j}) \end{aligned} \tag{2.5}$$

When infinitesimal deformations are considered, the above equation is simplified as follows:

$$\begin{aligned} \frac{d(\rho\mathbf{v})}{dt} &= \rho\frac{\partial \mathbf{v}}{\partial t} = \rho\frac{\partial^2 \mathbf{u}}{\partial t^2} = \rho\ddot{\mathbf{u}} \quad or \\ &= \rho\frac{\partial v_i}{\partial t} = \rho\frac{\partial^2 u_i}{\partial t^2} = \rho\ddot{u}_i \end{aligned} \tag{2.6}$$

where the first partial derivative of velocity $\mathbf{v}$ with respect to time has been replaced by the second derivative $\ddot{\mathbf{u}}$ of the displacement vector $\mathbf{u}$. In addition, equation (2.4) is written for this case as follows:

$$\int_V \rho\mathbf{f}dV + \int_S \mathbf{n}\cdot\boldsymbol{\sigma}dS = \int_V \rho\ddot{\mathbf{u}}dV \tag{2.7}$$

Next, the divergence theorem is used to transform the surface integral of equation (2.7) into volume integral and the following equation is obtained:

$$\begin{aligned} &\int_V \rho\mathbf{f}dV + \int_V \nabla\cdot\boldsymbol{\sigma}dV = \int_V \rho\ddot{\mathbf{u}}dV \Rightarrow \\ &\int_V (\rho\mathbf{f} + \nabla\cdot\boldsymbol{\sigma} - \rho\ddot{\mathbf{u}})dV = \mathbf{0} \end{aligned} \tag{2.8}$$

Relation (2.8), which expresses the global form of equation of motion, results in the following local form:

$$\rho\mathbf{f} + \nabla\cdot\boldsymbol{\sigma} - \rho\ddot{\mathbf{u}} = \mathbf{0} \Rightarrow \nabla\cdot\boldsymbol{\sigma} + \rho\mathbf{f} = \rho\ddot{\mathbf{u}} \tag{2.9}$$

Equation (2.9) is written in terms of governing differential equations, which are applied to any material point of the continuum. For this reason, it is said that it states a problem in *strong form*. In a cartesian coordinate system, equation (2.9) can be written as follows:

$$\sigma_{ji,j} + \rho f_i = \rho\ddot{u}_i \quad or \quad \frac{\partial \sigma_{ji}}{\partial x_j} + \rho f_i = \rho\frac{\partial^2 u_i}{\partial t^2} \tag{2.10}$$

When a static problem is considered indicating that inertial effects are neglected, equation (2.10) is further simplified according to the following equation:

$$\frac{\partial \sigma_{ji}}{\partial x_j} + \rho f_i = 0 \Rightarrow \frac{\partial \sigma_{ji}}{\partial x_j} = -\rho f_i \tag{2.11}$$

Equation (2.11) represents the *equilibrium* condition for a material point. This is the strong form of the equilibrium state, since it is expressed in terms of governing differential equations representing equilibrium for any material point of the continuum. It is noted that if the body force $\mathbf{f}$ is given as the force per unit volume (instead of force per unit mass which was initially considered), then the body force for an infinitesimal volume $dV$ is $\mathbf{f}dV$ and the strong form of the equilibrium equation becomes:

$$\frac{\partial \sigma_{ji}}{\partial x_j} + f_i = 0 \quad or \quad \nabla \cdot \boldsymbol{\sigma} + \mathbf{f} = \mathbf{0} \tag{2.12}$$

### 2.2.2 Energy principles and weak form

The strong form of the equilibrium state, which was presented in the previous section, needs to be further elaborated in order to provide a convenient framework for the implementation of the finite element method. This convenient framework arises, when the differential equations are implicitly included in integral expressions. Then, all conditions must be satisfied in an average or integral sense which is named *weak form*, contrary to the strong form where conditions were met at every material point of the continuum. According to the formulation of the weak form, integrals are used to cover the domain of a problem and integrations of the underlying expressions, in combination with a finite element discretization, result in a finite number of algebraic equations providing the solution of the structural analysis problem.

The most significant weak formulations are represented by the *principle of virtual work* and by the *principle of stationary potential energy.* These principles are also called *variational methods.*

A preliminary condition of these formulations is the *compatibility condition.* A body satisfies compatibility if the displacement field is continuous and presents a single-value function of position. Practically this means that the structure does not present any cracks which would violate the displacement continuity, kinks due to bending or interpenetration between structural parts.

#### 2.2.2.1 Principle of virtual work

The relation between strains and displacements given by equation (1.77) for the Green strain tensor under small displacement gradients, is expressed here as follows:

$$\begin{bmatrix} \varepsilon_{11} \\ \varepsilon_{22} \\ \varepsilon_{33} \\ \gamma_{23} \\ \gamma_{13} \\ \gamma_{12} \end{bmatrix} = \begin{bmatrix} \frac{\partial}{\partial x_1} & 0 & 0 \\ 0 & \frac{\partial}{\partial x_2} & 0 \\ 0 & 0 & \frac{\partial}{\partial x_3} \\ 0 & \frac{\partial}{\partial x_3} & \frac{\partial}{\partial x_2} \\ \frac{\partial}{\partial x_3} & 0 & \frac{\partial}{\partial x_1} \\ \frac{\partial}{\partial x_2} & \frac{\partial}{\partial x_1} & 0 \end{bmatrix} \begin{bmatrix} u_1 \\ u_2 \\ u_3 \end{bmatrix} \quad or \quad \boldsymbol{\varepsilon} = \boldsymbol{\partial}\mathbf{u} \tag{2.13}$$

The matrix $\boldsymbol{\partial}$ can be used to provide the divergence operator of the second-order stress tensor, as follows:

$$\nabla \cdot \boldsymbol{\sigma} = \boldsymbol{\partial}^T \boldsymbol{\sigma} \tag{2.14}$$

where the stress tensor $\boldsymbol{\sigma}$ is considered in the vector format of relations (1.127) with dimensions $6 \times 1$. Using equation (2.14), relation (2.9) expressing the strong form of the equation of motion is written as:

$$\rho\mathbf{f} + \boldsymbol{\partial}^T\boldsymbol{\sigma} - \rho\ddot{\mathbf{u}} = \mathbf{0} \tag{2.15}$$

Next, the concept of *virtual displacement* is introduced. This is an imaginary, small displacement vector which is *admissible*, indicating that it satisfies compatibility and the boundary conditions of the structure. Both parts of equation (2.15) are multiplied by the virtual displacement vector field $\delta\mathbf{u}^T$ and integrated over the volume domain $V$ of the body:

$$\int_V \delta\mathbf{u}^T(\rho\mathbf{f} + \boldsymbol{\partial}^T\boldsymbol{\sigma} - \rho\ddot{\mathbf{u}})dV = \mathbf{0} \tag{2.16}$$

The following formulation is then developed, using summation notation.

$$\int_V \frac{\partial(\sigma_{ij}\delta u_i)}{\partial x_j}dV = \int_V \sigma_{ij}\frac{\partial \delta u_i}{\partial x_j}dV + \int_V \frac{\partial \sigma_{ij}}{\partial x_j}\delta u_i dV \tag{2.17}$$

The volume integral which appears in the left-hand part of equation (2.17) is replaced by a surface integral using the divergence theorem:

$$\begin{aligned} &\int_S \sigma_{ij}n_j\delta u_i dS = \int_V \sigma_{ij}\frac{\partial \delta u_i}{\partial x_j}dV + \int_V \frac{\partial \sigma_{ij}}{\partial x_j}\delta u_i dV \Rightarrow \\ &\int_V \frac{\partial \sigma_{ij}}{\partial x_j}\delta u_i dV = \int_S \sigma_{ij}n_j\delta u_i dS - \int_V \sigma_{ij}\frac{\partial \delta u_i}{\partial x_j}dV \Rightarrow \\ &\int_V \frac{\partial \sigma_{ij}}{\partial x_j}\delta u_i dV = \int_S t_i\delta u_i dS - \int_V \sigma_{ij}\frac{\partial \delta u_i}{\partial x_j}dV \end{aligned} \tag{2.18}$$

where Cauchy theorem was used according to $t_i = \sigma_{ij} n_j$. The last of equations (2.18) is rewritten using matrix-vector notation:

$$\int_V \delta\mathbf{u}^T \boldsymbol{\partial}^T \boldsymbol{\sigma} dV = \int_S \delta\mathbf{u}^T \mathbf{t} dS - \int_V (\boldsymbol{\partial}\delta\mathbf{u})^T \boldsymbol{\sigma} dV \tag{2.19}$$

By substituting the left-hand part of equation (2.19) into equation (2.16), the following equation, which is known as the *principle of virtual work*, arises:

$$\begin{gathered}\int_V \rho\delta\mathbf{u}^T \ddot{\mathbf{u}} dV + \int_V (\boldsymbol{\partial}\delta\mathbf{u})^T \boldsymbol{\sigma} dV = \int_V \rho\delta\mathbf{u}^T \mathbf{f} dV + \int_S \delta\mathbf{u}^T \mathbf{t} dS \Rightarrow \\ \int_V \rho\delta\mathbf{u}^T \ddot{\mathbf{u}} dV + \int_V \delta\boldsymbol{\varepsilon}^T \boldsymbol{\sigma} dV = \int_V \rho\delta\mathbf{u}^T \mathbf{f} dV + \int_S \delta\mathbf{u}^T \mathbf{t} dS\end{gathered} \tag{2.20}$$

where equation (2.13) relating the strain vector and the displacement field is introduced in the second equation. When inertial effects are neglected, the following form arises:

$$\int_V \delta\boldsymbol{\varepsilon}^T \boldsymbol{\sigma} dV = \int_V \rho\delta\mathbf{u}^T \mathbf{f} dV + \int_S \delta\mathbf{u}^T \mathbf{t} dS \tag{2.21}$$

#### 2.2.2.2 Principle of stationary potential energy

The potential energy will be expressed in the form of an integral, which is called *functional*. The principle of stationary potential energy will be defined for *conservative* systems. A structural system is conservative when the works of the internal and the external forces, between an initial and a current configuration, are each independent of the path from the initial to the current configuration.

The potential energy $\Pi$ is defined by the summation of the strain energy of internal forces $U$ and the potential energy of external forces $\Omega$, expressed as their potential to produce work when a displacement arises:

$$\Pi = U + \Omega \tag{2.22}$$

Then, the principle of stationary potential energy defines that for a conservative system, the admissible configuration (among several) which satisfies equilibrium, is the one which corresponds to stationary value of the potential energy, when small, admissible variations of displacement are applied to the system. In addition, the equilibrium state is stable, when the stationary potential energy becomes minimum. The principle of stationary potential energy is given by the following relation:

$$\delta\Pi = 0 \tag{2.23}$$

The *strain energy density* is defined for a unit volume, by the term:

$$U_0 = \int \boldsymbol{\sigma}^T d\boldsymbol{\varepsilon} \tag{2.24}$$

Then, by substituting the generalized Hooke's law given below in equation (2.24), the strain energy density $U_0$ is provided by (2.26).

$$\boldsymbol{\sigma} = \mathbf{C}\boldsymbol{\varepsilon} \tag{2.25}$$

$$U_0 = \int \boldsymbol{\varepsilon}^T \mathbf{C} d\boldsymbol{\varepsilon} = \frac{1}{2}\boldsymbol{\varepsilon}^T \mathbf{C}\boldsymbol{\varepsilon} \tag{2.26}$$

The total strain energy of the system is then defined as follows:

$$U = \int_V U_0 dV = \frac{1}{2}\int_V \boldsymbol{\varepsilon}^T \mathbf{C}\boldsymbol{\varepsilon} dV \tag{2.27}$$

Next, the total potential energy can be defined, as the sum of the strain energy and the potential energy of the external forces:

$$\Pi = \frac{1}{2}\int_V \boldsymbol{\varepsilon}^T \mathbf{C}\boldsymbol{\varepsilon} dV - \int_V \rho \mathbf{u}^T \mathbf{f} dV - \int_S \mathbf{u}^T \mathbf{t} dS \tag{2.28}$$

The negative sign for the potential energy of the external forces which is assigned in equation (2.28), indicates that as the external forces are producing work due to displacement, the capacity of external forces to produce new work is equally reduced and thus, the negative sign denotes the lost potential.

By applying the principle of stationary potential energy given in equation (2.23) into (2.28), the following relation is obtained:

$$\delta\Pi = 0 \Rightarrow \int_V \delta\boldsymbol{\varepsilon}^T \boldsymbol{\sigma} dV = \int_V \rho \delta\mathbf{u}^T \mathbf{f} dV + \int_S \delta\mathbf{u}^T \mathbf{t} dS \tag{2.29}$$

It is noted that when the virtual strain $\delta\boldsymbol{\varepsilon}$ and the virtual displacement $\delta\mathbf{u}$ are considered, equation (2.29) is identical to the principle of virtual work given in (2.21).

---

## 2.3 The finite element method

In the framework of the finite element method, the overall domain representing a structural system is divided into several sub-domains, the *finite elements*. Finite elements are interconnected at points, which are called *nodes*. The process of using finite elements in order to approximate the structural response of a body is called *discretization* and the assembly of all elements in a structure is the *mesh*. An example of the mesh for a fibre reinforced composite material is provided in figure 2.1.

A core concept of the finite element method is that within each element, a spatial variation for a field variable is decided. This spatial variation is usually

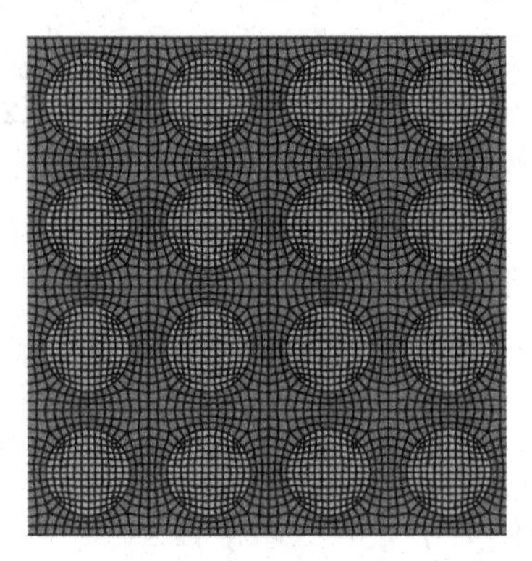

**FIGURE 2.1**: Mesh for a fibre reinforced composite material.

polynomial and represents an approximation of the actual variation for the field, indicating that the finite element method is an approximate method. The most widely used field is the displacement field, constituting the *displacement-based* finite element method, which is elaborated in this book. There are, however, alternative considerations where force or stress field variations are adopted within *force-based* methods. When both the displacement field and the stress field are simultaneously approximated, *mixed* finite element solutions are obtained. When the fields on a mixed finite element and on its boundaries are independently assumed, a *hybrid* formulation is defined [32].

For the implementation of the solution, the weak formulations presented in Section 2.2.2 are used, resulting in equilibrium equations involving integrals, with unknowns the values of the field quantity at the nodes. For the approximation of the integrals, numerical integration schemes are adopted, indicating that eventually a system of algebraic equations needs to be solved. This solution can nicely be implemented within programming codes and be included in structural analysis software, highlighting the advantage of adopting weak formulations, comparing to the strong form where partial differential equations should be elaborated.

In this section the main steps of the finite element method are presented. Then, it will be indicated how non-linear finite element analysis is formulated.

### 2.3.1 Discretization

Within the displacement-based finite element analysis, the field variable which is approximated is the displacement field. Thus, an interpolating polynomial distribution of the displacement field $u$ of an element in respect to the independent variable $x$ representing cartesian coordinates, is given by the following relation:

$$u = \sum_{i=0}^{n} \alpha_i x^i, \quad or \quad u = \mathbf{x}\boldsymbol{\alpha} \tag{2.30}$$

An example for the form of vectors $\mathbf{x}$ and $\boldsymbol{\alpha}$ which implement the polynomial distribution is given in the equation (2.31) below. This could be the case for a one-dimensional type finite element, representing for instance rod or beam elements.

$$\mathbf{x} = \begin{bmatrix} 1 & x & x^2 & \dots & x^n \end{bmatrix}, \quad \boldsymbol{\alpha} = \begin{bmatrix} \alpha_0 & \alpha_1 & \alpha_2 & \dots & \alpha_n \end{bmatrix}^T \tag{2.31}$$

A linear interpolation is obtained in this case for $n = 1$, a quadratic for $n = 2$ and a higher-order interpolation is obtained for higher values of $n$.

Next, known values of the components $x$ are considered at the nodes of the element and equation (2.30) is used to provide the corresponding nodal displacements:

$$\mathbf{u}_e = \mathbf{A}\boldsymbol{\alpha} \Rightarrow \boldsymbol{\alpha} = \mathbf{A}^{-1}\mathbf{u}_e \tag{2.32}$$

where $\mathbf{u}_e$ represents the vector of nodal displacements of the element and $\mathbf{A}$ is the matrix with components given in terms of the known values of $x$ at the nodes. By substituting (2.32) into (2.30), the following relation arises:

$$u = \mathbf{x}\mathbf{A}^{-1}\mathbf{u}_e \Rightarrow u = \mathbf{N}\mathbf{u}_e \tag{2.33}$$

with

$$\mathbf{N} = \mathbf{x}\mathbf{A}^{-1}, \quad \text{where} \quad \mathbf{N} = \begin{bmatrix} N_1 & N_2 & \dots & N_{n+1} \end{bmatrix} \tag{2.34}$$

Equation (2.33) indicates that the relation between the displacement field and the nodal displacements is expressed in terms of the vector $\mathbf{N}$ of dimensions $1 \times (n+1)$ and constitutes a fundamental concept of the finite element method. Each component of $\mathbf{N}$ is an interpolation function, which is also called *shape function* or *basis function*.

Some fundamental properties of the shape functions are now discussed. According to equation (2.33), the nodal displacement $u_i$ at $x = x_i$ can be obtained if the component $N_i(x_i)$ of the shape function vector $\mathbf{N}$ is equal to 1 and the remaining components of $\mathbf{N}$ at $x = x_j$ for $i \neq j$ are equal to 0. This is expressed as follows:

$$N_i(x_j) = \delta_{ij} = \begin{cases} 1 & \text{if } i = j \\ 0 & \text{if } i \neq j \end{cases} \tag{2.35}$$

where $\delta_{ij}$ is also known as Kronecker-delta. Another significant property of the shape functions is the *partition of unity*, given by the following equation:

$$\sum_{j=1}^{n} N_j(x) = 1 \tag{2.36}$$

This property is properly elaborated when damage of composite materials is studied, in the framework of the Extended Finite Element Method (XFEM).

In the given descriptions, the displacement field is interpolated in piecewise mode within each finite element, indicating a smooth variation of the field within the element. However, the variation of the displacement field between elements may not be smooth, highlighting the need for the definition of the *degree of continuity*. A displacement field is $C^m$ continuous if both the field and its derivatives up to (and including) degree $m$ are continuous. In particular, $C^0$ continuity denotes that the displacement field $u$ is continuous. Equivalently, $C^1$ continuity indicates that the displacement field $u$ and its first derivative $\frac{\partial u}{\partial x}$ are continuous.

### 2.3.2 Equilibrium within the finite element method

In the framework of the finite element method, equilibrium can be defined using the weak formulations of section 2.2.2 and the finite element discretization presented in section 2.3.1. Equation (2.33) is rewritten here to express the interpolation of the displacement field $\mathbf{u}$, using matrix $\mathbf{N}$ with components the shape functions $N_i$ and the vector of nodal displacements $\mathbf{u}_e$.

$$\mathbf{u} = \mathbf{N}\mathbf{u}_e \quad \text{where} \quad \mathbf{u} = \begin{bmatrix} u & v & w \end{bmatrix}^T \tag{2.37}$$

When the displacement field $\mathbf{u}$ includes three displacement degrees of freedom, the vector $\mathbf{u}$ is of dimensions $3 \times 1$ and in every node of the element three displacement degrees of freedom are also assigned. Thus, the dimensions of the nodal displacement vector $\mathbf{u}_e$ is $3n_e \times 1$ and the dimensions of matrix $\mathbf{N}$ is $3 \times 3n_e$, with $n_e$ denoting the number of nodes for the element. It is noted that according to this description, one shape function $N_i$ is introduced for every node of the element. It is also pointed out that the unknown quantity of the finite element formulation, is the vector of nodal displacements $\mathbf{u}_e$.

Then, the displacement field $\mathbf{u}$ of equation (2.37) is substituted into the strain-displacement relation (2.13) and the following strain-nodal displacement equation is obtained:

$$\boldsymbol{\varepsilon} = \boldsymbol{\partial}\mathbf{u} \Rightarrow \boldsymbol{\varepsilon} = \boldsymbol{\partial}\mathbf{N}\mathbf{u}_e \tag{2.38}$$

with matrix $\boldsymbol{\partial}$ given in (2.13). Equation (2.38) is rewritten as follows:

$$\boldsymbol{\varepsilon} = \mathbf{B}\mathbf{u}_e \quad \text{where} \quad \mathbf{B} = \boldsymbol{\partial}\mathbf{N} \tag{2.39}$$

Matrix $\mathbf{B}$ is called *strain-displacement* matrix and is of dimensions $6 \times 3n_e$. Next, the variations of the virtual strain $\delta\boldsymbol{\varepsilon}$ and virtual displacement $\delta\mathbf{u}$, which appear in the principle of virtual work equation (2.20) representing equilibrium, are formulated using equations (2.39) and (2.37), respectively:

$$\boldsymbol{\varepsilon} = \mathbf{B}\mathbf{u}_e \Rightarrow \delta\boldsymbol{\varepsilon} = \mathbf{B}\delta\mathbf{u}_e \Rightarrow \delta\boldsymbol{\varepsilon}^T = \delta\mathbf{u}_e^T\mathbf{B}^T \tag{2.40}$$

$$\mathbf{u} = \mathbf{N}\mathbf{u}_e \Rightarrow \delta\mathbf{u} = \mathbf{N}\delta\mathbf{u}_e \Rightarrow \delta\mathbf{u}^T = \delta\mathbf{u}_e^T\mathbf{N}^T \tag{2.41}$$

In addition, the second derivative over time of the displacement field can be expressed as follows, using equation (2.37):

$$\mathbf{u} = \mathbf{N}\mathbf{u}_e \Rightarrow \ddot{\mathbf{u}} = \mathbf{N}\ddot{\mathbf{u}}_e \tag{2.42}$$

To build the equilibrium equation in the framework of the finite element method, equations (2.40), (2.41) and (2.42) denoting the virtual strain, the virtual displacement and the second derivative of displacement over time are substituted in the weak formulation (2.20) representing the principle of virtual work, as shown in (2.43). In addition, summation over the volume $V_e$ and surface $S_e$ of each element, for the overall number of elements $N_e$ of the structure is considered, to indicate that an assembly process over all finite elements of the structural model is conducted.

$$\begin{aligned} &\sum_{i=1}^{N_e} \int_{V_e} \rho \delta \mathbf{u}_e^T \mathbf{N}^T \mathbf{N} \ddot{\mathbf{u}}_e dV + \sum_{i=1}^{N_e} \int_{V_e} \delta \mathbf{u}_e^T \mathbf{B}^T \boldsymbol{\sigma} dV = \\ &\sum_{i=1}^{N_e} \int_{V_e} \rho \delta \mathbf{u}_e^T \mathbf{N}^T \mathbf{f} dV + \sum_{i=1}^{N_e} \int_{S_e} \delta \mathbf{u}_e^T \mathbf{N}^T \mathbf{t} dS \end{aligned} \tag{2.43}$$

The assembly of the virtual displacement vector $\delta \mathbf{u}_e$ of each element leads to the virtual displacement vector $\delta \mathbf{U}$ of the whole structure:

$$\begin{aligned} &\delta \mathbf{U}^T \sum_{i=1}^{N_e} \int_{V_e} \rho \mathbf{N}^T \mathbf{N} \ddot{\mathbf{u}}_e dV + \delta \mathbf{U}^T \sum_{i=1}^{N_e} \int_{V_e} \mathbf{B}^T \boldsymbol{\sigma} dV = \\ &\delta \mathbf{U}^T \sum_{i=1}^{N_e} \int_{V_e} \rho \mathbf{N}^T \mathbf{f} dV + \delta \mathbf{U}^T \sum_{i=1}^{N_e} \int_{S_e} \mathbf{N}^T \mathbf{t} dS \end{aligned} \tag{2.44}$$

Since $\delta \mathbf{U}$ takes discrete values, it is left out of the integrals and the summation in equation (2.44) and eventually vanishes. Furthermore, the second derivative $\ddot{\mathbf{U}}$ over time of the total, nodal displacement vector $\mathbf{U}$ of the whole structure is introduced. The principle of the virtual work is then written as follows:

$$\sum_{i=1}^{N_e} \int_{V_e} \rho \mathbf{N}^T \mathbf{N} dV \ddot{\mathbf{U}} + \sum_{i=1}^{N_e} \int_{V_e} \mathbf{B}^T \boldsymbol{\sigma} dV = \sum_{i=1}^{N_e} \int_{V_e} \rho \mathbf{N}^T \mathbf{f} dV + \sum_{i=1}^{N_e} \int_{S_e} \mathbf{N}^T \mathbf{t} dS \tag{2.45}$$

Since the virtual nodal displacement vector of equation (2.43) has been eliminated in (2.45), the right-hand part in (2.45) represents the discretized, nodal external force vector $\mathbf{F}_{ext}$ for the overall structure, which was initially defined in equation (2.2). It is observed that the same shape functions used to interpolate the displacement field, have been considered for the discretization of the external forces. In addition, the second term of the left-hand part of (2.45) represents the nodal internal force vector $\mathbf{F}_{int}$ given in the form of integration of internal stresses. Considering these two statements, equation (2.45) can be rewritten in the following form, which is suitable for the implementation of the solution in the framework of non-linear finite element analysis:

$$\mathbf{M} \ddot{\mathbf{U}} = \mathbf{F}_{ext} - \mathbf{F}_{int} \tag{2.46}$$

where the *mass matrix* $\mathbf{M}$ is defined as:

$$\mathbf{M} = \sum_{i=1}^{N_e} \int_{V_e} \rho \mathbf{N}^T \mathbf{N} dV \tag{2.47}$$

The external and internal nodal force vectors, are given by:

$$\mathbf{F}_{ext} = \sum_{i=1}^{N_e} \int_{V_e} \rho \mathbf{N}^T \mathbf{f} dV + \sum_{i=1}^{N_e} \int_{S_e} \mathbf{N}^T \mathbf{t} dS \tag{2.48}$$

$$\mathbf{F}_{int} = \sum_{i=1}^{N_e} \int_{V_e} \mathbf{B}^T \boldsymbol{\sigma} dV \tag{2.49}$$

To further elaborate equation (2.45), the stress vector $\boldsymbol{\sigma}$ expressed as:

$$\boldsymbol{\sigma} = \mathbf{C}\boldsymbol{\varepsilon} \Rightarrow \boldsymbol{\sigma} = \mathbf{C}\mathbf{B}\mathbf{u}_e \tag{2.50}$$

is substituted in (2.45). By considering also the assembly of all element nodal displacement vectors $\mathbf{u}_e$ into the total, nodal displacement vector $\mathbf{U}$, the following equation arises:

$$\sum_{i=1}^{N_e} \int_{V_e} \rho \mathbf{N}^T \mathbf{N} dV \ddot{\mathbf{U}} + \sum_{i=1}^{N_e} \int_{V_e} \mathbf{B}^T \mathbf{C}\mathbf{B} dV \mathbf{U} = \sum_{i=1}^{N_e} \int_{V_e} \rho \mathbf{N}^T \mathbf{f} dV + \sum_{i=1}^{N_e} \int_{S_e} \mathbf{N}^T \mathbf{t} dS \tag{2.51}$$

It is noted that the total nodal displacement vector $\mathbf{U}$ is the unknown quantity of the finite element formulation and it appears out of the integral, since it is not a function of the coordinates but instead, it includes discrete displacement values.

By observing equation (2.51), the global *stiffness matrix* $\mathbf{K}$ of the structure is defined as the sum of the stiffness matrix $\mathbf{k}$ of each element. Both the global and the element stiffness matrices are given below:

$$\mathbf{K} = \sum_{i=1}^{N_e} \int_{V_e} \mathbf{B}^T \mathbf{C}\mathbf{B} dV, \quad \mathbf{k} = \int_{V_e} \mathbf{B}^T \mathbf{C}\mathbf{B} dV \tag{2.52}$$

When inertial effects are neglected, equation (2.51) can be rewritten in the form:

$$\mathbf{K}\mathbf{U} = \mathbf{F}_{ext} \tag{2.53}$$

In the given descriptions, the assembly of the stiffness matrix and force vector of each element into the global stiffness matrix and total, nodal force vector of the structure, should be considered. This assembly is expressed in these equations by the sum of the quantities in the integrals. For the practical implementation of the assembly in a finite element code, the degrees of freedom corresponding to each component of the element stiffness matrix and force vector are initially assigned. Then, summation of these components takes place, to the positions of the overall stiffness matrix and nodal force vector corresponding to the same degrees of freedom. This approach for the assembly can be implemented within loops in a programming code, and it is found in Matlab finite element codes discussed in the appendix of the book. Relevant technical details can also be found in introductory finite element textbooks [17, 50].

Finally, it should be noted that the general formulation which is described in this section although it clearly highlights the main steps of the finite element method, it presents significant disadvantages that need to be addressed. Thus, this formulation is not appropriate when non-rectangular, curve-sided elements are used. In addition, the numerical implementation of the method can be very difficult, when global (cartesian) coordinates are adopted. For instance, the analytical calculation of the integrals over the element domain, is a quite demanding task if not impossible for several cases, such as when higher-order interpolation is adopted or when geometric/material non-linearities are involved. For these reasons, an alternative formulation called *isoparametric* is usually adopted for the implementation of the finite element analysis method, as presented in section 2.3.4.

### 2.3.3 Boundary conditions

Equation (2.53) can only be solved and nodal displacements can be calculated when *boundary* or *support conditions* are considered, since otherwise the stiffness matrix $\mathbf{K}$ is singular (has no inverse) due to arising rigid body movements. Boundary conditions usually involve prescribed nodal displacements, which can be either zero or non-zero. When non-zero prescribed displacements are considered, it is stated that a *displacement loading* is included in the formulation. This type of loading is often preferred in respect to force loading, as explained in section 2.4.3.

There are two main approaches for imposing boundary conditions within finite element analysis. According to the first approach, the system of equilibrium equations (2.53) is partitioned into parts corresponding to free, "$f$" and supported, "$s$" degrees of freedom. Free degrees of freedom are related to unknown nodal displacements, which are about to be calculated. Supported degrees of freedom correspond to nodes with prescribed (zero or non-zero) displacements.

The partition of equation (2.53) takes place as follows:

$$\begin{bmatrix} \mathbf{K}_{ff} & \mathbf{K}_{fs} \\ \mathbf{K}_{sf} & \mathbf{K}_{ss} \end{bmatrix} \begin{bmatrix} \mathbf{U}_f \\ \mathbf{U}_s \end{bmatrix} = \begin{bmatrix} \mathbf{F}_{ext,f} \\ \mathbf{F}_{ext,s} \end{bmatrix} \tag{2.54}$$

By developing (2.54), the following relations are derived:

$$\begin{aligned} \mathbf{K}_{ff}\mathbf{U}_f + \mathbf{K}_{fs}\mathbf{U}_s &= \mathbf{F}_{ext,f} \\ \mathbf{K}_{sf}\mathbf{U}_f + \mathbf{K}_{ss}\mathbf{U}_s &= \mathbf{F}_{ext,s} \end{aligned} \tag{2.55}$$

The first of equations (2.55) is used to determine the unknown nodal displacements $\mathbf{U}_f$, corresponding to the free degrees of freedom:

$$\mathbf{U}_f = \mathbf{K}_{ff}^{-1}(\mathbf{F}_{ext,f} - \mathbf{K}_{fs}\mathbf{U}_s) \tag{2.56}$$

When boundary conditions are properly considered in the prescribed degrees of freedom such that no rigid body motions are developed, the submatrix $\mathbf{K}_{ff}$ is non-singular and can be inverted. It is noted that the external forces $\mathbf{F}_{ext,f}$ at free degrees of freedom as well as the prescribed displacements $\mathbf{U}_s$ are known. If zero displacements are considered in the supports, the term $\mathbf{K}_{fs}\mathbf{U}_s$ at the right-hand part of (2.56) vanishes.

The second of equations (2.55) is then used for the calculation of the external forces at the supported degrees of freedom, $\mathbf{F}_{ext,s}$, by substituting $\mathbf{U}_f$ from equation (2.56). These are the reaction forces, developed at the prescribed nodes.

According to the second approach that can be used to impose boundary conditions, no partitioning of the system of equations is needed. Instead, the overall size of the stiffness matrix, the displacement and the force vector are used. To implement this approach, the degrees of freedom which correspond to prescribed displacements are identified. Then, the stiffness matrix is properly formulated such that all its members in the row and column corresponding to prescribed degrees of freedom receive a zero value, except the main diagonal component which is set equal to one.

The displacement vector includes all the nodal displacements and is not changed, while the force vector should also be reformulated. In particular, the components of the force vector corresponding to prescribed degrees of freedom, receive the value of the prescribed displacements (zero or non-zero). For the remaining components of the force vector, the quantity $K_{is}\Delta_s$ is subtracted by each external nodal force. With $K_{is}$ is denoted the component of the stiffness matrix at row $i$ and column $s$, with $s$ being the prescribed degree of freedom. Values of $i$ are all row numbers, except the row corresponding to the prescribed degree of freedom.

This is better illustrated by developing the relevant formulations for a structural system, assuming that the nodal displacement $U_2$ of a structure is prescribed, thus, $U_2 = \Delta_2$. For this case, $i = 1, 3, \ldots, n$ and $s = 2$. Then, it can be proved that the initial system of equilibrium equations is equivalent to the following system, which is built according to the above descriptions:

$$\begin{bmatrix} K_{11} & 0 & K_{13} & \dots & K_{1n} \\ 0 & 1 & 0 & \dots & 0 \\ K_{31} & 0 & K_{33} & \dots & K_{3n} \\ \vdots & \vdots & \vdots & \ddots & \vdots \\ K_{n1} & 0 & K_{n3} & \dots & K_{nn} \end{bmatrix} \begin{bmatrix} U_1 \\ U_2 \\ U_3 \\ \vdots \\ U_n \end{bmatrix} = \begin{bmatrix} F_{ext,1} - K_{12}\Delta_2 \\ \Delta_2 \\ F_{ext,3} - K_{32}\Delta_2 \\ \vdots \\ F_{ext,n} - K_{n2}\Delta_2 \end{bmatrix} \tag{2.57}$$

From (2.57), it is clear that the second equation imposes the prescribed displacement condition $U_2 = \Delta_2$. For the above example, the reaction force $F_{ext,2}$ at the node of the prescribed displacement is unknown. This reaction force can be determined using the initial scheme given in equation (2.53), after all nodal displacements have been calculated by (2.57). It is also noted that if the prescribed displacement $\Delta_2$ is zero, then relations derived in (2.57) are further simplified.

This second approach does not need any re-arranging of the stiffness matrix, displacement and force vectors, which is needed when the first approach is adopted. On the other hand, the first approach can be useful in multi-scale methods, which are adopted to depict the non-linear response of composite materials. More relevant details can be found in chapter 7 of the book.

Finally, when the prescribed displacements are considered as additional equality constraints, more methods can be deployed to include the boundary conditions, such as the *penalty* or *Lagrange multipliers* methods.

### 2.3.4 Isoparametric formulation

The *isoparametric formulation* is an approach which presents some advantages comparing to the general discretization description provided in previous sections. According to this approach, interpolation can be achieved not only for rectangular but also for quadrilateral or even curve-sided finite elements. In addition, the isoparametric formulation allows for the calculation of the integrals using numerical integration, for the solution of the equilibrium equations represented by the weak form.

Within the isoparametric formulation, the finite element is described by the *natural* coordinate system, which is a coordinate system defined by the geometry of the element. Thus, natural coordinates are attached to the element and do not depend on the orientation of the element in respect to the global coordinate system. A mapping between natural and global coordinates is then introduced and a relevant transformation should be considered in the subsequent steps of the finite element method. In figure 2.2 of section 2.3.5, a quadrilateral element is depicted in cartesian and natural coordinates.

Another significant property of the isoparametric formulation is that the same shape functions which are used to express the relation between the displacement field and the nodal displacements, are also used to express the position on the element in respect to the nodal coordinates. Thus, the same matrix $\mathbf{N}$ with shape functions $N_i$ is used to interpolate both the displacement field and the position of a point within the element in respect to nodal displacements and nodal positions (coordinates), respectively. It is noted that within this formulation, every shape function $N_i$ is provided in terms of the natural coordinates, usually expressed by $\xi$, $\eta$, $\zeta$ for a three-dimensional problem.

The general isoparametric formulation is initially presented. Next, an example of the formulation for a plane, quadrilateral element is highlighted. For a three-dimensional problem, the natural coordinates are given by the vector $\boldsymbol{\xi}$:

$$\boldsymbol{\xi} = \begin{bmatrix} \xi & \eta & \zeta \end{bmatrix}^T \tag{2.58}$$

The displacement field $\mathbf{u} = [u \quad v \quad w]^T$ and the position $\mathbf{x} = [x \quad y \quad z]^T$ of the element are then interpolated by the same shape functions included in $\mathbf{N}$:

$$\mathbf{u} = \mathbf{N}\mathbf{u}_e, \quad \mathbf{x} = \mathbf{N}\mathbf{x}_e \tag{2.59}$$

where vector $\mathbf{x}_e$ denotes the nodal cartesian coordinates and vector $\mathbf{u}_e$ the nodal displacements. Each shape function $N_i$ is expressed in terms of the natural coordinates found in $\boldsymbol{\xi}$. To proceed with the weak formulation, matrix $\mathbf{B}$ which contains the partial derivatives of the shape functions over the cartesian coordinates, needs to be calculated. Contrary to the general formulation presented in section 2.3.2, where the shape functions were defined in terms of cartesian coordinates and matrix $\mathbf{B}$ could directly be calculated, within the isoparametric formulation this straightforward calculation is not feasible, since shape functions are given in terms of the natural coordinates. This issue is resolved using the chain rule according to the following relation:

$$\frac{\partial N_i}{\partial \boldsymbol{\xi}} = \frac{\partial \mathbf{x}}{\partial \boldsymbol{\xi}} \frac{\partial N_i}{\partial \mathbf{x}} \tag{2.60}$$

From equation (2.60), the *Jacobian* matrix $\mathbf{J}$ is defined:

$$\mathbf{J} = \frac{\partial \mathbf{x}}{\partial \boldsymbol{\xi}} \tag{2.61}$$

By substituting the right equation of (2.59) into (2.61), it is obtained that:

$$\mathbf{J} = \sum_{i=1}^{n_e} \frac{\partial N_i}{\partial \boldsymbol{\xi}} \mathbf{x}_{ei}^T \tag{2.62}$$

where $n_e$ is the number of nodes for the element and $\mathbf{x}_{ei} = [x_{ei} \quad y_{ei} \quad z_{ei}]^T$ is the nodal position vector of node $i$. Then, from equations (2.60) and (2.61):

$$\frac{\partial N_i}{\partial \boldsymbol{\xi}} = \mathbf{J} \frac{\partial N_i}{\partial \mathbf{x}} \Rightarrow \frac{\partial N_i}{\partial \mathbf{x}} = \mathbf{J}^{-1} \frac{\partial N_i}{\partial \boldsymbol{\xi}} \tag{2.63}$$

The strain-displacement matrix $\mathbf{B} = \boldsymbol{\partial}\mathbf{N}$ (equation (2.39)) which includes the terms $\frac{\partial N_i}{\partial \mathbf{x}}$ with $N_i$ being functions of natural coordinates, can now be derived by substituting these terms using equation (2.63). This is illustrated in the next section, for an isoparametric quadrilateral element.

Finally, the element stiffness matrix is provided within isoparametric formulation, by the following relation:

$$\mathbf{k} = \int_{V_e} \mathbf{B}^T\mathbf{C}\mathbf{B}dV \Rightarrow \mathbf{k} = \int_{-1}^{1}\int_{-1}^{1}\int_{-1}^{1} \mathbf{B}^T\mathbf{C}\mathbf{B}\det(\mathbf{J})d\xi d\eta d\zeta \tag{2.64}$$

For the calculation of the element stiffness matrix, numerical integration is used as it is presented in section 2.3.6.

### 2.3.5 Quadrilateral isoparametric element

The representation of a plane, quadrilateral isoparametric element in cartesian and natural coordinates is shown in figure 2.2. The natural coordinates $\xi$, $\eta$ in the cartesian system, may not be orthogonal or parallel to cartesian coordinates $x$, $y$. The centre of the element is assigned natural coordinates ($\xi$, $\eta$)=(0,0) and may not be the centroid of the element. Values of the natural coordinates on the edges of the element vary between $-1$ to 1 irrespective of the shape, the size or the orientation of the element in regard with the cartesian system. In addition, the displacements $u$, $v$ are directed parallel to cartesian coordinates $x$, $y$. The mapping of the element on the natural system results in the square shaped element which is shown in figure 2.2.

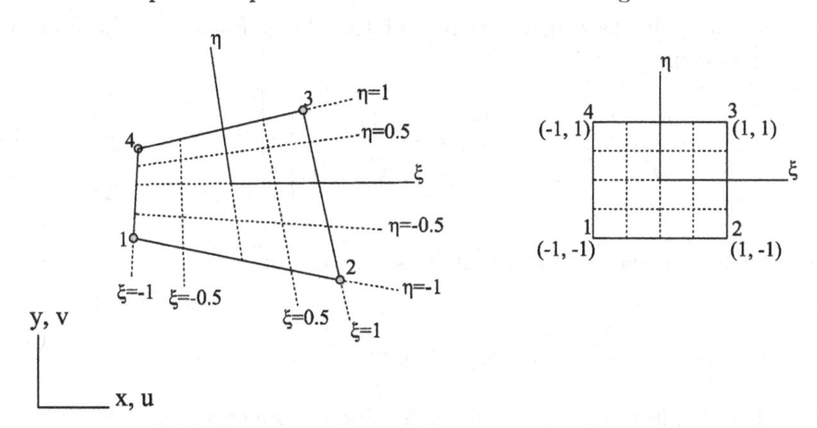


**FIGURE 2.2**: Quadrilateral element in cartesian (left) and in natural coordinate system (right).

The nodal displacement vector $\mathbf{u}_e$, the nodal position vector $\mathbf{x}_e$ and the shape function matrix $\mathbf{N}$ which appear in equation (2.59), are given below for a quadrilateral element with four nodes and two displacement degrees of freedom per node:

$$
\begin{aligned}
\mathbf{u}_e &= \begin{bmatrix} u_1 & v_1 & u_2 & v_2 & u_3 & v_3 & u_4 & v_4 \end{bmatrix}^T \\
\mathbf{x}_e &= \begin{bmatrix} x_1 & y_1 & x_2 & y_2 & x_3 & y_3 & x_4 & y_4 \end{bmatrix}^T \\
\mathbf{N} &= \begin{bmatrix} N_1 & 0 & N_2 & 0 & N_3 & 0 & N_4 & 0 \\ 0 & N_1 & 0 & N_2 & 0 & N_3 & 0 & N_4 \end{bmatrix}
\end{aligned}
\tag{2.65}
$$

The shape functions are provided for this element in terms of the natural coordinates $\xi$, $\eta$:

$$
\begin{aligned}
N_1 &= \frac{1}{4}(1-\xi)(1-\eta) \\
N_2 &= \frac{1}{4}(1+\xi)(1-\eta) \\
N_3 &= \frac{1}{4}(1+\xi)(1+\eta) \\
N_4 &= \frac{1}{4}(1-\xi)(1+\eta)
\end{aligned}
\tag{2.66}
$$

Then, equation (2.62) is rewritten in a matrix notation and the Jacobian $\mathbf{J}$ is calculated for the isoparametric element:

$$
\mathbf{J} = \begin{bmatrix} \frac{\partial N_1}{\partial \xi} & \frac{\partial N_2}{\partial \xi} & \frac{\partial N_3}{\partial \xi} & \frac{\partial N_4}{\partial \xi} \\ \frac{\partial N_1}{\partial \eta} & \frac{\partial N_2}{\partial \eta} & \frac{\partial N_3}{\partial \eta} & \frac{\partial N_4}{\partial \eta} \end{bmatrix} \begin{bmatrix} x_1 & y_1 \\ x_2 & y_2 \\ x_3 & y_3 \\ x_4 & y_4 \end{bmatrix}
\tag{2.67}
$$

By calculating the partial derivatives of the shape functions, the Jacobian matrix is determined:

$$
\mathbf{J} = \frac{1}{4} \begin{bmatrix} -(1-\eta) & (1-\eta) & (1+\eta) & -(1+\eta) \\ -(1-\xi) & -(1+\xi) & (1+\xi) & (1-\xi) \end{bmatrix} \begin{bmatrix} x_1 & y_1 \\ x_2 & y_2 \\ x_3 & y_3 \\ x_4 & y_4 \end{bmatrix} = \begin{bmatrix} J_{11} & J_{12} \\ J_{21} & J_{22} \end{bmatrix}
\tag{2.68}
$$

The inverse $\mathbf{D}$ of the Jacobian matrix is then calculated:

$$
\mathbf{D} = \mathbf{J}^{-1} = \frac{1}{\det(\mathbf{J})} \begin{bmatrix} J_{22} & -J_{12} \\ -J_{21} & J_{11} \end{bmatrix}
\tag{2.69}
$$

with $\det(\mathbf{J})$ being the determinant of the Jacobian matrix.

To build the strain-displacement matrix $\mathbf{B}$ for the isoparametric quadrilateral element, equation (2.13) is properly reformulated:

$$
\boldsymbol{\varepsilon} = \begin{bmatrix} \varepsilon_{xx} \\ \varepsilon_{yy} \\ \gamma_{xy} \end{bmatrix} = \begin{bmatrix} \frac{\partial}{\partial x} & 0 \\ 0 & \frac{\partial}{\partial y} \\ \frac{\partial}{\partial y} & \frac{\partial}{\partial x} \end{bmatrix} \begin{bmatrix} u \\ v \end{bmatrix} \quad \text{with} \quad \boldsymbol{\partial} = \begin{bmatrix} \frac{\partial}{\partial x} & 0 \\ 0 & \frac{\partial}{\partial y} \\ \frac{\partial}{\partial y} & \frac{\partial}{\partial x} \end{bmatrix}
\tag{2.70}
$$

where the terms $\varepsilon_{11}$, $\varepsilon_{22}$ and $\gamma_{12}$ of (2.13) are written here as $\varepsilon_{xx}$, $\varepsilon_{yy}$ and $\gamma_{xy}$. Then, equality $\mathbf{B} = \boldsymbol{\partial}\mathbf{N}$ presented in (2.39) is used, considering matrix $\mathbf{N}$ from (2.65) and $\boldsymbol{\partial}$ from (2.70):

$$\mathbf{B} = \boldsymbol{\partial}\mathbf{N} = \begin{bmatrix} \frac{\partial}{\partial x} & 0 \\ 0 & \frac{\partial}{\partial y} \\ \frac{\partial}{\partial y} & \frac{\partial}{\partial x} \end{bmatrix} \begin{bmatrix} N_1 & 0 & N_2 & 0 & N_3 & 0 & N_4 & 0 \\ 0 & N_1 & 0 & N_2 & 0 & N_3 & 0 & N_4 \end{bmatrix}$$

$$= \begin{bmatrix} \frac{\partial N_1}{\partial x} & 0 & \frac{\partial N_2}{\partial x} & 0 & \frac{\partial N_3}{\partial x} & 0 & \frac{\partial N_4}{\partial x} & 0 \\ 0 & \frac{\partial N_1}{\partial y} & 0 & \frac{\partial N_2}{\partial y} & 0 & \frac{\partial N_3}{\partial y} & 0 & \frac{\partial N_4}{\partial y} \\ \frac{\partial N_1}{\partial y} & \frac{\partial N_1}{\partial x} & \frac{\partial N_2}{\partial y} & \frac{\partial N_2}{\partial x} & \frac{\partial N_3}{\partial y} & \frac{\partial N_3}{\partial x} & \frac{\partial N_4}{\partial y} & \frac{\partial N_4}{\partial x} \end{bmatrix} \tag{2.71}$$

Every component of (2.71) is calculated using (2.63):

$$\frac{\partial N_i}{\partial \mathbf{x}} = \mathbf{J}^{-1}\frac{\partial N_i}{\partial \boldsymbol{\xi}} \Rightarrow \begin{bmatrix} \frac{\partial N_i}{\partial x} \\ \frac{\partial N_i}{\partial y} \end{bmatrix} = \begin{bmatrix} D_{11} & D_{12} \\ D_{21} & D_{22} \end{bmatrix} \begin{bmatrix} \frac{\partial N_i}{\partial \xi} \\ \frac{\partial N_i}{\partial \eta} \end{bmatrix} \tag{2.72}$$

where terms $D_{ij}$ of the inverse of the Jacobian matrix have been calculated in (2.68) and (2.69). By conducting the described calculations, the strain-displacement matrix is rewritten as follows:

$$\mathbf{B} = \begin{bmatrix} D_{11} & D_{12} & 0 & 0 \\ 0 & 0 & D_{21} & D_{22} \\ D_{21} & D_{22} & D_{11} & D_{12} \end{bmatrix} \begin{bmatrix} \frac{\partial N_1}{\partial \xi} & 0 & \frac{\partial N_2}{\partial \xi} & 0 & \frac{\partial N_3}{\partial \xi} & 0 & \frac{\partial N_4}{\partial \xi} & 0 \\ \frac{\partial N_1}{\partial \eta} & 0 & \frac{\partial N_2}{\partial \eta} & 0 & \frac{\partial N_3}{\partial \eta} & 0 & \frac{\partial N_4}{\partial \eta} & 0 \\ 0 & \frac{\partial N_1}{\partial \xi} & 0 & \frac{\partial N_2}{\partial \xi} & 0 & \frac{\partial N_3}{\partial \xi} & 0 & \frac{\partial N_4}{\partial \xi} \\ 0 & \frac{\partial N_1}{\partial \eta} & 0 & \frac{\partial N_2}{\partial \eta} & 0 & \frac{\partial N_3}{\partial \eta} & 0 & \frac{\partial N_4}{\partial \eta} \end{bmatrix}$$

$$= \begin{bmatrix} D_{11} & D_{12} & 0 & 0 \\ 0 & 0 & D_{21} & D_{22} \\ D_{21} & D_{22} & D_{11} & D_{12} \end{bmatrix} \frac{1}{4} \begin{bmatrix} -(1-\eta) & 0 & (1-\eta) & 0 & (1+\eta) & 0 & -(1+\eta) & 0 \\ -(1-\xi) & 0 & -(1+\xi) & 0 & (1+\xi) & 0 & (1-\xi) & 0 \\ 0 & -(1-\eta) & 0 & (1-\eta) & 0 & (1+\eta) & 0 & -(1+\eta) \\ 0 & -(1-\xi) & 0 & -(1+\xi) & 0 & (1+\xi) & 0 & (1-\xi) \end{bmatrix} \tag{2.73}$$

where the partial derivatives of the shape functions over natural coordinates given in (2.68) have been used. Finally, the stiffness matrix of the isoparametric quadrilateral element is given below:

$$\mathbf{k} = \int_{V_e} \mathbf{B}^T\mathbf{C}\mathbf{B}dV \Rightarrow \mathbf{k} = \int_{-1}^{1}\int_{-1}^{1} \mathbf{B}^T\mathbf{C}\mathbf{B}t\det(\mathbf{J})d\xi d\eta \tag{2.74}$$

where $t$ is the thickness of the element and $dV = tdxdy = t\det(\mathbf{J})d\xi d\eta$ [40]. More formulations for different types of elements such as trusses, beams, plates, shells and hexagonal elements can be found in literature [50]. For each element type, appropriate shape functions are chosen and the process presented in this section is repeated.

### 2.3.6 Numerical integration

As mentioned in previous sections, the analytical calculation of the integrals which appear in the element stiffness matrix, in the external and in the internal force vectors is quite demanding, indicating that a numerical integration technique should be adopted. The general concept which is applied for the numerical integration (or *quadrature*), is to calculate a definite integral by developing the summation of the integrand evaluated at specific, known points, multiplied by also known weights.

The method which is widely adopted for numerical integration and is used in this book is the *Gauss integration* rule. A different numerical integration approach may also be chosen, such as the Newton-Cotes or Simpson rule. Gauss integration is of the highest accuracy for continuum elements which are elaborated in this book. For different element types such as beam or plate elements, which require integration along the thickness, the Simpson or Newton-Cotes rule may be more accurate [54]. The Gauss integration rule is briefly presented below.

For an one-dimension case, integration of function $f(x)$ from $x_1$ to $x_2$ is substituted by integration of function $\phi(\xi)$ from $-1$ to 1, considering that a mapping between coordinate $x$ and $\xi$ appears and that function $\phi(\xi)$ incorporates the Jacobian $J = \frac{\partial x}{\partial \xi}$ of this transformation. This could be for instance, the case of integration for an one-dimensional element from the cartesian to natural coordinate system as shown below:

$$\int_{x_1}^{x_2} f(x)dx \rightarrow \int_{-1}^{1} \phi(\xi)d\xi \tag{2.75}$$

Then, function $\phi$ is evaluated in known discrete values $\xi_i$ and multiplied by also known weighting factors $\alpha_i$. The terms which are obtained are added and the integral is calculated:

$$\int_{-1}^{1} \phi(\xi)d\xi = \alpha_1\phi(\xi_1) + \alpha_2\phi(\xi_2) + \cdots + \alpha_n\phi(\xi_n) \tag{2.76}$$

Each discrete value $\xi_i$ (sample point) represents an integration point or *Gauss* point and the number $n$ of Gauss points considered in equation (2.76) defines the *order of integration*. At each Gauss point, a weight factor $\alpha_i$ is assigned.

The physical interpretation of numerical integration can be derived when one value $\xi = 0$ at the middle of the interval between $-1$ and 1 is used to evaluate the function $\phi(\xi)$. The integral in this case is approximated as the rectangular area under $\phi$, by multiplying $\phi(\xi = 0)$ and the length of the interval which is equal to 2, as shown in relation (2.77). When the function $\phi$ is a straight line, the numerical integration described here is accurate.

$$\int_{-1}^{1} \phi(\xi)d\xi = 2\phi(\xi = 0) \tag{2.77}$$

The precision of the Gauss integration is related to the order of integration and the order of the polynomial integrand. Gauss quadrature of order $n$ is generally accurate, when the integrand is a polynomial of degree $2n - 1$. If the integrand is not a polynomial, Gauss quadrature is not exact, but the accuracy is improved when more Gauss points are considered. In table 2.1 Gauss points locations and corresponding weights are given, for integration between $-1$ and 1.

For the two-dimensional case, Gauss integration is implemented according to the following relation:

$$\int_{-1}^{1}\int_{-1}^{1} \phi(\xi,\eta)d\xi d\eta = \int_{-1}^{1} \sum_{i} \alpha_i \phi(\xi_i,\eta)d\eta = \sum_{i}\sum_{j} \alpha_i \alpha_j \phi(\xi_i,\eta_j) \quad (2.78)$$

In figure 2.3 the Gauss points locations for a quadrilateral plane element are given for the case of second and third-order of integration along each direction.

Finally, Gauss integration is applied for the calculation of the stiffness matrix and external force vector within linear finite element analysis as well as for the calculation of the internal force vector within non-linear analysis.

**TABLE 2.1:** Gauss points locations and corresponding weights for integration in the interval between $\xi = -1$ and $\xi = 1$.

| Order of integration $n$ | Gauss points locations $\xi_i$ | Corresponding weights $\alpha_i$ |
|---|---|---|
| 1 | 0 | 2 |
| 2 | ±0.57735 | 1 |
| 3 | ±0.77459 | 0.55555 |
| | 0 | 0.88888 |

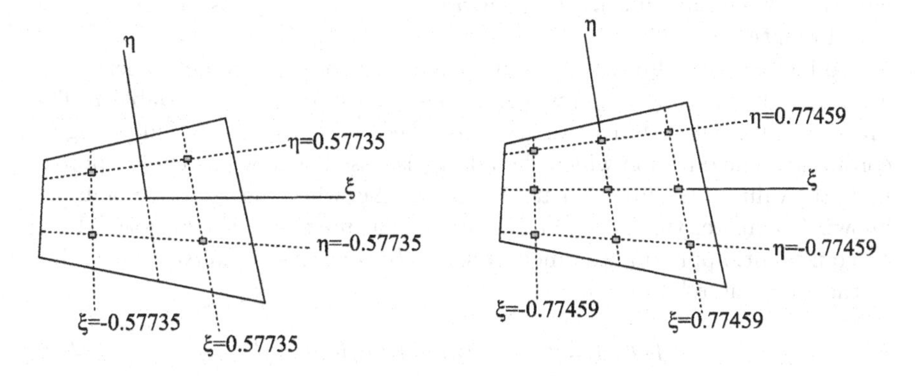


**FIGURE 2.3**: Gauss points locations for second-order integration along each direction (left) and for third-order integration along each direction (right).

Towards implementation within programming codes, the Gauss point locations and weights are identified for every Gauss point and the matrices or vectors corresponding to each Gauss point are built. Then, summation into overall matrices or vectors takes place using the concept of the assembly which was discussed in section 2.3.2.

---

## 2.4 Non-linear finite element analysis

The presented description of the finite element method can properly be extended to address non-linear problems arising on composite heterogeneous structures. Among the types of non-linear analysis which are presented at the beginning of chapter 2, emphasis in this book is given in the cases of large displacement analysis, material non-linearity and boundary non-linearity involving contact mechanics applications. The first and second types are discussed in sections 2.4.4 and 2.4.5 of this chapter while the third is presented in chapter 4.

### 2.4.1 The general framework of non-linear analysis

A non-linear problem arises in mechanics when the governing equilibrium equations contain components which are non-linear. Such problems appear when a non-linear stress-strain material law is adopted to provide the structural response or in case a structure undergoes large displacements. The main action which is needed in those cases, is the implementation of a numerical procedure allowing for the solution of the non-linear governing equations, within an iterative scheme. The most widely used approach, relies on the transformation of the initial governing equations into equivalent, *approximate* equations which are linear. Thus, a *linearization* process is needed for this transformation.

To highlight the linearization process, an example is presented here using a scalar valued function $f(x)$ which is non-linear in respect to variable $x$. This function and its first derivative are considered to be continuous indicating $C^1$ continuity. The question which should be discussed, is how one could approximate the value of the function at a point $x_0+\Delta u$, when the value at point $x_0$ is known. To answer this question, the non-linear function $f$ is transformed into an equivalent, approximate function using *Taylor series expansion*, according to the following relation:

$$f(x_0 + \Delta u) = f(x_0) + f'(x_0)\Delta u + R \tag{2.79}$$

The term $R$ includes additional components which are functions of the increment $\Delta u$. When $\Delta u$ is sufficiently small, the $R$ components are also small and $R$ is neglected. In this case, the right-hand part of the above equation is linear, since $f(x_0)$ and $f'(x_0)$ represent discrete values, indicating that all components are linear.

Following this procedure, it can be stated that the non-linear function $f$ is linearized at point $x_0$ by considering the transformation of $f$ at $x_0 + \Delta u$ for a small increment $\Delta u$, as given by the right-hand part of equation (2.79). It is noted that for the implementation of the linearization, the discrete value of function $f(x_0)$ at $x_0$ as well as the discrete value of the first derivative $f'(x_0)$ should be known. This is an indication that an incremental process is needed, which uses the known solution calculated in a previous step.

The graphical representation of equation (2.79) is given in figure 2.4. According to this figure, the tangent line to the curve $f(x)$ at point $x_0$ is used to define the linear part of $f(x)$ at $x_0$. Since $f(x_0)$ and $f'(x_0)$ are known, this linear part is a function of the increment $\Delta u$.

For the more general case of N-dimensional space, linearization of the non-linear function $\mathbf{G}$ at point $\mathbf{x}_0$ can be implemented using the above descriptions, as follows:

$$\mathbf{G}(\mathbf{x}_0 + \Delta\mathbf{u}) = \mathbf{G}(\mathbf{x}_0) + \left(\frac{\partial \mathbf{G}}{\partial \mathbf{x}}\right)_{\mathbf{x}=\mathbf{x}_0} \Delta\mathbf{u} + \mathbf{R} \tag{2.80}$$

Like before, for small increment $\Delta\mathbf{u}$ the higher-order terms $\mathbf{R}$ are neglected, and equation (2.80) provides the linear part of $\mathbf{G}$ at $\mathbf{x}_0$. Equation (2.80) can then be rewritten in the following form:

$$\mathbf{G}(\mathbf{x}_0 + \Delta\mathbf{u}) = \mathbf{G}(\mathbf{x}_0) + \Delta\mathbf{G} \tag{2.81}$$

where

$$\Delta\mathbf{G} = \left(\frac{\partial \mathbf{G}}{\partial \mathbf{x}}\right)_{\mathbf{x}=\mathbf{x}_0} \Delta\mathbf{u} \tag{2.82}$$

According to this formulation, the non-linear function $\mathbf{G}$ is linearized at $\mathbf{x}_0$, after is decomposed to the known $\mathbf{G}(\mathbf{x}_0)$ and the unknown $\Delta\mathbf{G}$. Linearization then takes place for $\Delta\mathbf{G}$ using equation (2.82). The presented scheme is widely

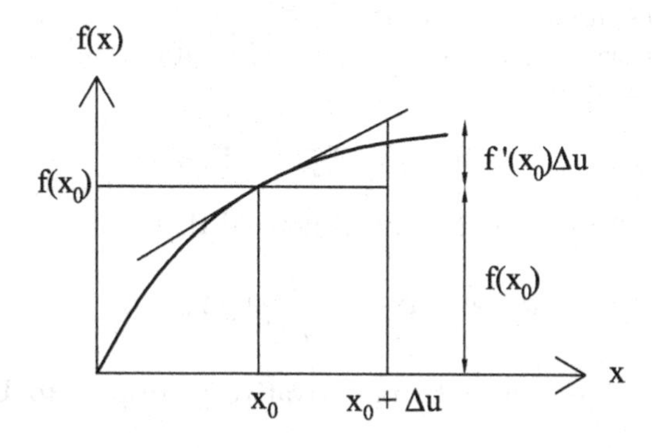


**FIGURE 2.4**: Linearization of the function $f$ at $x_0$.

used to linearize non-linear terms which appear in non-linear finite element analysis, as it will be shown in the next sections of this chapter.

### 2.4.2 Newton-Raphson incremental-iterative process

The method presented in section 2.4.1 for approximating non-linear functions can be implemented for the solution of a non-linear finite element analysis problem, within an incremental-iterative process. By neglecting inertial effects, equilibrium is expressed in a general form by the following relation, emanating from equation (2.46):

$$\mathbf{F}_{ext}^{t+\Delta t} - \mathbf{F}_{int}^{t+\Delta t} = 0 \Rightarrow \mathbf{G}(\mathbf{U}^{t+\Delta t}) = \mathbf{0} \tag{2.83}$$

Equation (2.83) defines the equilibrium of a structure, as the balance between the external and the internal forces at time $t+\Delta t$. Next, within an incremental framework, it is considered that the solution is known at time $t$ and a solution at $t+\Delta t$ needs to be found, with $\Delta t$ being a suitably chosen time or loading increment. As noticed previously in this book, for static problems without time effects (e.g. without creep), time $t$ is only used to assign an order in the sequence of events, within an incremental loading process provided by load steps.

Within these descriptions, the nodal point displacement vector $\mathbf{U}$ is decomposed as follows:

$$\mathbf{U}^{t+\Delta t} = \mathbf{U}^t + \Delta\mathbf{U} \tag{2.84}$$

where $\mathbf{U}^t$ is the known, nodal displacement vector at time $t$ and $\Delta\mathbf{U}$ the unknown, incremental nodal displacement vector. Then, the non-linear equilibrium equation (2.83) can be linearized as shown in section 2.4.1:

$$\mathbf{G}(\mathbf{U}^t + \Delta\mathbf{U}) = \mathbf{0} \Rightarrow \mathbf{G}(\mathbf{U}^t) + \frac{\partial \mathbf{G}(\mathbf{U}^t)}{\partial \mathbf{U}^t}\Delta\mathbf{U} = \mathbf{0} \tag{2.85}$$

The vector $\mathbf{G}(\mathbf{U}^t)$ is defined in equation (2.86) as the difference between the external force vector $\mathbf{F}_{ext}^{t+\Delta t}$ at time $t+\Delta t$ and the internal force vector $\mathbf{F}_{int}^{t}$ at time $t$. Within iterative analysis, $\mathbf{F}_{ext}^{t+\Delta t}$ is considered independent of the deformation history and hence constant, while iterations are conducted for the internal force vector.

$$\mathbf{G}(\mathbf{U}^t) = \mathbf{F}_{ext}^{t+\Delta t} - \mathbf{F}_{int}^{t} \tag{2.86}$$

By substituting (2.86) into (2.85), it is derived that

$$\mathbf{F}_{ext}^{t+\Delta t} - \mathbf{F}_{int}^{t} - \frac{\partial \mathbf{F}_{int}^{t}}{\partial \mathbf{U}^t}\Delta\mathbf{U} = \mathbf{0} \tag{2.87}$$

where since $\mathbf{F}_{ext}^{t+\Delta t}$ is constant, its derivative in respect to $\mathbf{U}^t$ is zero. By assigning:

$$\mathbf{K}_T = \frac{\partial \mathbf{F}_{int}^{t}}{\partial \mathbf{U}^t} \tag{2.88}$$

equation (2.87) is written as follows:

$$\mathbf{K}_T \Delta \mathbf{U} = \mathbf{F}_{ext}^{t+\Delta t} - \mathbf{F}_{int}^{t} \tag{2.89}$$

The matrix $\mathbf{K}_T$ is known as the *tangent stiffness matrix* and according to equation (2.88) is provided by the derivative of the internal force vector $\mathbf{F}_{int}^{t}$ given in (2.49), in respect to nodal displacements.

Using equation (2.89), the *out of balance vector* or *residual vector* $\mathbf{G}$ is defined as shown below:

$$\mathbf{G}(\mathbf{U}^{t+\Delta t}) = \mathbf{F}_{ext}^{t+\Delta t} - \mathbf{F}_{int}^{t} - \mathbf{K}_T \Delta \mathbf{U} \tag{2.90}$$

As it will be shown, goal of this numerical procedure is to determine the solution for which the residual vector becomes sufficiently small.

The presented description becomes more specific, when iterations are introduced. As given above, the solution at time $t$ is known indicating that the internal force vector $\mathbf{F}_{int}^{t}$ and the nodal displacement vector $\mathbf{U}^{t}$ are also known. Next, the nodal displacement vector $\mathbf{U}^{t+\Delta t}$ will be defined by demanding equilibrium at time $t + \Delta t$. This will occur by solving the following equations, within iterations $i$:

$$\mathbf{K}_{T,i} \Delta \mathbf{U}_{i+1} = \mathbf{F}_{ext}^{t+\Delta t} - \mathbf{F}_{int,i}^{t+\Delta t} \tag{2.91}$$

$$\mathbf{U}_{i+1}^{t+\Delta t} = \mathbf{U}_{i}^{t+\Delta t} + \Delta \mathbf{U}_{i+1} \tag{2.92}$$

This numerical approach is known as *Newton-Raphson* process and is schematically shown in figure 2.5 for an one-dimensional case. Equilibrium at time $t$ on the curve shown in the figure has been achieved, indicating that the internal force $\mathrm{F}_{\mathrm{int}}^{\mathrm{t}} = \mathrm{F}_{\mathrm{int,0}}^{\mathrm{t+\Delta t}}$ and the nodal displacement $\mathrm{U}^{\mathrm{t}} = \mathrm{U}_{0}^{\mathrm{t+\Delta t}}$ are known. The next equilibrium point which corresponds to external load $\mathrm{F}_{\mathrm{ext}}^{\mathrm{t+\Delta t}}$ will then be found, using iterations:

$$\begin{aligned}
\mathrm{K}_{\mathrm{T,0}} \Delta \mathrm{U}_1 &= \mathrm{F}_{\mathrm{ext}}^{\mathrm{t+\Delta t}} - \mathrm{F}_{\mathrm{int,0}}^{\mathrm{t+\Delta t}}, \quad \mathrm{U}_1^{\mathrm{t+\Delta t}} = \mathrm{U}_0^{\mathrm{t+\Delta t}} + \Delta \mathrm{U}_1 \\
\mathrm{K}_{\mathrm{T,1}} \Delta \mathrm{U}_2 &= \mathrm{F}_{\mathrm{ext}}^{\mathrm{t+\Delta t}} - \mathrm{F}_{\mathrm{int,1}}^{\mathrm{t+\Delta t}}, \quad \mathrm{U}_2^{\mathrm{t+\Delta t}} = \mathrm{U}_1^{\mathrm{t+\Delta t}} + \Delta \mathrm{U}_2 \\
\mathrm{K}_{\mathrm{T,2}} \Delta \mathrm{U}_3 &= \mathrm{F}_{\mathrm{ext}}^{\mathrm{t+\Delta t}} - \mathrm{F}_{\mathrm{int,2}}^{\mathrm{t+\Delta t}}, \quad \mathrm{U}_3^{\mathrm{t+\Delta t}} = \mathrm{U}_2^{\mathrm{t+\Delta t}} + \Delta \mathrm{U}_3
\end{aligned} \tag{2.93}$$

According to figure 2.5 iterations are held until an approximate solution, which is close enough to the real solution, is obtained. Thus, a convergence criterion is needed for the termination of the iterations, once an adequate solution close enough to the bold line of figure 2.5 is found. This convergence criterion naturally arises according to figure 2.5, when the residual force defined in (2.90) is sufficiently small and the nodal displacement increment is also small.

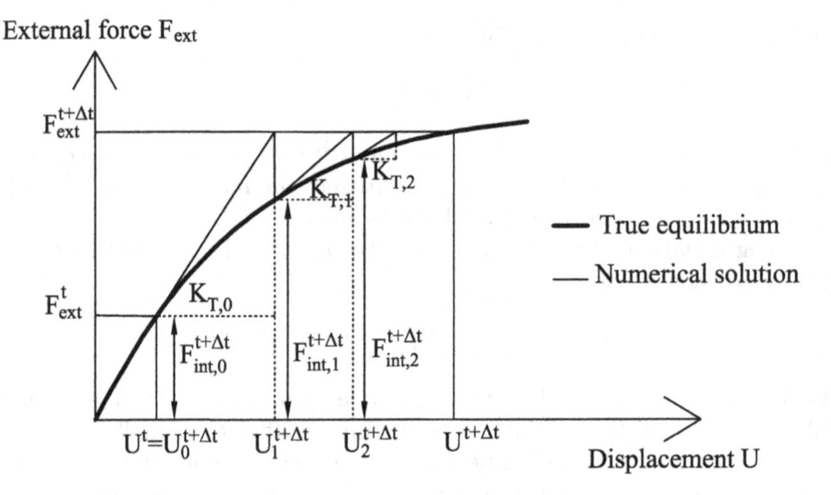


**FIGURE 2.5**: Newton-Raphson incremental process.

For the general case of higher dimensions, convergence of Newton-Raphson process is provided by the following relations:

$$\left\|\mathbf{G}(\mathbf{U}_{i+1}^{t+\Delta t})\right\| \leq \text{tolerance}, \quad \text{or} \quad \begin{aligned} \left\|\mathbf{F}_{ext}^{t+\Delta t} - \mathbf{F}_{int,i+1}^{t+\Delta t}\right\| &\leq \text{tolerance} \\ \left\|\mathbf{U}_{i+1}^{t+\Delta t} - \mathbf{U}_{i}^{t+\Delta t}\right\| &\leq \text{tolerance} \end{aligned} \tag{2.94}$$

Some significant points, which are related to the implementation of the Newton-Raphson numerical scheme, are discussed below.

**Convergence rate:** It defines how fast the convergence criterion given in (2.94) is achieved. The convergence rate is *quadratic* when for the error $e_i$ and $e_{i+1}$ emanating from non-balancing residuals after two subsequent iterations, the following inequality applies:

$$\frac{e_{i+1}}{e_i^2} \leq \mathrm{c} \tag{2.95}$$

with $c$ being a constant. Quadratic convergence often arises near the solution point, indicating its local character. This convergence rate is an advantage of the presented numerical scheme, since a relatively small number of iterations is needed until a solution within the given accuracy is achieved.

**Incremental-iterative scheme:** Quite often, equation (2.91) is slightly modified in order to define load increments, according to the following form:

$$\mathbf{K}_{T,i}\Delta\mathbf{U}_{i+1} = (\lambda\mathbf{F}_{ext}^{t} + \mathbf{F}_{ext}^{t}) - \mathbf{F}_{int,i}^{t+\Delta t} \tag{2.96}$$

The scalar $\lambda$ is used to define a fraction of the overall external load and hence, to introduce load increments. A load increment $\lambda \mathrm{F}_{\mathrm{ext}}^{\mathrm{t}}$ appears between $\mathrm{F}_{\mathrm{ext}}^{\mathrm{t}}$ and $\mathrm{F}_{\mathrm{ext}}^{\mathrm{t}+\Delta\mathrm{t}}$ in figure 2.5, within which iterations are held in the framework of the Newton-Raphson process. The benefit of introducing increments instead of using the total external load, is that with increments it becomes more possible for the algorithm to converge faster to a solution. It is noted that in modern commercial finite element packages, proper algorithms that automatically adjust $\lambda$ are implemented, targeting in improving the computational time. For completeness, it must also be noted that the parameter $\lambda$ may involve any action imposed on the structure, like prescribed displacements at the boundaries. Mathematically, one has an one-dimensional parametrization and the solution path along this parametrization is calculated, including any possible non-linear effect that may arise (bifurcation of solutions, snap-through, etc.).

**Internal force vector:** This is a very important parameter because it influences the accuracy of the method. An incorrect calculation of $\mathbf{F}_{int,i}^{t+\Delta t}$ may still result in convergence of the scheme, however, the corresponding solution may be mistaken. It is noticed that according to equation (2.49) integration of the element stresses is needed for the calculation of the internal force vector, indicating that a prediction for the stress should be used. Therefore, a constitutive stress versus strain description is needed to provide the stress at every iteration of the method.

**Tangent stiffness matrix:** This is another key parameter for the method, since it significantly influences the convergence rate. An appropriate calculation of the tangent stiffness may result in faster convergence. However, even a less accurate tangent stiffness may result in a correct solution with a slower convergence. Within the presented scheme, it is calculated at every iteration and for this reason the method is known as the *Full Newton-Raphson.* Although the calculation of the tangent stiffness and its inverse (which is needed for the solution of (2.91)) at every iteration may increase the convergence rate, it is a quite time consuming process. Hence, alternative approaches can also be used, adopting a different concept for the calculation of the tangent stiffness. Namely, these are the *Modified*, the *Initial Stiffness* and the *Quasi-* Newton-Raphson schemes.

The Modified Newton-Raphson states that the tangent stiffness matrix is built only once per load increment and thus, it is kept constant during each increment. The Initial Stiffness scheme suggests that the initial stiffness matrix, build for linear analysis, is used for every load increment. The Quasi- Newton-Raphson relies on the calculation of the tangential stiffness matrix using a secant scheme. It is noticed that adopting one of these alternative approaches may result in the reduction of the convergence rate, indicating that the quadratic convergence which characterizes the Full Newton-Raphson may be lost and more (or many more) iterations within the increment shall be needed. On the other hand, their usage could save some computational time when mild non-linearities are studied.

**Range of applicability:** The Newton-Raphson method provides solution to a wide range of non-linear structural problems, including material, geometric and boundary non-linearities and thus, it is the most commonly used approach in finite element analysis. There are, however, specific cases which cannot be solved using the Newton-Raphson method indicating that advanced numerical schemes are required. One of this case arises for the *snap-back* response shown in figure 2.6. This response appears for geometric non-linearity of thin shells as well as for strain-softening relations. An alternative method called *path-following* or *arc-length*, is suggested for this case [232, 54]. The main idea within the arc-length method is that the load increment is considered as an additional unknown of the problem. Another advanced technique which is used to increase the limited radius of convergence arising from the Newton-Raphson approach, is the *line search* [232, 54].

**Softening:** Quite often the structural behaviour of composite materials is characterized by a softening response, as shown in figure 2.6. When the Newton-Raphson scheme is used, which relies on the application of external forces and for this reason is known as a *force control* scheme, analysis shall not converge to a solution for this softening response. An alternative approach in this case, is to use a *displacement control* scheme within Newton-Raphson, where prescribed nodal displacements instead of forces are applied to enforce loading on the structure. A more detailed description of the displacement control scheme is provided in the next section of this chapter. It should be noted that softening is accompanied by localization phenomena. As it is discussed in the following chapters, advanced numerical methods are required in this case, to depict the mechanical behaviour of the structures under investigation.

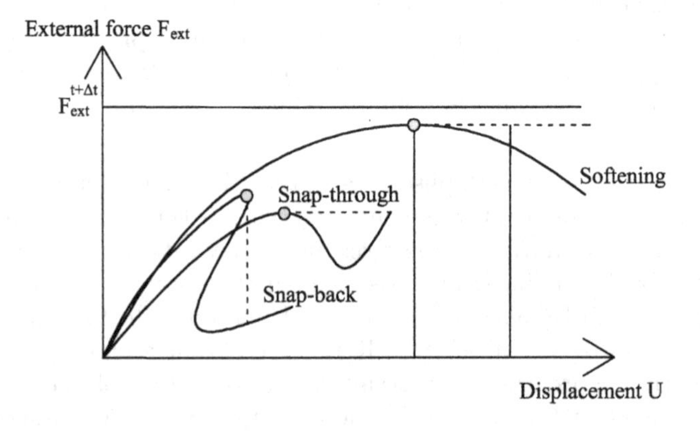


**FIGURE 2.6**: Different types of non-linear response.

### 2.4.3 Newton-Raphson under displacement control

The main concept of the displacement control, is that the loading of the structural system consists of prescribed nodal displacements which are applied incrementally. The external forces in this case are zero and the reaction forces are developed at the nodes with prescribed displacements.

Some significant benefits of using displacement control instead of force control can be identified. First, the tangent stiffness matrix is better conditioned under displacement control than under force control, resulting in a faster convergence. In addition, at the peak (limit) point of the softening (or snap-through) force-displacement graph shown in figure 2.6 for an one-dimensional problem, the tangent is parallel to the horizontal axis and the $\mathrm{F}_{\mathrm{ext}}^{\mathrm{t+\Delta t}}$ axis. Thus, the tangent at this point cannot intersect the $\mathrm{F}_{\mathrm{ext}}^{\mathrm{t+\Delta t}}$ axis, which is required for the load control process shown in figure 2.5. This results in convergence difficulties after the peak, when softening behaviour is studied using load control. It is noted that at the peak of the softening curve shown in figure 2.6, the tangent stiffness matrix becomes singular.

When displacement control is used indicating that prescribed displacements are applied, the vertical line from the loading prescribed displacement on the horizontal axis to the force-displacement diagram, should be crossed by the tangent line at every previous point of the diagram. This is feasible for the softening case, as shown in figure 2.6. However, it is still not feasible when the snap-back response shown in figure 2.6 is studied. For this case the arc-length method should be adopted.

To implement displacement control within non-linear finite element analysis, equation (2.91) is rewritten, after partitioning the tangent stiffness matrix, the nodal displacement vector as well as the internal and external force vectors into parts corresponding to free, "*f*" and supported, "*s*" degrees of freedom, the latter for nodes with prescribed displacements:

$$\begin{bmatrix} \mathbf{K}_{ff} & \mathbf{K}_{fs} \\ \mathbf{K}_{sf} & \mathbf{K}_{ss} \end{bmatrix} \begin{bmatrix} \Delta\mathbf{U}_f \\ \Delta\mathbf{U}_s \end{bmatrix} = - \begin{bmatrix} \mathbf{F}_{int,f} \\ \mathbf{F}_{int,s} \end{bmatrix} \tag{2.97}$$

In the above equation no external forces have been considered indicating that $\mathbf{F}_{ext} = 0$. By developing (2.97), the following relations are derived:

$$\begin{aligned} \mathbf{K}_{ff}\Delta\mathbf{U}_f + \mathbf{K}_{fs}\Delta\mathbf{U}_s &= -\,\mathbf{F}_{int,f} \\ \mathbf{K}_{sf}\Delta\mathbf{U}_f + \mathbf{K}_{ss}\Delta\mathbf{U}_s &= -\,\mathbf{F}_{int,s} \end{aligned} \tag{2.98}$$

The first of equations (2.98) is used for the calculation of the incremental nodal displacements, as follows:

$$\Delta\mathbf{U}_f = -\mathbf{K}_{ff}^{-1}(\mathbf{K}_{fs}\Delta\mathbf{U}_s + \mathbf{F}_{int,f}) \tag{2.99}$$

Then, the second of equations (2.98) is used for the calculation of the internal force vector $\mathbf{F}_{int,s}$ at nodes with prescribed displacements. This is equal to the vector providing the reaction forces developed at these nodes.

It is noted that the prescribed displacement increment $\Delta\mathbf{U}_s$ is considered during the first iteration within each increment. Then, relation (2.92) is used to determine the overall nodal displacement vector. Within this vector, the prescribed nodal displacements are considered fixed for the subsequent iterations of the increment, in the same sense that external forces were fixed within a force increment. Thus, from the second to the last iteration within the increment, no $\Delta\mathbf{U}_s$ is considered and the nodal displacements are determined using the following relation:

$$\Delta\mathbf{U}_f = -\mathbf{K}_{ff}^{-1}(\mathbf{F}_{int,f}) \tag{2.100}$$

### 2.4.4 Geometrical non-linearity

For the formulation of the finite element method in the framework of large displacement gradients which is the case of geometrical non-linearity, equilibrium will be adopted using the principle of virtual work. The second Piola-Kirchhoff stress tensor and the Green strain tensor, which both refer to the reference configuration which is known will be used, noticing that these tensors are energetically conjugate.

According to the general process which is followed, the state of the material at the current configuration $t + \Delta t$ is decomposed into the state at the configuration $t$ plus an increment $\Delta t$. The state at current configuration $t$ is known and the state at $\Delta t$ is unknown. This incremental decomposition takes place for the stress and strain tensors as well as for the displacement vector. Next, the variation of the strain tensor representing the virtual strain in the principle of virtual work, is determined. Finally, the non-linearity which is attributed to non-linear terms arising in the governing equilibrium equation is resolved, using a linearization process which relies on the Taylor series expansion presented in section 2.4.1. According to this process, the non-linear components are substituted by equivalent, approximately linear components neglecting remaining non-linear terms, within an incremental framework.

The descriptions which follow rely on the *total Lagrangian formulation*, indicating that every quantity refers to the initial, undeformed configuration at time 0. An alternative *updated Lagrangian formulation* which refers to the previous known configuration $t$ at the beginning of the load step can also be used [17].

#### 2.4.4.1 Continuum mechanics formulations

The principle of virtual work given in (2.21) for the case that inertial effects are neglected, is initially adopted using matrix notation:

$$\int_{V_0} (\delta\mathbf{E}^{t+\Delta t})^T \mathbf{S}^{t+\Delta t} dV = \int_{V_0} \rho_0 (\delta\mathbf{u}^{t+\Delta t})^T \mathbf{f}^{t+\Delta t} dV + \int_{S_0} (\delta\mathbf{u}^{t+\Delta t})^T \mathbf{t}_0^{t+\Delta t} dS \tag{2.101}$$

Vector $\delta\mathbf{E}^{t+\Delta t}$ represents the variation of the virtual Green strain tensor given in equation (1.72) and vector $\mathbf{S}^{t+\Delta t}$ the second Piola-Kirchhoff stress tensor defined in (1.111), both referring to the original, undeformed configuration. With $\mathbf{t}_0^{t+\Delta t}$ is denoted the traction vector as the force over the undeformed area.

At configuration $t + \Delta t$ the stress and the strain tensors are unknown, indicating that it is necessary to further elaborate these tensors. The unknown second Piola-Kirchhoff stress tensor $\mathbf{S}^{t+\Delta t}$, is decomposed into the known stress tensor $\mathbf{S}^t$ and the increment of the stress tensor $\Delta\mathbf{S}$ as follows:

$$\mathbf{S}^{t+\Delta t} = \mathbf{S}^t + \Delta\mathbf{S} \tag{2.102}$$

By substituting (2.102) into (2.101), it is obtained that:

$$\begin{aligned} &\int_{V_0} (\delta\mathbf{E}^{t+\Delta t})^T \mathbf{S}^t dV + \int_{V_0} (\delta\mathbf{E}^{t+\Delta t})^T \Delta\mathbf{S} dV = \\ &\int_{V_0} \rho_0 (\delta\mathbf{u}^{t+\Delta t})^T \mathbf{f}^{t+\Delta t} dV + \int_{S_0} (\delta\mathbf{u}^{t+\Delta t})^T \mathbf{t}_0^{t+\Delta t} dS \end{aligned} \tag{2.103}$$

The first component in the left-hand part of the governing equation (2.103) is linear. However, the second component in the left-hand part is non-linear, indicating that a linearization process is needed. Both statements are discussed in the following lines of this section.

Similar to the stress tensor, the Green strain tensor is decomposed as follows:

$$\mathbf{E}^{t+\Delta t} = \mathbf{E}^t + \Delta\mathbf{E} \tag{2.104}$$

where $\mathbf{E}^t$ is the known strain tensor at configuration $t$ and $\Delta\mathbf{E}$ the unknown incremental strain. Using the definition of the Green strain tensor given in (1.74) adopting summation notation, the strain tensors at configurations $t$ and $t + \Delta t$ are expressed as:

$$E_{ij}^t = \frac{1}{2}\left(\frac{\partial u_i^t}{\partial X_j} + \frac{\partial u_j^t}{\partial X_i} + \frac{\partial u_k^t}{\partial X_i}\frac{\partial u_k^t}{\partial X_j}\right) \tag{2.105}$$

$$E_{ij}^{t+\Delta t} = \frac{1}{2}\left(\frac{\partial u_i^{t+\Delta t}}{\partial X_j} + \frac{\partial u_j^{t+\Delta t}}{\partial X_i} + \frac{\partial u_k^{t+\Delta t}}{\partial X_i}\frac{\partial u_k^{t+\Delta t}}{\partial X_j}\right) \tag{2.106}$$

The increment $\Delta\mathbf{E}$ of the Green strain tensor is then derived according to (2.104), as the difference $\Delta\mathbf{E} = \mathbf{E}^{t+\Delta t} - \mathbf{E}^t$ and is provided by:

$$\Delta E_{ij} = \frac{1}{2}\left(\frac{\partial \Delta u_i}{\partial X_j} + \frac{\partial \Delta u_j}{\partial X_i} + \frac{\partial u_k^t}{\partial X_i}\frac{\partial \Delta u_k}{\partial X_j} + \frac{\partial \Delta u_k}{\partial X_i}\frac{\partial u_k^t}{\partial X_j}\right) + \frac{1}{2}\left(\frac{\partial \Delta u_k}{\partial X_i}\frac{\partial \Delta u_k}{\partial X_j}\right) \tag{2.107}$$

where the decomposition of the displacement vector:

$$\mathbf{u}^{t+\Delta t} = \mathbf{u}^t + \Delta\mathbf{u} \tag{2.108}$$

has been used. The first term of the right-hand part of equation (2.107) is linear in respect to displacement increment $\Delta u$, since it is composed of four linear components. It is noted that the components $\frac{\partial u_k^t}{\partial X_i}\frac{\partial \Delta u_k}{\partial X_j}$ and $\frac{\partial \Delta u_k}{\partial X_i}\frac{\partial u_k^t}{\partial X_j}$ are linear in respect to $\Delta u$, since $u_k^t$ is the known displacement at configuration $t$ indicating that its partial derivatives are known. The second term of the right-hand part of equation (2.107) is non-linear in $\Delta u$ as it consists of the product of the unknown partial derivatives of displacement increments.

According to this description, the increment of the Green strain tensor $\Delta\mathbf{E}$ can be decomposed into a linear part $\Delta\mathbf{E}_L$ and a non-linear part $\Delta\mathbf{E}_{NL}$ as follows:

$$\Delta\mathbf{E} = \Delta\mathbf{E}_L + \Delta\mathbf{E}_{NL} \tag{2.109}$$

where the linear and non-linear parts are provided in summation notation:

$$\begin{aligned} \Delta E_{ij}^{L} &= \frac{1}{2}\left(\frac{\partial \Delta u_i}{\partial X_j} + \frac{\partial \Delta u_j}{\partial X_i} + \frac{\partial u_k^t}{\partial X_i}\frac{\partial \Delta u_k}{\partial X_j} + \frac{\partial \Delta u_k}{\partial X_i}\frac{\partial u_k^t}{\partial X_j}\right) \\ \Delta E_{ij}^{NL} &= \frac{1}{2}\left(\frac{\partial \Delta u_k}{\partial X_i}\frac{\partial \Delta u_k}{\partial X_j}\right) \end{aligned} \tag{2.110}$$

In addition, the variation of the Green strain tensor $\mathbf{E}^{t+\Delta t}$ given in equation (2.104) results in the following relation:

$$\delta\mathbf{E}^{t+\Delta t} = \delta\mathbf{E}^{t} + \delta\Delta\mathbf{E} \tag{2.111}$$

The Green strain tensor $\mathbf{E}^t$ at time $t$ is known (and constant) indicating that $\delta\mathbf{E}^t = 0$. Thus, equation (2.111) becomes:

$$\delta\mathbf{E}^{t+\Delta t} = \delta\Delta\mathbf{E} \tag{2.112}$$

By substituting (2.109) into (2.112), it is derived that:

$$\delta\mathbf{E}^{t+\Delta t} = \delta\Delta\mathbf{E}_L + \delta\Delta\mathbf{E}_{NL} \tag{2.113}$$

The variation of the non-linear part of the strain increment $\delta\Delta\mathbf{E}_{NL}$ can then be determined by using the second of equations (2.110), as follows:

$$\delta\Delta E_{ij}^{NL} = \frac{1}{2}\left(\delta\frac{\partial \Delta u_k}{\partial X_i}\frac{\partial \Delta u_k}{\partial X_j} + \frac{\partial \Delta u_k}{\partial X_i}\delta\frac{\partial \Delta u_k}{\partial X_j}\right) \tag{2.114}$$

The components $\delta\frac{\partial \Delta u_k}{\partial X_i}$ and $\delta\frac{\partial \Delta u_k}{\partial X_j}$ in (2.114) are constants, indicating that the variation of the non-linear part of the strain increment $\delta\Delta E_{ij}^{NL}$ is linear in respect to the displacement increment. In addition, the variation of the linear part of the strain increment $\delta\Delta E_{ij}^{L}$ given in the first of equations (2.110) is constant. These statements will be considered in the following section, for the linearization of any non-linear terms arising in the governing equation.

### 2.4.4.2 Linearization of the governing equation

As mentioned in section 2.4.4.1 the first component of the left-hand part of the governing equilibrium equation (2.103) is linear while the second is non-linear. Thus, a linearization process needs to be applied, to allow for the numerical implementation.

One reason for the non-linearity of the second component of (2.103) is the potential material non-linearity resulting in non-linear stress versus strain response. Another reason which is discussed below further, is that the terms $\delta\mathbf{E}^{t+\Delta t}$ and $\Delta\mathbf{S}$ are expanded to equivalent terms, the product of which is non-linear in respect to the displacement increment.

This second component of the governing equation (2.103) is linearized here, by using a Taylor series expansion presented in equation (2.82):

$$\int_{V_0} (\delta\mathbf{E}^{t+\Delta t})^T \Delta\mathbf{S} dV = \int_{V_0} (\delta\mathbf{E}^{t+\Delta t})^T \left(\frac{\partial \mathbf{S}^t}{\partial \mathbf{E}^t}\Delta\mathbf{E} + \mathbf{R}\right) dV \qquad (2.115)$$

with $\mathbf{R}$ representing higher-order terms. By substituting in (2.115), the increment of the Green strain tensor $\Delta\mathbf{E}$ given in (2.109) and the variation $\delta\mathbf{E}^{t+\Delta t}$ given in (2.113), equation (2.115) is transformed as follows:

$$\int_{V_0} (\delta\mathbf{E}^{t+\Delta t})^T \Delta\mathbf{S} dV = \int_{V_0} (\delta\Delta\mathbf{E}_L + \delta\Delta\mathbf{E}_{NL})^T \left(\frac{\partial \mathbf{S}^t}{\partial \mathbf{E}^t}(\Delta\mathbf{E}_L + \Delta\mathbf{E}_{NL}) + \mathbf{R}\right) dV \qquad (2.116)$$

According to the descriptions presented in section 2.4.4.1, the component $\delta\Delta\mathbf{E}_{NL}$ of (2.116) is linear and the component $\Delta\mathbf{E}_{NL}$ is non-linear in respect to the displacement increment $\Delta\mathbf{u}$. Thus, their product in equation (2.116) will result in non-linear terms, indicating that these two terms must be neglected in the linearized form given in the right-hand part of (2.116). For small values on $\Delta\mathbf{E}$ the higher-order terms $\mathbf{R}$ can also be neglected. Moreover, $\delta\Delta\mathbf{E}_L$ is constant and $\Delta\mathbf{E}_L$ is linear in $\Delta\mathbf{u}$ as presented in section 2.4.4.1, indicating that their product is linear and it can be kept.

Next, the relation between the stress and the strain is provided below:

$$\mathbf{D} = \frac{\partial \mathbf{S}^t}{\partial \mathbf{E}^t} \qquad (2.117)$$

with $\mathbf{D}$ denoting the *instantaneous* stiffness modulus. When linear elasticity is considered, $\mathbf{D}$ represents the elasticity tensor. It should be noticed that even in the case that no material non-linearity appears, the non-linear character of equation (2.116) remains, due to the non-linear terms arising after the products of the Green strain tensor are considered according to previous explanations. This is attributed to the large displacement analysis which has been assumed.

By considering the above statements, equation (2.116) is written in the final, linearized form which follows:

$$\begin{aligned}\int_{V_0} (\delta \mathbf{E}^{t+\Delta t})^T \Delta \mathbf{S} dV &= \int_{V_0} (\delta \Delta \mathbf{E}_L)^T \frac{\partial \mathbf{S}^t}{\partial \mathbf{E}^t} \Delta \mathbf{E}_L dV \\ &= \int_{V_0} (\delta \Delta \mathbf{E}_L)^T \mathbf{D} \Delta \mathbf{E}_L dV\end{aligned} \tag{2.118}$$

The first component of the governing equation (2.103) can be rewritten, by substituting $\delta \mathbf{E}^{t+\Delta t}$ from relation (2.113):

$$\begin{aligned}\int_{V_0} (\delta \mathbf{E}^{t+\Delta t})^T \mathbf{S}^t dV &= \int_{V_0} (\delta \Delta \mathbf{E}_L + \delta \Delta \mathbf{E}_{NL})^T \mathbf{S}^t dV \\ &= \int_{V_0} (\delta \Delta \mathbf{E}_L)^T \mathbf{S}^t dV + \int_{V_0} (\delta \Delta \mathbf{E}_{NL})^T \mathbf{S}^t dV\end{aligned} \tag{2.119}$$

The first component on the right-hand part of the last equation in (2.119) is a constant, known quantity in displacement increment, since the term $\delta \Delta \mathbf{E}_L$ is constant and the second Piola-Kirchhoff stress tensor $\mathbf{S}^t$ is known at configuration $t$. This component represents the internal force vector. The second component on the right-hand part is linear in respect to the displacement increment, since the term $\delta \Delta \mathbf{E}_{NL}$ is also linear (and multiplied by the constant $\mathbf{S}^t$). Hence, equation (2.119) does not include any non-linear term and does not need any linearization.

The final version of the governing equation is obtained, when the linearized equations (2.118) and (2.119) are substituted in the governing equation (2.103):

$$\begin{aligned}&\int_{V_0} (\delta \Delta \mathbf{E}_{NL})^T \mathbf{S}^t dV + \int_{V_0} (\delta \Delta \mathbf{E}_L)^T \mathbf{D} \Delta \mathbf{E}_L dV = \\ &\int_{V_0} \rho_0 (\delta \mathbf{u}^{t+\Delta t})^T \mathbf{f}^{t+\Delta t} dV + \int_{S_0} (\delta \mathbf{u}^{t+\Delta t})^T \mathbf{t}_0^{t+\Delta t} dS - \int_{V_0} (\delta \Delta \mathbf{E}_L)^T \mathbf{S}^t dV\end{aligned} \tag{2.120}$$

where the internal force vector has been transferred to the right-hand part. It is noted that this is the final linearized equation in the framework of continuum mechanics. However, finite element discretization has not yet been introduced. This is the next step towards the numerical implementation of large displacement analysis and is presented in the following section.

#### 2.4.4.3 Discretization of the governing equation

The above formulations have been developed within continuum mechanics. For the implementation in the framework of the finite element method, the terms of the governing equation (2.120) should be discretized. This will be done by introducing the unknown nodal displacement vector in this equation, using the strain-nodal displacement relation presented in section 2.3.2. For the linear increment $\Delta\mathbf{E}_L$ of the Green strain tensor, this is expressed as follows:

$$\Delta\mathbf{E}_L = \boldsymbol{\partial}\Delta\mathbf{u} \tag{2.121}$$

where matrix $\boldsymbol{\partial}$ includes the partial derivatives in respect to the reference coordinates, according to the first of equations (2.110). The displacement field is expressed in terms of the nodal displacement vector $\mathbf{u}_e$, using the interpolation functions $\mathbf{N}$:

$$\mathbf{u} = \mathbf{N}\mathbf{u}_e \Rightarrow \Delta\mathbf{u} = \mathbf{N}\Delta\mathbf{u}_e \tag{2.122}$$

By substituting $\Delta\mathbf{u}$ from (2.122) into (2.121), the linear increment of the Green strain tensor is given by:

$$\begin{aligned} \Delta\mathbf{E}_L = \boldsymbol{\partial}\mathbf{N}\Delta\mathbf{u}_e &\Rightarrow \Delta\mathbf{E}_L = \mathbf{B}_L\Delta\mathbf{u}_e \\ &\Rightarrow \delta\Delta\mathbf{E}_L = \mathbf{B}_L\delta\Delta\mathbf{u}_e \end{aligned} \tag{2.123}$$

where matrix $\mathbf{B}_L$ is defined as follows:

$$\mathbf{B}_L = \boldsymbol{\partial}\mathbf{N} \tag{2.124}$$

The components of matrix $\mathbf{B}_L$ are derived by substituting the discretized displacement field given in (2.122) into the first of equations (2.110). This will be presented in the next section, for a quadrilateral element.

The second component of the left-hand part of the governing equation (2.120) can then be rewritten using (2.123) as follows:

$$\int_{V_0} (\delta\Delta\mathbf{E}_L)^T\mathbf{D}\Delta\mathbf{E}_L dV = \sum_{i=1}^{N_e}\int_{V_e}(\delta\Delta\mathbf{u}_e)^T\mathbf{B}_L^T\mathbf{D}\mathbf{B}_L\Delta\mathbf{u}_e dV \tag{2.125}$$

Similar to linear finite element analysis, summation over the initial volume $V_e$ and surface $S_e$ of each element has been considered in (2.125), to indicate the assembly process over the number $N_e$ of all finite elements. By considering the nodal displacement vector $\mathbf{U}$ of the whole structure as the assembly of the nodal displacement vector $\mathbf{u}_e$ of each element, equation (2.125) results in:

$$\int_{V_0} (\delta\Delta\mathbf{E}_L)^T\mathbf{D}\Delta\mathbf{E}_L dV = (\delta\Delta\mathbf{U})^T\sum_{i=1}^{N_e}\int_{V_e}\mathbf{B}_L^T\mathbf{D}\mathbf{B}_L dV\Delta\mathbf{U} \tag{2.126}$$

where $\delta\Delta\mathbf{U}$ and $\Delta\mathbf{U}$ have been placed out of the integral and the summation, since they receive discrete values. From equation (2.126), the contribution $\mathbf{K}_L$ to the tangent stiffness matrix is defined as follows:

$$\mathbf{K}_L = \sum_{i=1}^{N_e}\int_{V_e}\mathbf{B}_L^T\mathbf{D}\mathbf{B}_L dV \tag{2.127}$$

To discretize the first component of the left-hand part of the governing equation (2.120), the variation of the non-linear increment of the Green strain tensor is expressed as:

$$\delta\Delta\mathbf{E}_{NL} = \mathbf{B}_{NL}\delta\Delta\mathbf{u}_e \tag{2.128}$$

where $\mathbf{B}_{NL}$ represents the non-linear strain-displacement matrix. This is determined using the second equation in (2.110) as it will be shown in the next section for a quadrilateral element.

Then, the first component of the left-hand part of the governing equation (2.120) is formulated as follows:

$$\int_{V_0} (\delta\Delta\mathbf{E}_{NL})^T\mathbf{S}^t dV = (\delta\Delta\mathbf{U})^T \sum_{i=1}^{N_e} \int_{V_e} \mathbf{B}_{NL}^T \widetilde{\mathbf{S}}^t \mathbf{B}_{NL} dV \Delta\mathbf{U} \tag{2.129}$$

It should be emphasized here, that the concept which is applied for the derivation of relation (2.129) is that the term inside the integral on the left-hand part is substituted by an equivalent term on the right-hand part, such that both terms lead to the same result [17]. To implement this concept, the second Piola-Kirchhoff stress tensor $\widetilde{\mathbf{S}}^t$ is introduced in a matrix form in the right-hand part of (2.129), contrary to the corresponding term $\mathbf{S}^t$ at the left-hand part which is given in a vector form. An example of building matrix $\widetilde{\mathbf{S}}^t$ will be given in the next section.

From the right-hand part of equation (2.129), the geometric contribution to the tangential stiffness matrix $\mathbf{K}_{NL}$ is introduced as:

$$\mathbf{K}_{NL} = \sum_{i=1}^{N_e} \int_{V_e} \mathbf{B}_{NL}^T \widetilde{\mathbf{S}}^t \mathbf{B}_{NL} dV \tag{2.130}$$

The final governing equation is derived, when the work terms of the body and surface forces, as well as the work term of the internal force vector in the right-hand part of (2.120), are discretized. First, the discretization of the variation of the displacement field takes place as follows:

$$\mathbf{u}^{t+\Delta t} = \mathbf{u}^t + \Delta\mathbf{u} \Rightarrow \delta\mathbf{u}^{t+\Delta t} = \delta\mathbf{u}^t + \delta\Delta\mathbf{u} \Rightarrow \delta\mathbf{u}^{t+\Delta t} = \delta\Delta\mathbf{u} = \mathbf{N}\delta\Delta\mathbf{u}_e \tag{2.131}$$

noticing that $\mathbf{u}^t$ is constant and known and thus, its variation is equal to zero. Then, the work terms for the external body and surface forces are expressed according to the following relations:

$$\begin{aligned}\int_{V_0} \rho_0(\delta\mathbf{u}^{t+\Delta t})^T\mathbf{f}^{t+\Delta t} dV &= \sum_{i=1}^{N_e} \int_{V_e} (\delta\Delta\mathbf{u}_e)^T \rho_0 \mathbf{N}^T \mathbf{f}^{t+\Delta t} dV \\ &= (\delta\Delta\mathbf{U})^T \sum_{i=1}^{N_e} \int_{V_e} \rho_0 \mathbf{N}^T \mathbf{f}^{t+\Delta t} dV\end{aligned} \tag{2.132}$$

$$\begin{aligned}\int_{S_0} (\delta \mathbf{u}^{t+\Delta t})^T \mathbf{t}_0^{t+\Delta t} dS =& \sum_{i=1}^{N_e} \int_{S_e} (\delta \Delta \mathbf{u}_e)^T \mathbf{N}^T \mathbf{t}_0^{t+\Delta t} dS \\ =& (\delta \Delta \mathbf{U})^T \sum_{i=1}^{N_e} \int_{S_e} \mathbf{N}^T \mathbf{t}_0^{t+\Delta t} dS\end{aligned} \tag{2.133}$$

Following a similar concept as above, the last component at the right-hand part of (2.120) representing the internal force vector, is discretized as follows:

$$\int_{V_0} \delta \Delta \mathbf{E}_L^T \mathbf{S}^t dV = (\delta \Delta \mathbf{U})^T \sum_{i=1}^{N_e} \int_{V_e} \mathbf{B}_L^T \mathbf{S}^t dV \tag{2.134}$$

By substituting equations (2.126), (2.129), (2.132), (2.133) and (2.134) into (2.120), the overall reformulation of the governing equation within non-linear finite element analysis takes place:

$$\begin{aligned}&(\delta \Delta \mathbf{U})^T \sum_{i=1}^{N_e} \int_{V_e} \mathbf{B}_L^T \mathbf{D} \mathbf{B}_L dV \Delta \mathbf{U} + (\delta \Delta \mathbf{U})^T \sum_{i=1}^{N_e} \int_{V_e} \mathbf{B}_{NL}^T \widetilde{\mathbf{S}}^t \mathbf{B}_{NL} dV \Delta \mathbf{U} = \\ &(\delta \Delta \mathbf{U})^T \sum_{i=1}^{N_e} \int_{V_e} \rho_0 \mathbf{N}^T \mathbf{f}^{t+\Delta t} dV + (\delta \Delta \mathbf{U})^T \sum_{i=1}^{N_e} \int_{S_e} \mathbf{N}^T \mathbf{t}_0^{t+\Delta t} dS - \\ &(\delta \Delta \mathbf{U})^T \sum_{i=1}^{N_e} \int_{V_e} \mathbf{B}_L^T \mathbf{S}^t dV\end{aligned} \tag{2.135}$$

By eliminating $(\delta \Delta \mathbf{U})^T$ from both parts, equation (2.135) is then rewritten in a final form as follows:

$$\begin{aligned}&\left( \sum_{i=1}^{N_e} \int_{V_e} \mathbf{B}_L^T \mathbf{D} \mathbf{B}_L dV + \sum_{i=1}^{N_e} \int_{V_e} \mathbf{B}_{NL}^T \widetilde{\mathbf{S}}^t \mathbf{B}_{NL} dV \right) \Delta \mathbf{U} = \\ &\sum_{i=1}^{N_e} \int_{V_e} \rho_0 \mathbf{N}^T \mathbf{f}^{t+\Delta t} dV + \sum_{i=1}^{N_e} \int_{S_e} \mathbf{N}^T \mathbf{t}_0^{t+\Delta t} dS - \sum_{i=1}^{N_e} \int_{V_e} \mathbf{B}_L^T \mathbf{S}^t dV\end{aligned} \tag{2.136}$$

or

$$(\mathbf{K}_L + \mathbf{K}_{NL}) \Delta \mathbf{U} = \mathbf{F}_{ext}^{t+\Delta t} - \mathbf{F}_{int}^t \tag{2.137}$$

with $\mathbf{F}_{ext}^{t+\Delta t}$ being the external force vector at configuration $t+\Delta t$ and $\mathbf{F}_{int}^{t}$ the internal force vector at configuration $t$, both provided by:

$$\mathbf{F}_{ext}^{t+\Delta t} = \sum_{i=1}^{N_e} \int_{V_e} \rho_0 \mathbf{N}^T \mathbf{f}^{t+\Delta t} dV + \sum_{i=1}^{N_e} \int_{S_e} \mathbf{N}^T \mathbf{t}_0^{t+\Delta t} dS \tag{2.138}$$

$$\mathbf{F}_{int}^{t} = \sum_{i=1}^{N_e} \int_{V_e} \mathbf{B}_L^T \mathbf{S}^t dV \tag{2.139}$$

Equation (2.137) is in the same form with equation (2.89) presented in section 2.4.2. Thus, the tangent stiffness matrix $\mathbf{K}_T$ of (2.89) is provided here by $\mathbf{K}_T = \mathbf{K}_L + \mathbf{K}_{NL}$, where the second term $\mathbf{K}_{NL}$ is called *geometric stiffness.* For the implementation of the solution, equation (2.137) is solved within the incremental-iterative Newton-Raphson process, using iterations depicted by equations (2.91) and (2.92). According to this scheme, the unknown nodal displacement increment $\Delta\mathbf{U}$ is determined at every iteration using (2.137) and the total displacement vector is then calculated using (2.92).

Finally, it is noted that the isoparametric formulation is also applied for geometrically non-linear analysis in a similar fashion applied for linear analysis. Thus, the derivatives of the shape functions inside $\mathbf{B}_L$ and $\mathbf{B}_{NL}$ matrices are calculated using equation (2.63).

#### 2.4.4.4 Formulations for a quadrilateral element

In this section, the matrices which are needed for the discretization of the geometrically non-linear finite element analysis will be presented, for a plane, quadrilateral element. These are the strain-displacement matrices $\mathbf{B}_L$ and $\mathbf{B}_{NL}$ as well as the second Piola-Kirchhoff stress tensor $\widetilde{\mathbf{S}}^t$ given in a matrix form and $\mathbf{S}^t$ given in a vector form, which appear in the governing equation (2.136).

The increment of the linear Green strain tensor components are given below, using the first of equations (2.110):

$$\begin{aligned}
\Delta E_{11}^L &= \frac{\partial \Delta u_1}{\partial X_1} + \frac{\partial u_1^t}{\partial X_1}\frac{\partial \Delta u_1}{\partial X_1} + \frac{\partial u_2^t}{\partial X_1}\frac{\partial \Delta u_2}{\partial X_1} \\
\Delta E_{22}^L &= \frac{\partial \Delta u_2}{\partial X_2} + \frac{\partial u_1^t}{\partial X_2}\frac{\partial \Delta u_1}{\partial X_2} + \frac{\partial u_2^t}{\partial X_2}\frac{\partial \Delta u_2}{\partial X_2} \\
\Delta E_{12}^L &= \frac{1}{2}\left(\frac{\partial \Delta u_1}{\partial X_2} + \frac{\partial \Delta u_2}{\partial X_1} + \frac{\partial u_1^t}{\partial X_1}\frac{\partial \Delta u_1}{\partial X_2} + \frac{\partial u_2^t}{\partial X_1}\frac{\partial \Delta u_2}{\partial X_2} \right. \\
&\quad \left. + \frac{\partial u_1^t}{\partial X_2}\frac{\partial \Delta u_1}{\partial X_1} + \frac{\partial u_2^t}{\partial X_2}\frac{\partial \Delta u_2}{\partial X_1}\right)
\end{aligned} \tag{2.140}$$

To provide the linear strain-displacement matrix $\mathbf{B}_L$, equation (2.122) is used to substitute the increment of the displacement field $\Delta u_i$ in (2.140), by its counterpart discretized component, according to $\Delta \mathbf{u} = \mathbf{N}\Delta \mathbf{u}_e$. Then, equations (2.140) are written in the following form:

$$\begin{bmatrix} \Delta E_{11}^L \\ \Delta E_{22}^L \\ 2\Delta E_{12}^L \end{bmatrix} = (\mathbf{B}_L)\mathbf{u}_e = (\mathbf{B}_{L0} + \mathbf{B}_{L1})\Delta \mathbf{u}_e \tag{2.141}$$

where $\Delta \mathbf{u}_e$ is the $8 \times 1$ incremental displacement nodal point vector of a four node quadrilateral element, with two degrees of freedom per node. According to (2.141), matrix $\mathbf{B}_L$ is determined by the summation of matrices $\mathbf{B}_{L0}$ and $\mathbf{B}_{L1}$, each of dimensions $3 \times 8$. Following this concept, the components of the two matrices are easily identified in equations (2.140), noticing that matrix $\mathbf{B}_{L0}$ incorporates the terms $\frac{\partial \Delta u}{\partial X}$ and matrix $\mathbf{B}_{L1}$ the terms $\frac{\partial u^t}{\partial X}\frac{\partial \Delta u}{\partial X}$.

Thus, it is found that matrix $\mathbf{B}_{L0}$ is identical to the strain-displacement matrix used in linear analysis (equation 2.71) and is given below:

$$\mathbf{B}_{L0} = \begin{bmatrix} \frac{\partial N_1}{\partial X_1} & 0 & \frac{\partial N_2}{\partial X_1} & 0 & \frac{\partial N_3}{\partial X_1} & 0 & \frac{\partial N_4}{\partial X_1} & 0 \\ 0 & \frac{\partial N_1}{\partial X_2} & 0 & \frac{\partial N_2}{\partial X_2} & 0 & \frac{\partial N_3}{\partial X_2} & 0 & \frac{\partial N_4}{\partial X_2} \\ \frac{\partial N_1}{\partial X_2} & \frac{\partial N_1}{\partial X_1} & \frac{\partial N_2}{\partial X_2} & \frac{\partial N_2}{\partial X_1} & \frac{\partial N_3}{\partial X_2} & \frac{\partial N_3}{\partial X_1} & \frac{\partial N_4}{\partial X_2} & \frac{\partial N_4}{\partial X_1} \end{bmatrix} \tag{2.142}$$

Then, matrix $\mathbf{B}_{L1}$ is built as follows:

$$\mathbf{B}_{L1} =$$

$$\left[\begin{matrix} \frac{\partial u_1^t}{\partial X_1}\frac{\partial N_1}{\partial X_1} & \frac{\partial u_2^t}{\partial X_1}\frac{\partial N_1}{\partial X_1} & \frac{\partial u_1^t}{\partial X_1}\frac{\partial N_2}{\partial X_1} & \frac{\partial u_2^t}{\partial X_1}\frac{\partial N_2}{\partial X_1} \cdots \\ \frac{\partial u_1^t}{\partial X_2}\frac{\partial N_1}{\partial X_2} & \frac{\partial u_2^t}{\partial X_2}\frac{\partial N_1}{\partial X_2} & \frac{\partial u_1^t}{\partial X_2}\frac{\partial N_2}{\partial X_2} & \frac{\partial u_2^t}{\partial X_2}\frac{\partial N_2}{\partial X_2} \cdots \\ \frac{\partial u_1^t}{\partial X_1}\frac{\partial N_1}{\partial X_2} + \frac{\partial u_1^t}{\partial X_2}\frac{\partial N_1}{\partial X_1} & \frac{\partial u_2^t}{\partial X_1}\frac{\partial N_1}{\partial X_2} + \frac{\partial u_2^t}{\partial X_2}\frac{\partial N_1}{\partial X_1} & \frac{\partial u_1^t}{\partial X_1}\frac{\partial N_2}{\partial X_2} + \frac{\partial u_1^t}{\partial X_2}\frac{\partial N_2}{\partial X_1} & \frac{\partial u_2^t}{\partial X_1}\frac{\partial N_2}{\partial X_2} + \frac{\partial u_2^t}{\partial X_2}\frac{\partial N_2}{\partial X_1} \cdots \end{matrix}\right.$$

$$\left.\begin{matrix} \frac{\partial u_1^t}{\partial X_1}\frac{\partial N_3}{\partial X_1} & \frac{\partial u_2^t}{\partial X_1}\frac{\partial N_3}{\partial X_1} & \frac{\partial u_1^t}{\partial X_1}\frac{\partial N_4}{\partial X_1} & \frac{\partial u_2^t}{\partial X_1}\frac{\partial N_4}{\partial X_1} \\ \frac{\partial u_1^t}{\partial X_2}\frac{\partial N_3}{\partial X_2} & \frac{\partial u_2^t}{\partial X_2}\frac{\partial N_3}{\partial X_2} & \frac{\partial u_1^t}{\partial X_2}\frac{\partial N_4}{\partial X_2} & \frac{\partial u_2^t}{\partial X_2}\frac{\partial N_4}{\partial X_2} \\ \frac{\partial u_1^t}{\partial X_1}\frac{\partial N_3}{\partial X_2} + \frac{\partial u_1^t}{\partial X_2}\frac{\partial N_3}{\partial X_1} & \frac{\partial u_2^t}{\partial X_1}\frac{\partial N_3}{\partial X_2} + \frac{\partial u_2^t}{\partial X_2}\frac{\partial N_3}{\partial X_1} & \frac{\partial u_1^t}{\partial X_1}\frac{\partial N_4}{\partial X_2} + \frac{\partial u_1^t}{\partial X_2}\frac{\partial N_4}{\partial X_1} & \frac{\partial u_2^t}{\partial X_1}\frac{\partial N_4}{\partial X_2} + \frac{\partial u_2^t}{\partial X_2}\frac{\partial N_4}{\partial X_1} \end{matrix}\right] \tag{2.143}$$

All the partial derivatives $\frac{\partial u^t}{\partial X}$ of the displacement field $u^t$ at the configuration $t$ which are found in equation (2.143) can be determined by introducing the discretized displacement $\mathbf{u}^t = \mathbf{N}\mathbf{u}_e^t$ into the partial derivatives. By doing this, the following relations are derived for those partial derivatives:

$$\begin{aligned} \frac{\partial u_1^t}{\partial X_1} &= \left(\sum_{k=1}^{n_e} \frac{\partial N_k}{\partial X_1}\right) {}^t u_1^k, \quad \frac{\partial u_1^t}{\partial X_2} = \left(\sum_{k=1}^{n_e} \frac{\partial N_k}{\partial X_2}\right) {}^t u_1^k \\ \frac{\partial u_2^t}{\partial X_1} &= \left(\sum_{k=1}^{n_e} \frac{\partial N_k}{\partial X_1}\right) {}^t u_2^k, \quad \frac{\partial u_2^t}{\partial X_2} = \left(\sum_{k=1}^{n_e} \frac{\partial N_k}{\partial X_2}\right) {}^t u_2^k \end{aligned} \tag{2.144}$$

where the nodal displacement vector at time $t$ of a quadrilateral element with $n_e = 4$ nodes, is provided below:

$$\mathbf{u}_e^t = \begin{bmatrix} {}^t u_1^1 & {}^t u_2^1 & {}^t u_1^2 & {}^t u_2^2 & {}^t u_1^3 & {}^t u_2^3 & {}^t u_1^4 & {}^t u_2^4 \end{bmatrix}^T \tag{2.145}$$

noticing that $\mathbf{u}_e^t$ at time $t$ is known. Those partial derivatives define the influence of the initial displacements (at time $t$) on the calculation of the strain-displacement matrix. When $\mathbf{u}_e^t = \mathbf{0}$, then $\mathbf{B}_{L1}$ vanishes.

The increment of the non-linear Green strain tensor components are given below for the quadrilateral element, using the second of equations (2.110):

$$\begin{aligned}
\Delta E_{11}^{NL} =& \frac{1}{2}\left( (\frac{\partial \Delta u_1}{\partial X_1})^2 + (\frac{\partial \Delta u_2}{\partial X_1})^2 \right) \\
\Delta E_{22}^{NL} =& \frac{1}{2}\left( (\frac{\partial \Delta u_1}{\partial X_2})^2 + (\frac{\partial \Delta u_2}{\partial X_2})^2 \right) \\
\Delta E_{12}^{NL} =& \Delta E_{21}^{NL} = \frac{1}{2}\left( \frac{\partial \Delta u_1}{\partial X_1}\frac{\partial \Delta u_1}{\partial X2} + \frac{\partial \Delta u_2}{\partial X_1}\frac{\partial \Delta u_2}{\partial X_2} \right)
\end{aligned} \tag{2.146}$$

Next, the non-linear strain-displacement matrix $\mathbf{B}_{NL}$ and the second Piola-Kirchhoff stress tensor $\widetilde{\mathbf{S}}^t$ which appear in the governing equation (2.136), shall be built. As mentioned in section 2.4.4.3 this will be done in such a way, that the left and the right hand-parts of equation (2.129) are equal. Using this concept, the strain-displacement matrix $\mathbf{B}_{NL}$ of dimensions $4 \times 8$ is provided below:

$$\mathbf{B}_{NL} = \begin{bmatrix}
\frac{\partial N_1}{\partial X_1} & 0 & \frac{\partial N_2}{\partial X_1} & 0 & \frac{\partial N_3}{\partial X_1} & 0 & \frac{\partial N_4}{\partial X_1} & 0 \\
\frac{\partial N_1}{\partial X_2} & 0 & \frac{\partial N_2}{\partial X_2} & 0 & \frac{\partial N_3}{\partial X_2} & 0 & \frac{\partial N_4}{\partial X_2} & 0 \\
0 & \frac{\partial N_1}{\partial X_1} & 0 & \frac{\partial N_2}{\partial X_1} & 0 & \frac{\partial N_3}{\partial X_1} & 0 & \frac{\partial N_4}{\partial X_1} \\
0 & \frac{\partial N_1}{\partial X_2} & 0 & \frac{\partial N_2}{\partial X_2} & 0 & \frac{\partial N_3}{\partial X_2} & 0 & \frac{\partial N_4}{\partial X_2}
\end{bmatrix} \tag{2.147}$$

The second Piola-Kirchhoff stress tensor $\widetilde{\mathbf{S}}^t$ of dimensions $4 \times 4$ can then be built in a matrix form, as follows:

$$\widetilde{\mathbf{S}}^t = \begin{bmatrix}
S_{11}^t & S_{12}^t & 0 & 0 \\
S_{21}^t & S_{22}^t & 0 & 0 \\
0 & 0 & S_{11}^t & S_{12}^t \\
0 & 0 & S_{21}^t & S_{22}^t
\end{bmatrix} \tag{2.148}$$

as well as in a vector form as below:

$$\mathbf{S}^t = \begin{bmatrix} S_{11}^t \\ S_{22}^t \\ S_{12}^t \end{bmatrix} \tag{2.149}$$

For a three-dimensional continuum element, the above formulation should properly be extended.

### 2.4.5 Material non-linearity

A structural problem presents material non-linearity when the constitutive stress versus strain relation is non-linear. For the descriptions which follow, small displacement gradients are considered. Due to this assumption, all the additional terms arising from the non-linear parts of the Green strain tensor which have been considered in geometrical non-linearity, now vanish. In addition, the current and the initial configuration coincide indicating that the Green strain tensor is equal to the Euler strain tensor and the second Piola-Kirchhoff stress tensor is equal to the Cauchy stress tensor. It is noted that both material and geometrical non-linearities may arise in a structural system. In this case, the formulations which are presented in section 2.4.4 should be adopted.

The problem of material non-linearity will initially be formulated within continuum mechanics. Then, discretization in the framework of finite element analysis will be implemented. For the descriptions which follow, it is assumed that a known equilibrium state is considered at iteration $i$ and the unknown state at iteration $i+1$ will be investigated.

The principle of virtual work given in (2.21) is adopted using matrix-vector notation, noticing that inertial effects have been neglected:

$$\int_V (\delta\boldsymbol{\varepsilon})^T \boldsymbol{\sigma} dV = \int_V \rho(\delta\mathbf{u})^T \mathbf{f} dV + \int_S (\delta\mathbf{u})^T \mathbf{t} dS \tag{2.150}$$

Next, the decomposition of the stress tensor $\boldsymbol{\sigma}$ and the strain tensor $\boldsymbol{\varepsilon}$ using the corresponding increments $\Delta\boldsymbol{\sigma}$ and $\Delta\boldsymbol{\varepsilon}$, respectively, is provided by:

$$\boldsymbol{\sigma}^{i+1} = \boldsymbol{\sigma}^i + \Delta\boldsymbol{\sigma} \tag{2.151}$$

$$\boldsymbol{\varepsilon}^{i+1} = \boldsymbol{\varepsilon}^i + \Delta\boldsymbol{\varepsilon} \tag{2.152}$$

The strain tensor presented in (1.76) for small displacement gradients, is given below at iterations $i$ and $i+1$:

$$\varepsilon_{mn}^i = \frac{1}{2}\left(\frac{\partial u_m^i}{\partial x_n} + \frac{\partial u_n^i}{\partial x_m}\right) \tag{2.153}$$

$$\varepsilon_{mn}^{i+1} = \frac{1}{2}\left(\frac{\partial u_m^{i+1}}{\partial x_n} + \frac{\partial u_n^{i+1}}{\partial x_m}\right) \tag{2.154}$$

The increment $\Delta\boldsymbol{\varepsilon}$ of the strain tensor can now be derived as the difference $\boldsymbol{\varepsilon}^{i+1} - \boldsymbol{\varepsilon}^i$:

$$\Delta\varepsilon_{mn} = \frac{1}{2}\left(\frac{\partial \Delta u_m}{\partial x_n} + \frac{\partial \Delta u_n}{\partial x_m}\right) \tag{2.155}$$

where the decomposition of the displacement vector:

$$\mathbf{u}^{i+1} = \mathbf{u}^{i} + \Delta\mathbf{u} \tag{2.156}$$

has been used. It is noticed that the incremental strain $\Delta\varepsilon$ given in equation (2.155) is linear in respect to displacement increment $\Delta u$, contrary to the corresponding relation for geometrical non-linearity given in (2.107), where non-linear terms in $\Delta u$ appear.

In addition, the variation of the strain tensor $\boldsymbol{\varepsilon}^{i+1}$ is obtained as follows:

$$\delta\boldsymbol{\varepsilon}^{i+1} = \delta\boldsymbol{\varepsilon}^{i} + \delta\Delta\boldsymbol{\varepsilon} \tag{2.157}$$

Since the strain tensor $\boldsymbol{\varepsilon}^{i}$ at iteration $i$ is known (and constant), it appears that $\delta\boldsymbol{\varepsilon}^{i} = 0$. Thus, equation (2.157) becomes:

$$\delta\boldsymbol{\varepsilon}^{i+1} = \delta\Delta\boldsymbol{\varepsilon} \tag{2.158}$$

By substituting relations (2.158) and (2.151) into (2.150), it is derived that:

$$\begin{gathered}\int_V (\delta\Delta\boldsymbol{\varepsilon})^T \boldsymbol{\sigma}^i dV + \int_V (\delta\Delta\boldsymbol{\varepsilon})^T \Delta\boldsymbol{\sigma} dV = \\ \int_V \rho(\delta\mathbf{u})^T \mathbf{f} dV + \int_S (\delta\mathbf{u})^T \mathbf{t} dS\end{gathered} \tag{2.159}$$

The first component of the left-hand part of equation (2.159) is a constant quantity in respect with the displacement increment, since the variation of the incremental strain $\delta\Delta\boldsymbol{\varepsilon}$ is constant and the stress tensor $\boldsymbol{\sigma}^i$ is known. The second component of the left-hand part is non-linear, due to the non-linear stress versus strain constitutive description. Linearization of this term is then applied, using the Taylor series expansion presented in equation (2.82):

$$\int_V (\delta\Delta\boldsymbol{\varepsilon})^T \Delta\boldsymbol{\sigma} dV = \int_V (\delta\Delta\boldsymbol{\varepsilon})^T \left(\frac{\partial\boldsymbol{\sigma}}{\partial\boldsymbol{\varepsilon}^i}\Delta\boldsymbol{\varepsilon}\right) dV \tag{2.160}$$

where the instantaneous stiffness modulus $\mathbf{D}$ derived from the constitutive description, also known in literature as *consistent tangent stiffness* operator, is defined here as:

$$\mathbf{D} = \frac{\partial\boldsymbol{\sigma}}{\partial\boldsymbol{\varepsilon}^i} \tag{2.161}$$

The linearized equilibrium equation within continuum mechanics is then obtained, after substituting (2.160) and (2.161) into (2.159):

$$\begin{gathered}\int_V (\delta\Delta\boldsymbol{\varepsilon})^T \boldsymbol{\sigma}^i dV + \int_V (\delta\Delta\boldsymbol{\varepsilon})^T \mathbf{D}\Delta\boldsymbol{\varepsilon} dV = \\ \int_V \rho(\delta\mathbf{u})^T \mathbf{f} dV + \int_S (\delta\mathbf{u})^T \mathbf{t} dS\end{gathered} \tag{2.162}$$

Next, discretization of the above equation will be conducted in the framework of finite element analysis. As already explained in previous sections, the incremental strain $\Delta\boldsymbol{\varepsilon}$ and its variation $\delta\Delta\boldsymbol{\varepsilon}$ are expressed in terms of the nodal displacements $\mathbf{u}_e$ as follows:

$$\Delta\boldsymbol{\varepsilon} = \mathbf{B}\Delta\mathbf{u}_e \Rightarrow \delta\Delta\boldsymbol{\varepsilon} = \mathbf{B}\delta\Delta\mathbf{u}_e \tag{2.163}$$

where the strain-displacement matrix $\mathbf{B}$ is the same with the one used also in linear finite element analysis. In addition, variation of (2.156) results in:

$$\delta\mathbf{u}^{i+1} = \delta\mathbf{u}^i + \delta\Delta\mathbf{u} \Rightarrow \delta\mathbf{u}^{i+1} = \delta\Delta\mathbf{u} = \mathbf{N}\delta\Delta\mathbf{u}_e \tag{2.164}$$

where matrix $\mathbf{N}$ represents the shape functions for the element. It is also noticed that $\mathbf{u}^i$ is constant and known and thus, its variation is equal to zero. By substituting $\delta\Delta\boldsymbol{\varepsilon}$, $\Delta\boldsymbol{\varepsilon}$ and $\delta\mathbf{u}^{i+1}$ from equations (2.163), (2.164) into (2.162), it is obtained that:

$$\begin{gathered}\sum_{i=1}^{N_e}\int_{V_e}(\delta\Delta\mathbf{u}_e)^T\mathbf{B}^T\boldsymbol{\sigma}^i dV + \sum_{i=1}^{N_e}\int_{V_e}(\delta\Delta\mathbf{u}_e)^T\mathbf{B}^T\mathbf{D}\mathbf{B}\Delta\mathbf{u}_e dV = \\ \sum_{i=1}^{N_e}\int_{V_e}(\delta\Delta\mathbf{u}_e)^T\rho\mathbf{N}^T\mathbf{f}dV + \sum_{i=1}^{N_e}\int_{S_e}(\delta\Delta\mathbf{u}_e)^T\mathbf{N}^T\mathbf{t}dS\end{gathered} \tag{2.165}$$

where summation over the volume $V_e$ and surface $S_e$ of each element has been considered, to indicate the assembly process over the number $N_e$ of all finite elements. By considering the nodal displacement vector $\mathbf{U}$ of the whole structure as the assembly of the nodal displacement vector $\mathbf{u}_e$ of each element, equation (2.165) becomes:

$$\begin{gathered}(\delta\Delta\mathbf{U})^T\sum_{i=1}^{N_e}\int_{V_e}\mathbf{B}^T\boldsymbol{\sigma}^i dV + (\delta\Delta\mathbf{U})^T\sum_{i=1}^{N_e}\int_{V_e}\mathbf{B}^T\mathbf{D}\mathbf{B}dV\Delta\mathbf{U} = \\ (\delta\Delta\mathbf{U})^T\sum_{i=1}^{N_e}\int_{V_e}\rho\mathbf{N}^T\mathbf{f}dV + (\delta\Delta\mathbf{U})^T\sum_{i=1}^{N_e}\int_{S_e}\mathbf{N}^T\mathbf{t}dS\end{gathered} \tag{2.166}$$

The above equation is written in a final form after eliminating all $(\delta\Delta\mathbf{U})^T$ terms and bringing the first component of the left-hand part of the equation, which represents the internal force vector, into the right-hand part:

$$\begin{gathered}\left(\sum_{i=1}^{N_e}\int_{V_e}\mathbf{B}^T\mathbf{D}\mathbf{B}dV\right)\Delta\mathbf{U} = \\ \sum_{i=1}^{N_e}\int_{V_e}\rho\mathbf{N}^T\mathbf{f}dV + \sum_{i=1}^{N_e}\int_{S_e}\mathbf{N}^T\mathbf{t}dS - \sum_{i=1}^{N_e}\int_{V_e}\mathbf{B}^T\boldsymbol{\sigma}^i dV\end{gathered} \tag{2.167}$$

or

$$\mathbf{K}_T \Delta \mathbf{U} = \mathbf{F}_{ext} - \mathbf{F}_{int} \tag{2.168}$$

where $\mathbf{K}_T$ is the tangent stiffness matrix, $\mathbf{F}_{ext}$ is the external force vector and $\mathbf{F}_{int}$ the internal force vector, both provided by:

$$\mathbf{F}_{ext} = \sum_{i=1}^{N_e} \int_{V_e} \rho \mathbf{N}^T \mathbf{f} dV + \sum_{i=1}^{N_e} \int_{S_e} \mathbf{N}^T \mathbf{t} dS \tag{2.169}$$

$$\mathbf{F}_{int} = \sum_{i=1}^{N_e} \int_{V_e} \mathbf{B}^T \boldsymbol{\sigma}^i dV \tag{2.170}$$

Equation (2.168) is solved and the unknown nodal displacement increment $\Delta \mathbf{U}$ is determined at every iteration, noticing that equation (2.168) is expressed using iterations by relation (2.91). The total displacement vector is then calculated using equation (2.92).

# Chapter 3

# Failure of heterogeneous materials using non-linear continuum laws

## 3.1 Introduction

Structural failure of materials may involve *creep*, which is time dependent damage, *fatigue* which is caused by repetitive or cyclic load, *ductile* damage in metals, including nucleation of cavities or microcracks and their propagation-coalescence as well as *brittle* or *quasi-brittle* damage in concrete, rocks, masonry and fibre reinforced composites including fibre debonding or breaking. When quasi-brittle damage appears, fracture is the result of nucleation, growth and coalescence of microscopic defects rather than the effect of one dominant crack. Thus, a gradual decrease of the deformation resistance is obtained, instead of the sudden loss of integrity observed in perfectly brittle fracture.

Non-linear continuum laws are widely used to depict some of the mentioned failure types on composite heterogeneous structures. Two of the most known numerical methods of this category are represented by *continuum damage mechanics* and *plasticity*. When during unloading the irreversible (or plastic) strain is negligible like in quasi-brittle materials, continuum damage laws can be adopted to simulate failure. However, when the irreversible strain is not negligible, which is the case for ducticle materials e.g. metals, then plasticity is adopted to simulate failure.

Within these methods, a constitutive description of failure is assigned in the structural length scale, which is also known as *macroscopic scale*. Experimental investigation, which incorporates interactive effects of the composite constituents, can be used to calibrate the parameters of the adopted laws. Furthermore, in some cases damage in lower length scales (*meso*, or *microscopic scale*) may be represented in these models by using internal damage variables. However, they cannot consider the interaction between the constituent materials taking place in lower length scales, as well as the influence of this interaction on failure. Thus, these models depict a phenomenological nature in the sense that they describe in the (homogeneous) structural scale, the failure which actually occurs in the constituent materials in lower length scales, without representing the interaction between those constituent materials.

DOI: 10.1201/9781003017240-3

On the other hand, the application of non-linear continuum laws in finite element models is relatively simple and the underlying computational cost manageable. Therefore, these methods are appropriate for practical applications, involving large structural systems.

---

## 3.2 Plasticity

A quite general framework for the description of non-linear material behaviour is the so-called *continuum theory with internal variables. Internal variables* describe, in a phenomenological way, every accumulated within the previous time or loading history change in the material behaviour [118]. Within the most common theory of elastoplasticity, the plastic flow is considered as an irreversible process characterized in terms of the history of the total strain tensor $\boldsymbol{\varepsilon}$ and two kinematic variables, the plastic strain tensor $\boldsymbol{\varepsilon}^p$ and a set of strain-like internal variables $\boldsymbol{\xi}$, which are second-order tensors or scalars describing internal irreversible phenomena like material hardening.

Following the theory of internal variables, the thermodynamically conjugates to the kinematic variables are the stress tensor $\boldsymbol{\sigma}$ and the stress-like internal variables called thermodynamic forces $\boldsymbol{\chi}$ [84], [87].

Plasticity may be expressed in a *strain-space* or a *stress-space* formulation. Within the strain-space formulation, the plastic flow is described in terms of the strain-like variables $(\boldsymbol{\varepsilon}, \boldsymbol{\varepsilon}^p, \boldsymbol{\xi})$. The stress-like variables $(\boldsymbol{\sigma}, \boldsymbol{\chi})$ are considered as dependent. This approach is commonly adopted in computational methods [129], [192]. It is further assumed that the total strain $\boldsymbol{\varepsilon}$ can be additively decomposed into the elastic strain and the plastic strain components:

$$\boldsymbol{\varepsilon} = \boldsymbol{\varepsilon}^e + \boldsymbol{\varepsilon}^p \tag{3.1}$$

For metals, plasticity is related to the deviatoric component of $\boldsymbol{\varepsilon}$. Then, the split of the second-order strain tensor into the *volumetric* and *deviatoric* components reads:

$$\boldsymbol{\varepsilon} = \frac{1}{3}\theta \mathbf{I} + \mathbf{e} \tag{3.2}$$

Relation (3.2) can in turn be used for both the elastic and the plastic components as follows:

$$\boldsymbol{\varepsilon}^e = \frac{1}{3}\theta^e \mathbf{I} + \mathbf{e}^e \tag{3.3}$$

$$\boldsymbol{\varepsilon}^p = \frac{1}{3}\theta^p \mathbf{I} + \mathbf{e}^p \tag{3.4}$$

noticing that $\theta^i = tr(\boldsymbol{\varepsilon}^i)$ for $i = e, p$. For many materials, including metals, the plastic deformation is assumed to be of deviatoric or shearing type, i.e.:

$$\boldsymbol{\varepsilon}^p = \mathbf{e}^p, \quad \theta^p = 0 \tag{3.5}$$

Furthermore, the *Helmholtz free energy* $\Psi$ is a function of the total strain, plastic strain and internal variables, i.e.:

$$\Psi = \Psi(\boldsymbol{\varepsilon}, \mathbf{e}^p, \boldsymbol{\xi}) = \Psi^e(\boldsymbol{\varepsilon}^e) + \Psi^p(\boldsymbol{\xi}) \tag{3.6}$$

derived due to the adopted additive decomposition. The *Gibbs free energy* $G$ is defined through a Legendre transformation of the Helmholtz free energy:

$$G = \boldsymbol{\sigma} : \boldsymbol{\varepsilon} - \Psi \tag{3.7}$$

In the case of isotropic linearized elasticity the previously introduced energy quantities are quadratic forms:

$$\Psi = \frac{1}{2}\boldsymbol{\varepsilon}^e : \mathcal{C}\boldsymbol{\varepsilon}^e + \frac{1}{2}\boldsymbol{\xi} \star \mathcal{D} \star \boldsymbol{\xi} \tag{3.8}$$

$$G = \frac{1}{2}\boldsymbol{\sigma} : \mathcal{C}^{-1}\boldsymbol{\sigma} + \frac{1}{2}\boldsymbol{\chi} \star \mathcal{D}^{-1} \star \boldsymbol{\chi} \tag{3.9}$$

Here the elastic tangent $\mathcal{C}$ is given by $\mathcal{C} = \frac{\partial^2 \Psi^e}{\partial(\boldsymbol{\varepsilon}^e)^2} = 3K\mathcal{I}_{vol} + 2\mu\mathcal{I}_{dev}$, the symbol $\star$ indicates the appropriate product between $\boldsymbol{\xi}$ and $\mathcal{D}$ and between $\boldsymbol{\chi}$ and $\mathcal{D}^{-1}$, respectively, and $\mathcal{D}$ is the matrix of generalized plastic moduli. In addition, $\mathcal{I}_{vol}, \mathcal{I}_{dev}$ are the volumetric and deviatoric components of the fourth-order identity tensor $\mathcal{I}$ and $\mu$, $K$ denote the shear and bulk modulus, respectively.

In elastoplasticity the evolution of stresses and corresponding plastic deformation is described by a convex scalar function $f$, the *yield criterion*. Admissible couples of $(\boldsymbol{\sigma}, \boldsymbol{\chi})$ are defined by:

$$E_{ad\sigma} = \{\boldsymbol{\sigma}, \boldsymbol{\chi} : f(\boldsymbol{\sigma}, \boldsymbol{\chi}) \leq 0\} \tag{3.10}$$

The elastic region is characterized by the interior of the set $E_{ad\sigma}$.

In addition, the stress $\boldsymbol{\sigma}$ can be expressed using the Helmholtz free energy, as provided by the following relation:

$$\boldsymbol{\sigma} = \frac{\partial \Psi}{\partial \boldsymbol{\varepsilon}^e} \tag{3.11}$$

and be split into its volumetric and deviatoric components, as follows:

$$\boldsymbol{\sigma} = p\mathbf{I} + \mathbf{s}, \quad p = \frac{1}{3}tr(\boldsymbol{\sigma}) \tag{3.12}$$

where $p$ and $\mathbf{s}$ denote the pressure and stress deviator, respectively.

At every point, time or loading increment for which the boundary of the *yield surface* is reached, plastic behaviour takes place. In metals, the yield criterion $f$ is independent from pressure, contrary to other materials like soils, concrete, rock or granular, where the yield criterion depends on pressure. For example, the pressure-insensitive *Tresca yield criterion* is defined by:

$$f = \tau_{max} - \tau_{yield} = (\sigma_{max} - \sigma_{min}) - \sigma_{yield} \tag{3.13}$$

where $\tau_{yield}$ is the shear yield stress, $\sigma_{yield} = 2\tau_{yield}$ is the uniaxial yield stress, $\tau_{max} = \frac{1}{2}(\sigma_{max} - \sigma_{min})$ is the maximum shear stress and $\sigma_{max}$, $\sigma_{min}$ denote the maximum and minimum principal stresses, respectively. The *von Mises yield criterion*, which is also pressure-insensitive, is written by using:

$$f = \sqrt{3J_2} - \sigma_{yield} \tag{3.14}$$

Here $\sigma_{yield}$ is the uniaxial yield stress and $\sqrt{3J_2}$ is the von Mises effective stress, with the second scalar invariant of the stress deviator $J_2 = \frac{1}{2}\mathbf{s} : \mathbf{s}$.

An example of a pressure-sensitive criterion is the *Mohr-Coulomb yield criterion* which is expressed by:

$$f = \tau - c + \sigma_n tan\phi = (\sigma_{max} - \sigma_{min}) + (\sigma_{max} + \sigma_{min})\, sin\phi - 2c(cos\phi) \tag{3.15}$$

In equation (3.15) $\tau$ is the shear stress, $c$ is the cohesion, $\phi$ is the angle of internal friction and $\sigma_n$ is the normal stress. The plasticity model is completed by introducing suitable *evolution laws* for the plastic strain and the strain-like internal variables. They are also known as plastic flow rule and hardening law, respectively, and take the following form:

$$\dot{\boldsymbol{\varepsilon}}^p = \dot{\gamma}\mathbf{n}, \quad \dot{\boldsymbol{\xi}} = \dot{\gamma}\mathbf{h} \tag{3.16}$$

where $\dot{\gamma}$ is the non-negative consistency parameter, $\mathbf{n} = \mathbf{n}(\boldsymbol{\sigma}, \boldsymbol{\chi})$ is the flow vector which describes the direction of the plastic flow and $\mathbf{h} = \mathbf{h}(\boldsymbol{\sigma}, \boldsymbol{\chi})$ is the generalized hardening modulus. The evolution laws are often defined in terms of a flow (or plastic) potential $Y = (\boldsymbol{\sigma}, \boldsymbol{\chi})$ such that:

$$\mathbf{n} = \frac{\partial Y}{\partial \boldsymbol{\sigma}}, \quad \mathbf{h} = \frac{\partial Y}{\partial \boldsymbol{\chi}} \tag{3.17}$$

In equation (3.17), $Y$ is a non-negative convex function of $\boldsymbol{\sigma}$ and $\boldsymbol{\chi}$ with zero value at the origin, which coincides with the flow potential when *associative plasticity* appears. In this case the evolution laws are called normality rules and the plastic strain rate is a tensor normal to the yield surface in the space of stresses. It is noted that for non-differentiable functions $Y$, the so-called pseudo- or generalized-potential and the flow rules are extended with the usage of the concept of the subdifferential of convex analysis [145], [165] and the treatment of contact problems (see relation (4.8) in chapter 4).

During loading in the plastic range, the yield surface of the material changes in shape and size. This change is defined by the *hardening rule*. In *perfect plasticity* no hardening is allowed. This mechanism can be described by a single internal variable, for instance $\boldsymbol{\varepsilon}^p$. *Isotropic hardening* leads to a uniform expansion of the initial yield surface without shifting. Stress-strain diagrams for perfect and strain-hardening plasticity are depicted in figure 3.1.

The set $\xi$ contains a scalar variable which determines the size of the yield surface. Strain-hardening is characterized by a scalar internal variable representing a measure of the accumulated plastic deformation. The case of linear hardening reads:

$$\sigma_{yield} = \sigma_{yield,0} + H_{isohard}\bar{\varepsilon}^p \tag{3.18}$$

where $\sigma_{yield,0}$ is the uniaxial initial yield stress, $H_{isohard}$ is the linear isotropic hardening modulus and $\bar{\varepsilon}^p$ is the accumulated plastic strain. Work-hardening adopts the dissipated plastic work $W^p$ as the scalar variable defining the state of hardening, where:

$$\dot{W}^p = \boldsymbol{\sigma} : \dot{\boldsymbol{\varepsilon}}^p \tag{3.19}$$

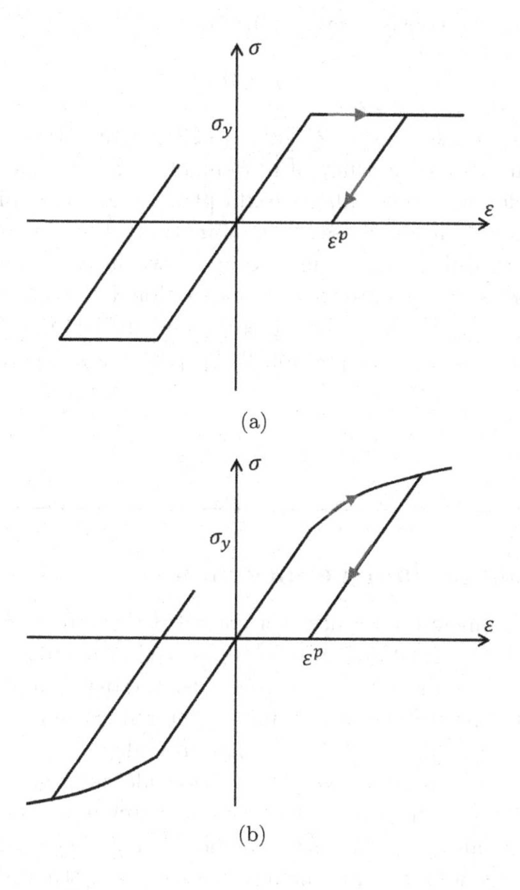


**FIGURE 3.1**: Stress-strain diagrams depicting a) perfect plasticity and b) strain-hardening plasticity.

A shifting of the initial yield surface is adopted in the *kinematic hardening* mechanism, and described by a single internal variable. For example, the von Mises yield function with shifting reads:

$$f = \sqrt{3J_2(\mathbf{s} - \boldsymbol{\alpha})} - \sigma_{yield} \tag{3.20}$$

where $\boldsymbol{\alpha}$ is the symmetric back-stress tensor that represents the translation of the yield surface in the space of stresses, $\mathbf{s} - \boldsymbol{\alpha}$ is the relative stress tensor, noticing that some evolution law for $\boldsymbol{\alpha}$ must be adopted.

The plasticity model is completed with the definition of the consistency parameter $\dot{\gamma}$ and the yield criterion $f$. Hence, the *Kuhn-Tucker complementarity conditions* are introduced:

$$\dot{\gamma} \geq 0, \quad f \leq 0, \quad \dot{\gamma} f = 0 \tag{3.21}$$

and coupled with the consistency condition:

$$\dot{\gamma} \dot{f} = 0 \tag{3.22}$$

The explanation of relations (3.21) and (3.22) representing the loading/unloading cases and the consistency of movement with the position of the yield surface can be found in specialized texts [92]. From the exploitation of the above conditions for various possible combinations, the more classical elasto-plastic tangent modulus can be derived as shown in [192]. Furthermore, the link to convex analysis and the theory of variational inequalities, analogously to the treatment of unilateral contact problems outlined in chapter 4, can be found in specialized texts [129], [148], [165], [184] and in review articles like [189].

---

## 3.3 Continuum damage mechanics

Continuum damage mechanics is a constitutive theory which depicts the progressive loss of material integrity taking place in the microscopic scale, due to propagation and coalescence of microcracks, microvoids and similar defects. This failure of the microstructure results in the reduction of the strength and stiffness on the macroscopic, structural length scale.

To characterize damage on the microscopic scale, *internal damage variables* are introduced. Depending on the required accuracy and type of application, different types of damage variables are defined. For the simplest representation of isotropic damage with no prominent orientation, one scalar damage variable $\omega$ with values between 0 and 1 is used. For $\omega = 0$ the material is intact while for $\omega = 1$ the material is totally damaged, indicating complete loss of integrity. This approach is very practical and for this reason is widely used. Two scalar-valued damage variables can also be adopted for isotropic damage of isotropic materials, one corresponding to Young's modulus and the other corresponding to Poisson's ratio. If Poisson's ratio is considered constant during the damage process, this reduces to the case of one scalar damage variable.

Several scalar damage variables may also be used, associated with predefined material directions, such as for fibre reinforced composites. For more complex cases of anisotropic damage, second-order (for initially isotropic materials) and fourth-order damage variables tensors (for the most general case of initially isotropic or anisotropic materials) can be adopted to capture damage.

For the structure shown in figure 3.2, the increase of the loading results in the gradual development of damage, represented by the scalar damage variable $\omega$. In region $\Omega_0$, no damage arises and $\omega = 0$. In region $\Omega_d$, some damage arises but not in a critical level noticing that $0 < \omega < 1$ while in region $\Omega_c$ the material is totally damaged indicating complete loss of integrity and $\omega = 1$.

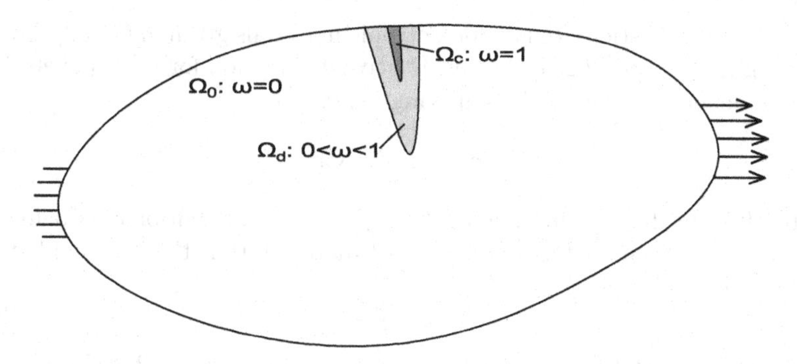


**FIGURE 3.2**: Distribution of damage in a structure.

### 3.3.1 Isotropic damage law

The constitutive stress-strain description is given within damage mechanics in the general form provided by the following equation:

$$\boldsymbol{\sigma} = \mathbf{D}^s(\omega, \boldsymbol{\omega}, \boldsymbol{\Omega})\boldsymbol{\varepsilon} \tag{3.23}$$

with $\omega$, $\boldsymbol{\omega}$ and $\boldsymbol{\Omega}$ denoting a scalar, second-order and fourth-order tensor representing the internal damage variables, $\boldsymbol{\sigma}$, $\boldsymbol{\varepsilon}$ being the stress and strain tensors in vector format and $\mathbf{D}^s$ expressing the *secant*, fourth-order elasticity tensor. As mentioned previously, when the degradation of stiffness is assumed to be isotropic and thus independent of the loading direction, two scalar damage variables $\omega_1$ and $\omega_2$ are introduced. The secant elasticity tensor $\mathbf{D}^s$ is then built from the elasticity tensor for isotropic materials given in relation (1.152), as follows:

$$\mathbf{D}^s = \frac{E^s}{(1+\nu^s)(1-2\nu^s)} \begin{bmatrix} 1-\nu^s & \nu^s & \nu^s & 0 & 0 & 0 \\ & 1-\nu^s & \nu^s & 0 & 0 & 0 \\ & & 1-\nu^s & 0 & 0 & 0 \\ & & & \frac{1-2\nu^s}{2} & 0 & 0 \\ & \text{Sym.} & & & \frac{1-2\nu^s}{2} & 0 \\ & & & & & \frac{1-2\nu^s}{2} \end{bmatrix} \tag{3.24}$$

where $E^s = (1 - \omega_1)E$ and $\nu^s = (1 - \omega_2)\nu$ are the secant elasticity modulus and Poisson's ratio, respectively. When Poisson's ratio is considered constant during damage process, then one scalar damage variable $\omega$ is used. In this case, the secant elasticity tensor is provided by:

$$\mathbf{D}^s = (1 - \omega)\mathbf{C} \tag{3.25}$$

where $\mathbf{C}$ is the elasticity tensor for isotropic materials given in (1.152). By substituting (3.25) into (3.23), the constitutive description for isotropic elasticity based-damage is given by the following expression:

$$\boldsymbol{\sigma} = (1 - \omega)\mathbf{C}\boldsymbol{\varepsilon} \tag{3.26}$$

Quite often in damage mechanics, the *effective stress* tensor $\hat{\boldsymbol{\sigma}}$ is defined as the stress tensor characterizing the intact material, thus the material between microcracks and cavities:

$$\hat{\boldsymbol{\sigma}} = \mathbf{C}\boldsymbol{\varepsilon} \tag{3.27}$$

Using equation (3.27), equation (3.26) can then be expressed as follows:

$$\boldsymbol{\sigma} = (1 - \omega)\hat{\boldsymbol{\sigma}} \tag{3.28}$$

To determine whether damage growth is possible, the *damage loading function* $f$ is defined in the general form of equation (3.29):

$$f = f(\tilde{\varepsilon}, \tilde{\sigma}, \kappa) \tag{3.29}$$

where the *equivalent* strain $\tilde{\varepsilon}$ and stress $\tilde{\sigma}$ are introduced in order to provide a strain and stress scalar measure, obtained using the components of the corresponding strain and stress tensors. With $\kappa$ is denoted an internal damage variable being a function of $\omega$ and thus, $\omega = \omega(\kappa)$. As it is explained in the following paragraphs, the reason that $\kappa$ is introduced, is to define damage when non-monotonic loading, e.g. unloading and reloading takes place and thus, $\kappa$ can be considered as a *history dependent* variable.

An assumption that is often considered, is to adopt a function $f$ which depends only on equivalent strain and not on equivalent stress:

$$f = \tilde{\varepsilon} - \kappa \tag{3.30}$$

Different expressions are found in literature for the equivalent strain. An appropriate expression for metal materials is provided below:

$$\tilde{\varepsilon} = \frac{1}{2}\boldsymbol{\varepsilon}^T\mathbf{C}\boldsymbol{\varepsilon} \tag{3.31}$$

A slightly modified expression of a dimensionless equivalent strain is given in (3.32), by dividing the numerator of (3.31) over Young's modulus $E$. Using

(3.32) for uniaxial conditions, the equivalent strain becomes equal to axial strain and for this reason it can be more convenient than (3.31).

$$\tilde{\varepsilon} = \sqrt{\frac{\boldsymbol{\varepsilon}^T \mathbf{C} \boldsymbol{\varepsilon}}{E}} \tag{3.32}$$

Relation (3.32) does not distinguish between tensile and compressive behaviour, indicating that it cannot properly represent quasi-brittle materials like concrete, which depict a significantly higher strength in compression than in tension. For this reason, an alternative expression for the equivalent strain is provided by the following relation [138]:

$$\tilde{\varepsilon} = \sqrt{\sum_{i=1}^{3} \langle \varepsilon_i \rangle^2} \tag{3.33}$$

where the principal strains $\varepsilon_i$ are introduced inside MacAulay brackets $\langle . \rangle$ defined such that $\langle \varepsilon_i \rangle = \varepsilon_i$ if $\varepsilon_i > 0$ and $\varepsilon_i = 0$ if $\varepsilon_i < 0$. Using this equation, the change of the equivalent strain is more sensitive under tension than under compression, noticing that increasing positive (tensile) principal strain components result in increasing equivalent strains while increasing negative (compressive) principal strains result in the same equivalent strain.

Another definition for the equivalent strain is introduced using a modified von Mises criterion [59], originating from plasticity models for polymers, initially formulated in terms of stresses. This definition for the equivalent strain is provided below:

$$\tilde{\varepsilon} = \frac{k-1}{2k(1-2\nu)} I_1 + \frac{1}{2k} \sqrt{\frac{(k-1)^2}{(1-2\nu)^2} I_1^2 - \frac{12k}{(1+\nu)^2} J_2} \tag{3.34}$$

where $I_1$ and $J_2$ denote the first invariant of the strain tensor and the second invariant of the deviatoric strain tensor, given by:

$$I_1 = \varepsilon_{ii} = tr(\boldsymbol{\varepsilon}), \quad J_2 = \frac{1}{6} I_1^2 - \frac{1}{2} \varepsilon_{ij} \varepsilon_{ij} \tag{3.35}$$

The parameter $k$ is used to express the sensitivity of the compression to tension relation and therefore, it is usually defined as the ratio between the compressive and the tensile strength.

After defining the damage loading function $f$ and the equivalent strain, relation $f = 0$ is applied to introduce a *loading surface* in the strain space. This compares to the elastoplastic *yield surface* defined in the stress space within plasticity. For strain states inside the loading surface, $f < 0$ and no damage growth arises. When $f \geq 0$ then damage arises and relation $f = 0$ is imposed, indicating that $\tilde{\varepsilon} = \kappa$.

It is also noticed that the scalar parameter $\kappa$ of equation (3.30) is a history dependent parameter, since it is influenced by the loading history, according to the descriptions which follow. At step 1 of the stress-strain diagram shown in figure 3.3, the strain increases and the material is elastic until the stress reaches the strength $\sigma_0$. After that point and during step 2 on the diagram, the strain increases further but the stress decreases, indicating that damage arises, $\omega > 0$ and $f = 0$. For every point on this step of the diagram, the secant stiffness is adopted. Next, loading is gradually removed and the unloading step 3 with decreasing strain (and equivalent strain) is obtained as shown in figure 3.3. During this step, no damage growth takes place and the internal damage variable calculated at the end of step 2 remains constant indicating that $f < 0$ and that the derivative over time for the parameter $\kappa$ is zero, thus, $\dot{\kappa} = 0$. In addition, the material behaves elastically with the secant stiffness provided by the reduced Young's modulus $(1 - \omega)E$.

Step 4 involves loading with the secant stiffness, noticing that the slope of step 4 is lower than slope of step 1. This is attributed to the damage which has been developed at the end of step 2, just before the unloading step 3 begins. At the end of step 4 of the diagram, strain as well as damage growth start increasing, at the point where the unloading of step 3 started. At this point $f = 0$ indicating that the parameter $\kappa$ is equal to the equivalent strain $\tilde{\varepsilon}$. According to this description, this equivalent strain is the one which had been calculated at the end of step 2 and since during steps 3 and 4 lower strains arise, it can be stated that $\kappa$ is equal to the largest value of the equivalent strain which was locally assigned during the loading history.

Mathematically these descriptions are expressed by the following, Karush-Kuhn-Tucker conditions:

$$f \leq 0, \quad \dot{\kappa} \geq 0, \quad f\dot{\kappa} = 0 \tag{3.36}$$

To complete the description of isotropic damage mechanics, a *damage evolution* law connecting the internal damage variables $\omega$ and $\kappa$ should be provided. Such relations can be found in literature for different materials and failure types. For quasi-brittle materials such as concrete, a linear or an exponential softening stress-strain law can be adopted, according to figure 3.4. For each stress-strain law, a corresponding damage evolution law is chosen, as shown in figure 3.4(a) for linear and in figure 3.4(b) for exponential softening, respectively [170].

By assuming a uniaxial response and that the equivalent strain $\tilde{\varepsilon}$ is equal to the axial strain $\varepsilon$, the linear softening law shown in figure 3.4(a) results in complete loss of integrity when the strain reaches the critical value $\varepsilon = \kappa_c$ and the damage variable $\omega$ becomes equal to 1. The exponential softening law shown in figure 3.4(b) depicts initially a drastic reduction of stress with increasing strain, followed by a more moderate decrease of stress. It is also noticed that for this case, the damage variable $\omega$ only asymptotically approaches the value 1.

The corresponding damage evolution relations are provided in (3.37) for linear softening and in (3.38) for exponential softening, respectively. Parameters $\alpha$ and $\beta$ in relation (3.38) adjust the shape of the stress-strain and the

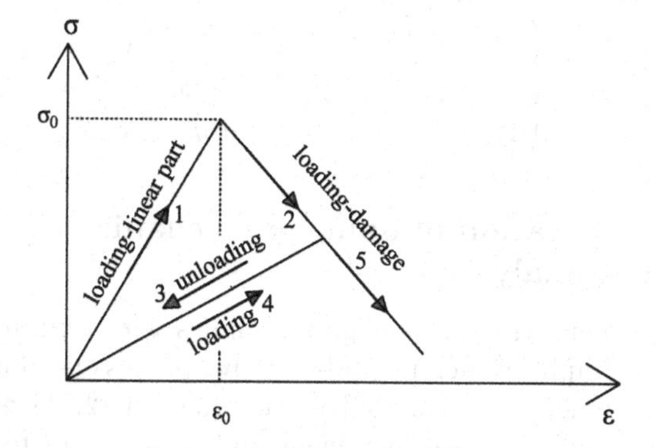


**FIGURE 3.3**: Stress-strain response representing a uniaxial continuum damage law.

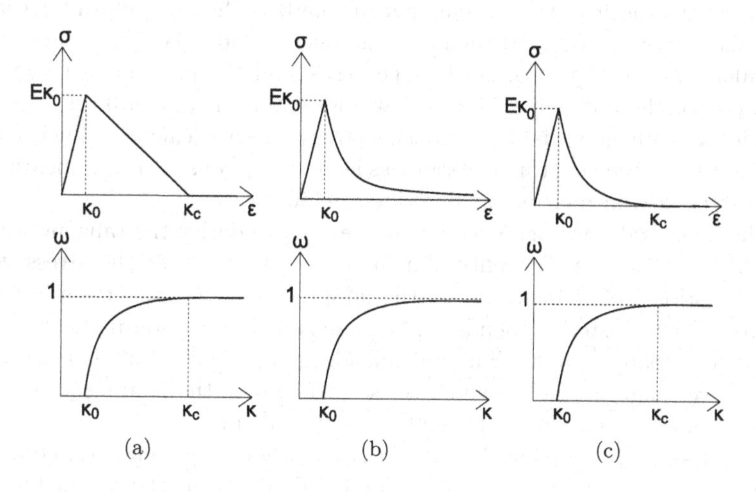


**FIGURE 3.4**: Uniaxial stress-strain and damage evolution laws for a) linear, b) exponential and c) modified exponential softening response.

evolution law. To describe failure of short glass-fibre reinforced polymers, a modified power law provided by figure 3.4(c) and equation (3.39), has been proposed in [80].

$$\omega = \begin{cases} 0 & \text{if } \kappa \leq \kappa_0 \\ \frac{\kappa_c}{\kappa}\frac{\kappa-\kappa_0}{\kappa_c-\kappa_0} & \text{if } \kappa_0 \leq \kappa \leq \kappa_c \\ 1 & \text{if } \kappa_c \leq \kappa \end{cases} \tag{3.37}$$

$$\omega = \begin{cases} 0 & \text{if } \kappa \leq \kappa_0 \\ 1 - \frac{\kappa_0}{\kappa}(1 - \alpha + \alpha e^{-\beta(\kappa-\kappa_0)} & \text{if } \kappa_0 \leq \kappa \end{cases} \tag{3.38}$$

$$\omega = \begin{cases} 0 & \text{if } \kappa \leq \kappa_0 \\ 1 - (\frac{\kappa_0}{\kappa})^\beta (\frac{\kappa_c - \kappa}{\kappa_c - \kappa_0})^\alpha & \text{if } \kappa_0 \leq \kappa \leq \kappa_c \\ 1 & \text{if } \kappa_c \leq \kappa \end{cases} \tag{3.39}$$

### 3.3.2 Implementation of damage mechanics within finite element analysis

For the implementation of damage mechanics within finite element analysis, the Newton-Raphson incremental-iterative process is adopted. After the displacement increment is calculated using equation (2.91) given in section 2.4.2, the corresponding strain increment and total strain tensor are calculated, for every Gauss point. Using the total strain, the equivalent strain is also calculated and the damage load function $f$ is evaluated using equation (3.30), where the history parameter $\kappa$ in (3.30) is the one determined in the previous iteration. If $f \geq 0$ then damage occurs and $\kappa$ is taken equal to the equivalent strain. If $f < 0$, no damage arises and the previous $\kappa$ is kept. Finally, one of the damage evolution laws presented in section 3.3.1 is used to provide the damage variable $\omega$, the new stress tensor is calculated using equation (3.26) and the new tangent stiffness is found by considering a linearization of the stress-strain relation, as described below.

The linearization process which will be used to derive the tangent stiffness matrix, relies on the differentiation in respect to time, of the stress versus strain relation (3.26) provided by the adopted damage law. Time is needed here to describe history dependent phenomena, involving non-monotonic loading, such as loading, unloading and reloading. Therefore, time is a monotonically increasing parameter which is used to control the loading process and thus, it does not generally represent the physical time.

As presented in section 3.3.1, $\omega = \omega(\kappa)$, where $\kappa$ depends on equivalent strain $\tilde{\varepsilon}$ according to relation (3.30) and $\tilde{\varepsilon}$ depends on the strain tensor $\boldsymbol{\varepsilon}$. Therefore, differentiation of the stress-strain relation (3.26) in respect to time, results in the following equation:

$$\dot{\boldsymbol{\sigma}} = (1 - \omega)\mathbf{C}\dot{\boldsymbol{\varepsilon}} - \dot{\omega}\mathbf{C}\boldsymbol{\varepsilon} \tag{3.40}$$

where $\dot{\boldsymbol{\sigma}}$ is the derivative of the stress over the time or the *stress rate* and $\dot{\boldsymbol{\varepsilon}}$ the derivative of the strain over the time or the *strain rate*. The derivative of the internal damage variable $\omega$ over time, $\dot{\omega}$, is then calculated, as follows:

$$\dot{\omega} = \frac{\partial \omega}{\partial \kappa} \frac{\partial \kappa}{\partial \tilde{\varepsilon}} \frac{\partial \tilde{\varepsilon}}{\partial \boldsymbol{\varepsilon}} \dot{\boldsymbol{\varepsilon}} \tag{3.41}$$

By substituting equation (3.41) into (3.40), it is derived that:

$$\dot{\boldsymbol{\sigma}} = \left( (1 - \omega)\mathbf{C} - \frac{\partial \omega}{\partial \kappa} \frac{\partial \kappa}{\partial \tilde{\varepsilon}} (\mathbf{C}\boldsymbol{\varepsilon}) (\frac{\partial \tilde{\varepsilon}}{\partial \boldsymbol{\varepsilon}})^T \right) \dot{\boldsymbol{\varepsilon}} \tag{3.42}$$

Equation (3.42) can be written alternatively, in the form:

$$\dot{\boldsymbol{\sigma}} = \left( (1-\omega)\mathbf{C} - \frac{\partial \omega}{\partial \kappa}\frac{\partial \kappa}{\partial \tilde{\varepsilon}}(\mathbf{C}\boldsymbol{\varepsilon}) \otimes \frac{\partial \tilde{\varepsilon}}{\partial \boldsymbol{\varepsilon}} \right) \dot{\boldsymbol{\varepsilon}} \tag{3.43}$$

where the dyadic product between vectors $\mathbf{C}\boldsymbol{\varepsilon}$ and $\frac{\partial \tilde{\varepsilon}}{\partial \boldsymbol{\varepsilon}}$ is used.

It is also noticed that during loading, the damage load function $f$ is zero which results in $\tilde{\varepsilon} = \kappa$ and thus, $\frac{\partial \kappa}{\partial \tilde{\varepsilon}} = 1$. In addition, during unloading $\kappa$ remains constant and equal to the one obtained in the last loading step, therefore $\frac{\partial \kappa}{\partial \tilde{\varepsilon}} = 0$. By substituting this expression into (3.42) or (3.43), it is derived that during unloading the second term of these equations cancels and the tangent stiffness becomes equal to $(1-\omega)\mathbf{C}$. This is the secant stiffness of the system, indicating that the tangent and secant stiffness coincide during unloading.

Finally, it is noticed that the tangent stiffness obtained in equations (3.42) and (3.43) is not symmetric. However, when relation (3.31) is adopted for the equivalent strain, then the tangent stiffness becomes symmetric.

### 3.3.3 Localization

*Localization* of deformation refers to the development of narrow regions within structures, where all further deformation concentrates (localizes), as the external loading increases monotonically. Outside these regions, the structure usually follows an unloading path. Localization is observed in several materials, such as rocks, concrete, soils, metals and polymers. However, the extend of relevant phenomena is different, noticing that the size of the narrow region may vary from less than one millimetre in metals to several millimetres in concrete or several centimetres/metres in rock.

Localization phenomena are related to *material instability* in the sense than when localization arises, the stability criterion given by inequality (3.44) is violated [89, 130]. Defined from a mechanical point of view, the stability criterion states that the inner product of the stress rate $\dot{\boldsymbol{\sigma}}$ and the strain rate $\dot{\boldsymbol{\varepsilon}}$ is positive. For a uniaxial tension or compression law, this condition is violated and becomes negative when for an increasing strain, the stress decreases. The corresponding stress-strain curve has a negative slope in this case, as it is shown in the diagrams presented in figure 3.4. Such a constitutive description depicts a *softening* and the phenomenon is called *strain softening.*

$$\dot{\boldsymbol{\varepsilon}}^T \dot{\boldsymbol{\sigma}} > 0 \tag{3.44}$$

A strain-softening behaviour is usually observed in non-metallic materials, such as concrete and rock under tension or unconfined compression and heavily consolidated soils under shear. During failure of these materials, potential flaws and local stress concentrations create strongly inhomogeneous deformations on the structure, defining the localization regions. Due to these localized, inhomogeneous deformations, the calculation of the stress as force per area and strain as the deformation over the initial length, that are usually adopted for the derivation of the stress-strain law, no longer represent the real conditions at a microscopic level. From a computational point of view, someone needs to enhance the given damage mechanics description, to properly address the effect of localization on the numerical simulation. Therefore, localization crucially influences the performance of continuum models adopted to describe failure of composite heterogeneous structures, indicating that relevant phenomena should carefully be taken into account before the selection of a proper constitutive description.

As it will be shown in the following lines, material instability may lead to structural instability. First, the linearized form of the stress-strain relation is considered, by differentiation in respect to time:

$$\dot{\boldsymbol{\sigma}} = \mathbf{D}\dot{\boldsymbol{\varepsilon}} \tag{3.45}$$

where $\mathbf{D}$ is the material tangent stiffness tensor. By substituting (3.45) into the material stability condition (3.44), it is obtained that:

$$\dot{\boldsymbol{\varepsilon}}^T \mathbf{D}\dot{\boldsymbol{\varepsilon}} > 0 \tag{3.46}$$

When inequality (3.46) becomes equality, the tangent stiffness tensor $\mathbf{D}$ is singular, it loses positive-definiteness (as it is also explained in section 1.2.1.2) and material instability is initiated. Then, the structural stability criterion given in (3.47) for a structure occupying a volume $V$ [89], may also be violated.

$$\int_V \dot{\boldsymbol{\varepsilon}}^T \dot{\boldsymbol{\sigma}} > 0 \tag{3.47}$$

In such a case, loss of positive-definiteness for the tangent stiffness tensor $\mathbf{D}$ results in the loss of positive-definiteness for the tangent stiffness matrix $\mathbf{K}_T$, for the calculation of which $\mathbf{D}$ is used as shown in relations (2.167) and (2.168) of section 2.4.5, in the framework of non-linear finite element analysis.

But the most crucial consequence of the loss of positive-definiteness of the material tangent stiffness tensor $\mathbf{D}$ is that it can result in *loss of ellipticity* of the governing equations. From a mathematical point of view, *ellipticity* is a necessary condition for the well-posedness of a boundary value problem, indicating that depending on the data, a finite number of linearly independent solutions are derived. Thus, loss of ellipticity results in infinite number of solutions for the boundary value problem. Within finite element analysis, this results in the pathological dependence of the solution on the discretization size. It is emphasized that this dependency goes beyond the expected correlation of the finite element solution and the mesh size and thus, it is not related to the finite element method. Loss of ellipticity therefore, leads to lack of a unique solution for the failure description, when strain softening appears. Practically, someone observes in this case a significant divergence of the force versus displacement post-peak response, for different mesh sizes.

In addition, when the governing differential equations expressing the strong form of equilibrium as presented in section 2.2.1, as well as all the kinematic and constitutive equations of a structural problem are elliptic, the solution of the set of governing equations cannot have discontinuous derivatives anywhere, indicating that no discontinuities can be described. On the contrary, when loss of ellipticity is observed, then discontinuous solutions may emerge. From a mathematical point of view, when the determinant of the *acoustic tensor* $\mathbf{A}$ vanishes:

$$\det(\mathbf{A}) = \det(\mathbf{n} \cdot \mathbf{D} \cdot \mathbf{n}) = 0 \tag{3.48}$$

then loss of ellipticity arises and the solution depicts a discontinuity, with a unit normal $\mathbf{n}$. To represent the discontinuity as a line crack with zero thickness, every numerical solution will attempt to reduce the width of the localization zone as much as possible indicating that the volume of the localization zone will also decrease and approach zero. But a zero volume leads to zero energy dissipation during failure, which is another consequence of the strain softening, not physically justified. For instance, in case a damage model for quasi-brittle materials is used, the described zero energy dissipation during failure results in immediate, complete, brittle failure after the elastic limit is reached, not representing the gradual loss of failure which was supposed to be depicted by the quasi-brittle damage model.

When the finite element method is adopted in strain softening problems, the solution will try to represent the discontinuity, by reducing the localization width as much as possible, reaching the size of one finite element. By gradually decreasing the size of elements, the localization region is further decreasing to approach the reduced size of elements and the mentioned dependence of the solution on the discretization is obtained.

A necessary condition for loss of ellipticity, is the loss of material stability which is expressed by the following condition [54, 57]:

$$\det(\mathbf{D}_{sym}) = 0 \tag{3.49}$$

where for the general case of a non-symmetric tensor $\mathbf{D}$, the determinant of its symmetric part $\mathbf{D}_{sym}$ has been used in (3.49).

It should be noticed that all these descriptions may apply not only to damage mechanics, but also to plasticity models, when strain softening behaviour appears. In addition, it is possible that the material stability condition expressed by inequality (3.44) is violated, even if no strain softening behaviour is involved. This is the case, when a non-symmetric tangent tensor $\mathbf{D}$ is adopted to describe the relation between the stress and strain rate, indicating that the non-symmetric nature of the tangent tensor leads to material instability. This behaviour applies to materials depicting frictional phenomena, resulting in a non-symmetric tangent tensor. The basic constitutive description provided in section 3.3.2 also leads to a non-symmetric tensor $\mathbf{D}$ in relation (3.42), resulting in material instability.

Finally, the details provided for strain softening and localization phenomena, refer to problems depicting small displacement gradients. When large displacement gradients are considered within geometric non-linearity, then loss of ellipticity may appear due to destabilizing geometric effects, even prior to material instability. This could be the case for buckling problems in metals following a stress-strain response with increasing post-peak slope (*strain hardening*) and for similar problems where geometric non-linearity leads to snap-back behaviour.

### 3.3.4 Enhancement of continuum damage mechanics to depict localization

Several approaches have been developed in the last years, to improve continuum damage mechanics, when strain softening and localization phenomena are present. These approaches are properly developed to simulate damage in problems involving discontinuities in the strain field, called *weak discontinuities* as well as discontinuities in the displacement field, called *strong discontinuities*. It is noted that all the developed methodologies present advantages and disadvantages indicating that the selection of the appropriate approach should carefully be made relying on the requirements of the problem under investigation.

A first group of models uses the *smeared crack* approach, where orthotropic constitutive laws are derived from the assumption that the cracks are smeared over an element [99, 186]. Thus, the nucleation of one or more cracks in a volume which corresponds to an integration point within an element, is translated into a deterioration of the current stiffness and strength at that integration point. The smeared crack models are often applied to simulate failure in concrete, where due to its heterogeneity and the presence of reinforcement, many small cracks nucleate and only in higher load levels, form one or more dominant cracks. This approach may result in sensitivity of the results in respect to discretization. On the other hand, is often adopted in commercial finite element analysis packages such as Abaqus, for practical applications.

A simple remedy may be obtained using the *crack band* approach [100, 172] which adopts an adjustment of the constitutive law according to the width of the localized band, related also to the finite element size. Objective is to preserve the overall energy dissipation expressed as the area under the stress-strain law, during the failure process. Following this approach, the direction of the crack growth is still sensitive to the orientation of the finite element mesh, indicating mesh alignment sensitivity of the results.

To avoid the dependency of the results on discretization, in another category of models, a *discrete interface* of zero thickness is introduced between adjacent finite elements in the continuum to simulate strong discontinuities (cracks). The constitutive response of each discontinuity is described in this case by a traction-relative displacement relation, noticing that the relative displacement represents the *displacement jump*, also called *separation*, between the two sides of the open discontinuity.

One of the methods of this category relies on the usage of the *cohesive zone model*, which introduces a discrete line interface in two-dimensional analysis or a discrete plane, in three-dimensional analysis to simulate strong discontinuities. Cohesive zone models were introduced in [16, 66] for ductile materials like metals, extended in [91] with the *fictitious crack model* for quasi-brittle materials and in [155] for description of inclusion debonding.

When the crack path is known a priori, which is the case for instance in laminate composites with interfaces depicting potential cracks, cohesive zone models can successfully be implemented by inserting interface elements between continuum elements, along the potential crack path. When multiple cracks arbitrarily oriented may appear, the method is not ideal and improvements are proposed, by introducing for instance interface elements within all continuum [237] or suggesting remeshing techniques [35]. Still, some mesh bias may arise due to the fact that the crack propagation direction is not totally free but depends on the interelement boundaries.

Another approach which is proposed to overcome the mentioned difficulties in localization problems, relies on the introduction of enhanced finite element interpolations to simulate a weak discontinuity as a band of fixed width [23, 163] or a strong discontinuity as a line (surface) [67, 125, 132, 162]. This is the *embedded discontinuity* method and it allows introducing a displacement or displacement gradient discontinuity into the finite element, independently of its boundaries. When weak discontinuities are introduced, similar disadvantages with the crack band approach, related to mesh orientation sensitivity, may arise [170].

When the size of the zone in front of the crack tip, which is called *fracture process zone* and shown in figure 3.5 of section 3.3.5, is sufficiently small in comparison with the structural dimensions, then principles from *linear fracture mechanics* can be adopted to simulate cracks in composite structures. In this case, potential inelastic phenomena are restricted to very small regions and can be neglected. Representative examples of materials depicting this behaviour, are metals or brittle materials. One of the methods which can be adopted to capture the response of these materials, is the *Extended Finite Element Method (XFEM)* as presented in section 5.2 of chapter 5. This method relies on the *partition of unity* property of finite element shape functions stating that the sum of the shape functions must be unity [22, 147]. Using this method, arbitrarily aligned cracks within the finite element mesh can be simulated, by introducing proper enrichment functions.

In an effort to further improve the performance of the mentioned constitutive descriptions when localization phenomena and strain softening arise, additional concepts have been introduced, combining these approaches. Among others, it has been proposed the extension of the partition of unity approach to cohesive fracture [229] as well as the formulation of the XFEM method to include cohesive cracks growth [146]. More relevant efforts can be found in [11, 56, 79, 228]. A basic description of the XFEM method is found in chapter

5, while advanced concepts involving the XFEM and multi-scale analysis are provided in section 7.

A final class of models adopted to provide a reliable numerical solution to localization phenomena is identified under the general title of *regularized damage models* [54]. These models suggest an enhancement of continuum theories which aims to avoid the loss of ellipticity, by considering a *characteristic length scale* used to prevent localization of strain into an arbitrarily small volume with zero energy dissipation. Examples of this category are the *non-local integral damage models*, the *gradient enhanced* as well as the *Cosserat continuum* models.

Non-local integral models [19, 59] substitute the concept of local action by introducing weighted volume average, thus, non-local quantities for certain variables such as the equivalent strain introduced in damage models. This non-locality has a smoothing effect on the deformation and damage which precludes localization in a very thin line or surface.

Gradient enhanced models adopt higher-order deformation gradients or higher-order gradients in the damage loading function, to create spatial interactions [5, 55]. These models can be formulated within non-local models, by expanding for instance the equivalent strain or internal variable $\kappa$ expressed as weighted volume average in Taylor series, or independently.

Cosserat continuum models [53, 150] introduce micro-rotations as degrees of freedom, in addition to conventional displacement degrees of freedom. Couples of stresses and gradients of these rotations are then added to the continuum description aiming in preventing the concentration of deformations in a line (surface). This approach is mainly appropriate when sliding type failure arises, but it may not work for opening type failure.

Other approaches may rely on the introduction of *viscous* or *time-dependent* terms in the constitutive description, to prevent loss of ellipticity [156]. However, adding viscosity may offer a solution to localization problems only when a sufficiently high rate sensitivity for the original material appears. Two representative models of this section, the cohesive zone model and the non-local damage model, are presented next.

### 3.3.5 Cohesive zone model

Cohesive zone models can be used to simulate damage due to crack initiation and propagation, when the size of the fracture process zone shown in figure 3.5 is not negligible in comparison to the structural dimensions, and the cohesive forces in this zone cannot be ignored. This behaviour appears when the fracture is not perfectly brittle but some ductility arises, after the maximum strength is reached.

Within these models, a softening *traction-separation* law is adopted to describe the response of the discontinuity. The area under this traction-separation curve denotes the energy dissipation needed for the separation and is called *work of separation* or *fracture energy*. Since energy dissipation is now

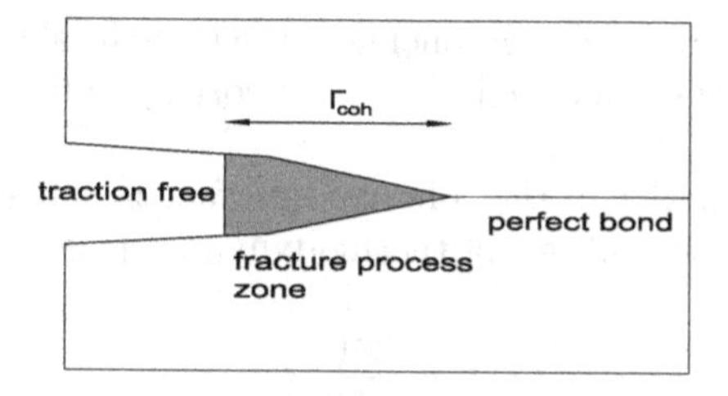


**FIGURE 3.5**: Cohesive zone interface.

non-zero, the quasi-brittle behaviour of materials like concrete can better be simulated and a relative objectivity in respect to the post-peak response of force-displacement curves is achieved.

For the implementation of the method within finite element analysis, cohesive zone elements representing cohesive forces, are placed between continuum (bulk) elements. When damage growth takes place, these elements open as shown in figure 3.6, simulating crack initiation and propagation. Thus, the direction of crack propagation depends on the presence of cohesive zone elements, indicating a mesh dependence. In addition and in connection with previous statements, a characteristic length scale is inserted in the formulation, associated with the traction-separation constitutive description of the cohesive interface.

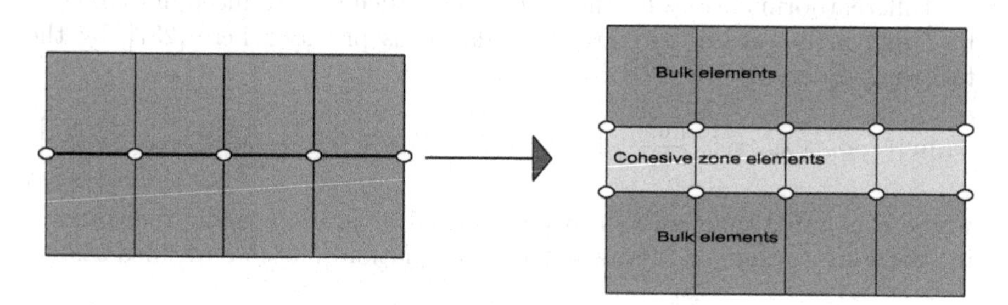


**FIGURE 3.6**: Representation of cohesive zone elements in continuum (bulk) elements before and after opening.

The first step towards establishing cohesive zone models is the formulation of the principle of virtual work, considering the cohesive interface which appears between finite elements. For such an interface, a traction vector $\mathbf{t}$ and a separation or displacement jump vector $||\mathbf{u}||$ are defined in two dimensions as follows:

$$\mathbf{t} = \begin{bmatrix} t_t \\ t_n \end{bmatrix} \tag{3.50}$$

$$||\mathbf{u}|| = \begin{bmatrix} ||u||_t \\ ||u||_n \end{bmatrix} \tag{3.51}$$

where $t_t$, $t_n$ are the traction components in the tangential and normal direction of the cohesive interface and $||u||_t$, $||u||_n$ the corresponding displacement jump components.

Then, the potential $\phi$ of the cohesive interface is defined as a consequence of the first and the second law of thermodynamics [128] and the cohesive law is expressed as:

$$\mathbf{t} = \frac{\partial \phi(||\mathbf{u}||)}{\partial ||\mathbf{u}||} \tag{3.52}$$

The principle of virtual work which was given in relation (2.21) can now be formulated by introducing an additional work term for the cohesive interface $\Gamma_{coh}$:

$$\int_V \delta \boldsymbol{\varepsilon}^T \boldsymbol{\sigma} dV + \int_{\Gamma_{coh}} \delta \phi dS = \int_V \rho \delta \mathbf{u}^T \mathbf{f} dV + \int_S \delta \mathbf{u}^T \mathbf{t} dS \tag{3.53}$$

By inserting the traction and displacement jump terms, the principle of virtual work becomes:

$$\int_V \delta \boldsymbol{\varepsilon}^T \boldsymbol{\sigma} dV + \int_{\Gamma_{coh}} \delta(||\mathbf{u}||)^T \mathbf{t} dS = \int_V \rho \delta \mathbf{u}^T \mathbf{f} dV + \int_S \delta \mathbf{u}^T \mathbf{t} dS \tag{3.54}$$

In equations (3.53) and (3.54) vector notation is used, inertial effects are neglected and small displacement gradients are considered.

Different formulations for the potential $\phi$ which is introduced in (3.52) can be found in literature. An early formulation is provided here [237], by the following equation:

$$\phi(||\mathbf{u}||) = \phi_n + \phi_n exp(-\frac{||u||_n}{\delta_n}) \left[(1 - r + \tfrac{||u||_n}{\delta_n})\tfrac{1-q}{r-1} - (q + \tfrac{r-q}{r-1}\tfrac{||u||_n}{\delta_n}) exp(-\tfrac{||u||_t^2}{\delta_t^2})\right] \tag{3.55}$$

where it is noted that both shear and normal separations along the cohesive interface are considered. Terms $q$ and $r$ of equation (3.55) are defined as:

$$q = \frac{\phi_t}{\phi_n}, \quad r = \frac{||u||_n^*}{\delta_n} \tag{3.56}$$

where $\phi_t$, $\phi_n$ represent the work of tangential and normal separation, as denoted by the areas under the shear and normal traction curves, respectively. With $||u||_n^*$ is provided the value of the displacement jump $||u||_n$ obtained after complete tangential separation with $t_n = 0$. The term $\delta_n$ of equations (3.55), (3.56) is a characteristic length, expressing the normal separation $||u||_n$ of the cohesive interface when the normal traction $t_n$ becomes equal to the corresponding strength $\sigma_{max}$ and thus:

$$||u||_n = \delta_n \rightarrow t_n(||u||_n = \delta_n) = \sigma_{max} \tag{3.57}$$

In a similar manner, the term $\delta_t$ of equation (3.55) represents a characteristic length which expresses the tangential separation $||u||_t = \frac{\sqrt{2}\delta_t}{2}$ when the shear traction $t_t$ becomes equal to the corresponding strength $\tau_{max}$ and thus:

$$||u||_t = \frac{\sqrt{2}\delta_t}{2} \rightarrow t_n(||u||_t = \frac{\sqrt{2}\delta_t}{2}) = \tau_{max} \quad (3.58)$$

The normal and tangential work of separation can then be written as follows:

$$\phi_n = exp(1)\sigma_{max}\delta_n, \quad \phi_t = \sqrt{\frac{exp(1)}{2}}\tau_{max}\delta_t \quad (3.59)$$

By substituting (3.55) into (3.52), the following relations are derived for the constitutive traction-separation laws in the normal and tangential direction of the cohesive interface:

$$t_n = -\frac{\phi_n}{\delta_n}exp(-\frac{||u||_n}{\delta_n})[\frac{||u||_n}{\delta_n}exp(-\frac{||u||_t^2}{\delta_t^2}) + \frac{1-q}{r-1}(1-exp(-\frac{||u||_t^2}{\delta_t^2}))(r - \frac{||u||_n}{\delta_n})] \quad (3.60)$$

$$t_t = -2(\frac{\phi_n||u||_t}{\delta_t^2})[q + \frac{r-q}{r-1}\frac{||u||_n}{\delta_n}]exp(-\frac{||u||_n}{\delta_n})exp(-\frac{||u||_t^2}{\delta_t^2}) \quad (3.61)$$

When the normal and the tangential behaviour of the cohesive interface are independent, indicating that for the normal behaviour no tangential separation and for the tangential behaviour no normal separation are considered, thus, $t_n = t_n(||u||_n, ||u||_t = 0)$, $t_t = t_t(||u||_t, ||u||_n = 0)$, the cohesive zone model is *uncoupled*. By substituting these conditions representing the uncoupled behaviour into equations (3.60), (3.61) and considering q=1, r=0 [237], the traction-separation curves given in figure 3.7 are obtained, for the normal and tangential direction, respectively.

In diagrams shown in figure 3.7, normalized, non-dimensional quantities are used for the traction and the separation. The normal traction-separation curve presents the classical softening diagram for positive values of traction and separation. When the separation becomes negative, the traction rapidly becomes negative, to prevent penetration. For the shear traction-separation curve, a change in the sign of separation indicates a different direction for the shear traction.

### 3.3.6 Implementation of cohesive zone models within finite element analysis

For the numerical implementation of cohesive zone models, discretization of the cohesive interface should initially be considered. Then, the solution of the governing equations, which include the non-linear traction-separation laws shown in figure 3.7, will be derived by linearization using the Newton-Raphson procedure.

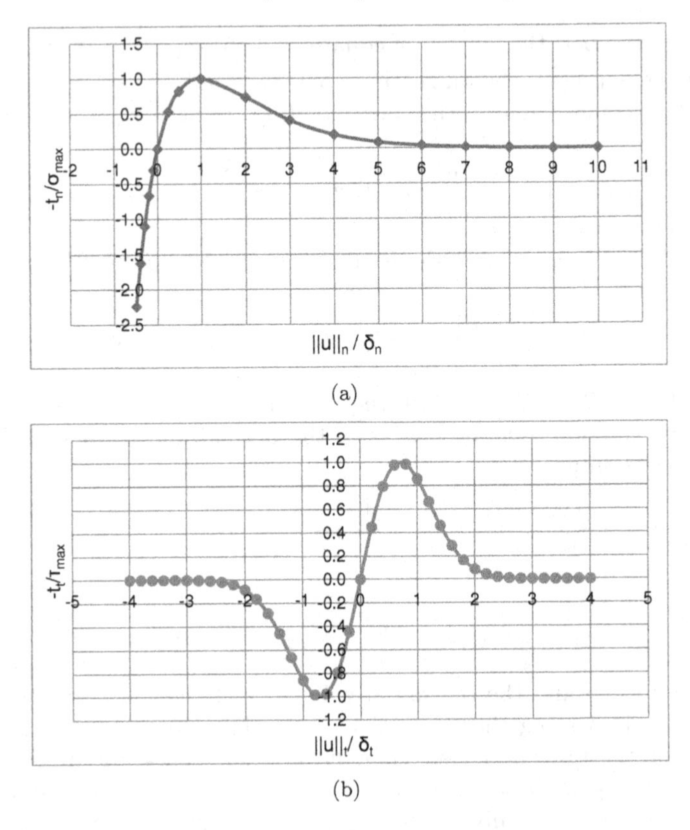


**FIGURE 3.7**: Normalized traction-separation curves for a) the normal behaviour with $||u||_t = 0$ and b) the tangential behaviour with $||u||_n = 0$.

In figure 3.8 a two-dimensional cohesive zone element with two translational degrees of freedom per node, is shown. The middle line A*- B* of the element is also depicted and a local coordinate system with origin at its centre O and axes along the tangential direction t and the normal direction n is introduced.

The first step towards discretizing the cohesive element, is to express the displacement jump in respect to nodal displacements. First, the nodal displacement vector for the cohesive element is provided below:

$$\mathbf{u}_e = \begin{bmatrix} u_1 & v_1 & u_2 & v_2 & u_3 & v_3 & u_4 & v_4 \end{bmatrix}^T \tag{3.62}$$

where $u_i$, $v_i$ are the displacement components along the global coordinate system.

As it is shown in figure 3.8(a), the middle line A*- B* may be rotated by an angle $\theta$ in respect to the global x axis. In this case, the displacement

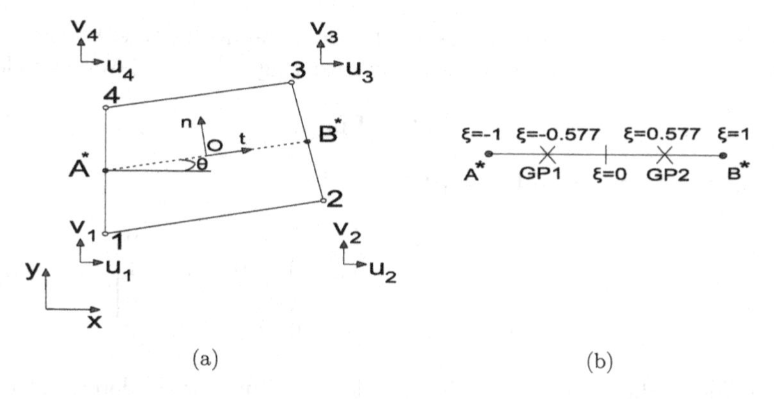


(a) (b)

**FIGURE 3.8**: a) Two-dimensional cohesive zone element and b) natural coordinates and Gauss points for the element.

components $\hat{u}_i$, $\hat{v}_i$ in the rotated, local coordinate system t-n are expressed in respect to the global displacements $u_i$, $v_i$ as shown below:

$$\begin{bmatrix} \hat{u}_i \\ \hat{v}_i \end{bmatrix} = \begin{bmatrix} cos(\theta) & sin(\theta) \\ -sin(\theta) & cos(\theta) \end{bmatrix} \begin{bmatrix} u_i \\ v_i \end{bmatrix} \Rightarrow \hat{\mathbf{u}}_i = \hat{\mathbf{T}}\mathbf{u}_i \tag{3.63}$$

where $\hat{\mathbf{T}}$ is the transformation matrix initially defined in section 1.2.2. Then, the nodal displacement vector $\hat{\mathbf{u}}_e$ is given below for the whole cohesive element in the local coordinate system:

$$\hat{\mathbf{u}}_e = \begin{bmatrix} \hat{u}_1 & \hat{v}_1 & \hat{u}_2 & \hat{v}_2 & \hat{u}_3 & \hat{v}_3 & \hat{u}_4 & \hat{v}_4 \end{bmatrix}^T \tag{3.64}$$

This vector is expressed in regard to the nodal displacement vector $\mathbf{u}_e$ defined in the global coordinate system, as follows:

$$\hat{\mathbf{u}}_e = \mathbf{T}\mathbf{u}_e \tag{3.65}$$

where:

$$\mathbf{T} = \begin{bmatrix} \hat{\mathbf{T}} & 0 & 0 & 0 \\ 0 & \hat{\mathbf{T}} & 0 & 0 \\ 0 & 0 & \hat{\mathbf{T}} & 0 \\ 0 & 0 & 0 & \hat{\mathbf{T}} \end{bmatrix} \tag{3.66}$$

Next, the vector $||\mathbf{u}||_e$ of the displacement jumps is defined for the cohesive element by the differences of the nodal displacements along the tangential and normal direction:

$$||\mathbf{u}||_e = \begin{bmatrix} \hat{u}_4 - \hat{u}_1 \\ \hat{v}_4 - \hat{v}_1 \\ \hat{u}_3 - \hat{u}_2 \\ \hat{v}_3 - \hat{v}_2 \end{bmatrix} \tag{3.67}$$

This vector can be expressed in respect to the nodal displacement vector $\hat{\mathbf{u}}_e$ in the local coordinate system, by properly rewriting relation (3.67), as follows:

$$||\mathbf{u}||_e = \mathbf{L}\hat{\mathbf{u}}_e \tag{3.68}$$

where matrix $\mathbf{L}$ is defined below [164]:

$$\mathbf{L} = \begin{bmatrix} -1 & 0 & 0 & 0 & 0 & 0 & 1 & 0 \\ 0 & -1 & 0 & 0 & 0 & 0 & 0 & 1 \\ 0 & 0 & -1 & 0 & 1 & 0 & 0 & 0 \\ 0 & 0 & 0 & -1 & 0 & 1 & 0 & 0 \end{bmatrix} \tag{3.69}$$

The vector of the displacement jump field $||\mathbf{u}||$ can now be determined at a point on the middle line A*- B* of the cohesive element, using interpolation of the nodal displacement jump vector $||\mathbf{u}||_e$:

$$||\mathbf{u}|| = \mathbf{N}_{coh}||\mathbf{u}||_e \tag{3.70}$$

where $\mathbf{N}_{coh}$ is the matrix which includes the shape functions $N_i$:

$$\mathbf{N}_{coh} = \begin{bmatrix} N_1 & 0 & N_2 & 0 \\ 0 & N_1 & 0 & N_2 \end{bmatrix} \tag{3.71}$$

Since interpolation for the displacement jump takes place on the line A*- B*, the shape functions which are adopted here are the ones used for a two-noded bar element, within isoparametric formulation. Thus:

$$N_1 = \frac{1-\xi}{2}, \quad N_1 = \frac{1+\xi}{2} \tag{3.72}$$

where the natural coordinate $\xi$, ranging between -1 and 1 as in standard two-noded isoparametric elements, is depicted in figure 3.8(b).

Using equations (3.65) and (3.68), the displacement jump $||\mathbf{u}||$ given in relation (3.70) can eventually be expressed in terms of the nodal displacement vector $\mathbf{u}_e$ as follows:

$$||\mathbf{u}|| = \mathbf{N}_{coh}\mathbf{L}\mathbf{T}\mathbf{u}_e \tag{3.73}$$

By introducing the following identity:

$$\mathbf{B}_{coh} = \mathbf{N}_{coh}\mathbf{L} \tag{3.74}$$

relation (3.73) can be rewritten as provided below:

$$||\mathbf{u}|| = \mathbf{B}_{coh}\mathbf{T}\mathbf{u}_e \tag{3.75}$$

where matrix $\mathbf{B}_{coh}$ is easily determined by substituting (3.71) and (3.69) into (3.74):

$$\mathbf{B}_{coh} = \begin{bmatrix} -N_1 & 0 & -N_2 & 0 & N_2 & 0 & N_1 & 0 \\ 0 & -N_1 & 0 & -N_2 & 0 & N_2 & 0 & N_1 \end{bmatrix} \tag{3.76}$$

The discretization of the principle of virtual work given in (3.54) can now be implemented. The cohesive work term at the left-hand part of this equation is discretized by substituting the transpose of the virtual displacement jump $\delta(||\mathbf{u}||)^T$ using relation (3.75). The remaining terms are discretized according to descriptions presented in previous sections, see for instance section 2.3.2:

$$\begin{aligned} &\delta\mathbf{u}_e^T \int_{V_e} \mathbf{B}^T \boldsymbol{\sigma} dV + \delta\mathbf{u}_e^T \int_{\Gamma_{coh_e}} \mathbf{T}^T \mathbf{B}_{coh}^T \mathbf{t} dS = \\ &\delta\mathbf{u}_e^T \int_{V_e} \rho \mathbf{N}^T \mathbf{f} dV + \delta\mathbf{u}_e^T \int_{S_e} \mathbf{N}^T \mathbf{t} dS \Rightarrow \\ &\int_{V_e} \mathbf{B}^T \boldsymbol{\sigma} dV + \int_{\Gamma_{coh_e}} \mathbf{T}^T \mathbf{B}_{coh}^T \mathbf{t} dS = \int_{V_e} \rho \mathbf{N}^T \mathbf{f} dV + \int_{S_e} \mathbf{N}^T \mathbf{t} dS \end{aligned} \tag{3.77}$$

For the non-linear traction-separation law, a linearization process will be followed. Thus, the traction at iteration $i+1$ is expanded in Taylor series, in the framework of the incremental-iterative Newton-Raphson process:

$$\mathbf{t}^{i+1} = \mathbf{t}^i + \Delta\mathbf{t} \tag{3.78}$$

where it is also considered that:

$$||\mathbf{u}||^{i+1} = ||\mathbf{u}||^i + \Delta||\mathbf{u}|| \tag{3.79}$$

By using relation (2.82) of section 2.4.1 as well as (3.75), the following equation is obtained for $\Delta\mathbf{t}$:

$$\begin{aligned} \Delta\mathbf{t} = \frac{\partial\mathbf{t}}{\partial||\mathbf{u}||^i}\Delta||\mathbf{u}|| \Rightarrow \Delta\mathbf{t} &= \frac{\partial\mathbf{t}}{\partial||\mathbf{u}||^i}\mathbf{B}_{coh}\mathbf{T}\Delta\mathbf{u}_e \\ &= \mathbf{D}_{coh}\mathbf{B}_{coh}\mathbf{T}\Delta\mathbf{u}_e \end{aligned} \tag{3.80}$$

For the calculation of the cohesive tangent stiffness $\mathbf{D}_{coh} = \frac{\partial\mathbf{t}}{\partial||\mathbf{u}||^i}$, the constitutive traction-separation law selected for the current application, like for instance the one presented in equations (3.60) and (3.61) of section 3.3.5, will be used. In general, this will result in the following identity:

$$\frac{\partial\mathbf{t}}{\partial||\mathbf{u}||^i} = \mathbf{D}_{coh} = \begin{bmatrix} \frac{\partial t_t}{\partial||u||_t} & \frac{\partial t_t}{\partial||u||_n} \\ \frac{\partial t_n}{\partial||u||_t} & \frac{\partial t_n}{\partial||u||_n} \end{bmatrix} \tag{3.81}$$

which stands for coupled traction-separation laws. When uncoupled traction-separation is adopted according to the descriptions of section 3.3.5, then the following relation applies:

$$\frac{\partial\mathbf{t}}{\partial||\mathbf{u}||^i} = \mathbf{D}_{coh} = \begin{bmatrix} \frac{\partial t_t}{\partial||u||_t} & 0 \\ 0 & \frac{\partial t_n}{\partial||u||_n} \end{bmatrix} \tag{3.82}$$

In case material non-linearity is considered for the bulk material, linearization of the corresponding stress-strain law is formulated as described in section 2.4.5:

$$\boldsymbol{\sigma}^{i+1} = \boldsymbol{\sigma}^i + \Delta\boldsymbol{\sigma}, \quad \text{with} \quad \Delta\boldsymbol{\sigma} = \mathbf{D}\Delta\boldsymbol{\varepsilon} = \mathbf{D}\mathbf{B}\Delta\mathbf{u}_e \tag{3.83}$$

where $\mathbf{B}$ is the strain-displacement matrix for the bulk elements and $\mathbf{D}$ the instantaneous stiffness modulus derived from the constitutive description as follows:

$$\mathbf{D} = \frac{\partial\boldsymbol{\sigma}}{\partial\boldsymbol{\varepsilon}^i} \tag{3.84}$$

By substituting (3.78), (3.80) and (3.83) into equation (3.77), it is obtained that:

$$\begin{gathered}
\int_{V_e} \mathbf{B}^T\boldsymbol{\sigma}^i dV + \int_{V_e} \mathbf{B}^T\Delta\boldsymbol{\sigma} dV + \int_{\Gamma_{coh_e}} \mathbf{T}^T\mathbf{B}_{coh}^T\mathbf{t}^i dS + \int_{\Gamma_{coh_e}} \mathbf{T}^T\mathbf{B}_{coh}^T\Delta\mathbf{t} dS = \\
\int_{V_e} \rho\mathbf{N}^T\mathbf{f} dV + \int_{S_e} \mathbf{N}^T\mathbf{t} dS \Rightarrow \\
\left(\int_{V_e} \mathbf{B}^T\mathbf{D}\mathbf{B} dV + \int_{\Gamma_{coh_e}} \mathbf{T}^T\mathbf{B}_{coh}^T\mathbf{D}_{coh}\mathbf{B}_{coh}\mathbf{T} dS\right)\Delta\mathbf{u}_e = \\
\int_{V_e} \rho\mathbf{N}^T\mathbf{f} dV + \int_{S_e} \mathbf{N}^T\mathbf{t} dS - \left(\int_{V_e} \mathbf{B}^T\boldsymbol{\sigma}^i dV + \int_{\Gamma_{coh_e}} \mathbf{T}^T\mathbf{B}_{coh}^T\mathbf{t}^i\right)
\end{gathered} \tag{3.85}$$

The tangent stiffness matrix and the internal force vector for the cohesive element are derived from the above equation as follows:

$$\mathbf{k}_{coh_e} = \int_{\Gamma_{coh_e}} \mathbf{T}^T\mathbf{B}_{coh}^T\mathbf{D}_{coh}\mathbf{B}_{coh}\mathbf{T} dS, \quad \mathbf{f}_{coh_e} = \int_{\Gamma_{coh_e}} \mathbf{T}^T\mathbf{B}_{coh}^T\mathbf{t}^i dS \tag{3.86}$$

Assembly for all elements of the structure is then considered for the tangent stiffness, external and internal force vectors and the equation (3.85) is written in the final, general form:

$$(\mathbf{K} + \mathbf{K}_{coh})\Delta\mathbf{U} = \mathbf{F}_{ext} - \mathbf{F}_{int} \tag{3.87}$$

It is noted that the cohesive terms of the stiffness matrix and the internal force vector contribute to the stiffness and internal force of bulk elements, only when nodes of these elements belong also to the cohesive interface.

For the solution of the integrals in the governing equation (3.85) numerical integration using the Gauss rule can be adopted. For the cohesive terms, numerical integration is formulated as follows:

$$\mathbf{k}_{coh_e} = \int_{\Gamma_{coh_e}} \mathbf{T}^T\mathbf{B}_{coh}^T\mathbf{D}_{coh}\mathbf{B}_{coh}\mathbf{T}dS = \int_{-1}^{1} \mathbf{T}^T\mathbf{B}_{coh}^T\mathbf{D}_{coh}\mathbf{B}_{coh}\mathbf{T}\det(\mathbf{J})d\xi$$
$$= \mathbf{T}^T\left(\sum_{j=1}^{2}\mathbf{B}_{cohj}^T\mathbf{D}_{cohj}\mathbf{B}_{cohj}\alpha_j\right)\mathbf{T}\frac{L}{2} \tag{3.88}$$

$$\mathbf{f}_{coh_e} = \int_{\Gamma_{coh_e}} \mathbf{T}^T\mathbf{B}_{coh}^T\mathbf{t}^i dS = \int_{-1}^{1} \mathbf{T}^T\mathbf{B}_{coh}^T\mathbf{t}^i\det(\mathbf{J})d\xi$$
$$= \mathbf{T}^T\left(\sum_{j=1}^{2}\mathbf{B}_{cohj}^T\mathbf{t}_j^i\alpha_j\right)\frac{L}{2} \tag{3.89}$$

where the determinant of the Jacobian matrix $\det(\mathbf{J})$ is equal to $L/2$ for a two-noded bar element, with $L$ being the length of the middle line of the cohesive element. Numerical integration for the cohesive element shown in figure 3.8 is implemented in 2 Gauss points, similar to the case of numerical integration for bar elements. The values for the Gauss points locations $\xi_j$ and corresponding weights $\alpha_j$ can be found in section 2.3.6, table 2.1.

The above description does not include history dependent phenomena, which arise when unloading takes place. For the incorporation of such phenomena in cohesive zone models, a history dependent variable needs to be introduced to the constitutive description, in the general framework of the discussions presented in section 3.3.1 for damage models.

Finally, when the initial stiffness of cohesive zone elements is high, some spurious oscillations may be obtained for the traction profile, without a physical meaning. To avoid this issue, Newton-Cotes numerical integration can be used instead of Gauss integration. Alternatively, a sufficient mesh refinement can be adopted.

### 3.3.7 Non-local damage model

Within non-local damage models an internal length scale is introduced, by considering the modification of a variable which controls damage growth, such as the equivalent strain, using a spatial average (integral). Thus, for the continuum damage mechanics scheme given in section 3.3.1, the damage growth at a point $\mathbf{x}$ is no longer defined by the local equivalent strain $\tilde{\varepsilon}(\mathbf{x})$, but by the average $\bar{\varepsilon}$ over the volume $V$ occupied by the microstructural element at $\mathbf{x}$, as shown below:

$$\bar{\varepsilon}(\mathbf{x}) = \frac{\int_V \psi(\mathbf{y},\mathbf{x})\tilde{\varepsilon}(\mathbf{y})dV}{\int_V \psi(\mathbf{y},\mathbf{x})dV} \tag{3.90}$$

The coordinate $\mathbf{x}$ expresses the local or "target" point at which the average equivalent strain $\bar{\varepsilon}(\mathbf{x})$ is evaluated, the coordinate $\mathbf{y}$ represents each "source" point where the corresponding strain $\tilde{\varepsilon}(\mathbf{y})$ is found and $\psi(\mathbf{y}, \mathbf{x})$ is a *weight function.* This function is often considered as homogeneous and isotropic, depending only on the distance between the target point $\mathbf{x}$ and the source point $\mathbf{y}$. In this case, equation (3.90) is rewritten as follows:

$$\bar{\varepsilon}(\mathbf{x}) = \frac{\int_V \psi(||\mathbf{x} - \mathbf{y}||)\tilde{\varepsilon}(\mathbf{y})dV}{\int_V \psi(||\mathbf{x} - \mathbf{y}||)dV} \tag{3.91}$$

Within this framework, $\bar{\varepsilon}(\mathbf{x})$ denoting the *non-local equivalent strain* depends on the strain state calculated in a finite volume $V$ indicating the non-local character of the constitutive description.

The weight function is often taken as the Gauss distribution function:

$$\psi(||\mathbf{x} - \mathbf{y}||) = \exp\left(-\frac{||\mathbf{x} - \mathbf{y}||^2}{2\ell^2}\right) \tag{3.92}$$

where $\ell$ represents the internal length of the non-local continuum and thus, it must be related to the scale of the microstructure. Its value is not a trivial choice and it should properly be selected for the material under investigation. For standard concrete, the internal length can be chosen to be equal to 3 times the maximum diameter of the aggregate [20].

Another option for the weight function is provided by the following equation:

$$\psi(||\mathbf{x} - \mathbf{y}||) = \begin{cases} \left(1 - \frac{||\mathbf{x}-\mathbf{y}||^2}{R^2}\right)^2 & \text{if } 0 \leq ||\mathbf{x} - \mathbf{y}|| \leq R \\ 0 & \text{if } R \leq ||\mathbf{x} - \mathbf{y}|| \end{cases} \tag{3.93}$$

with $R$ being a parameter which is proportional to the internal length, expressing the largest distance of point $\mathbf{y}$ that affects the non-local average at point $\mathbf{x}$. Parameter R is called *interaction radius.* When relation (3.92) is used for the weight function, it is assumed that $R = \infty$.

After these descriptions, the damage loading function $f$ defined in section 3.3.1 is formulated using the non-local equivalent strain, as follows:

$$f = \bar{\varepsilon} - \kappa \tag{3.94}$$

For the implementation of the non-local damage law in a finite element analysis code, an additional step should be added to the descriptions given in sections 3.3.1 and 3.3.2, related to the calculation of the non-local equivalent strains. These will be evaluated on integration points, using numerical integration which provides the solution to the integral quantities given in equations (3.90) and (3.91):

$$\bar{\varepsilon}(\mathbf{x}_i) = \frac{\sum\limits_{\mathbf{y}_j \in V} \omega(\mathbf{y}_j)|J(\mathbf{y}_j)|\psi(||\mathbf{x}_i - \mathbf{y}_j||)\tilde{\varepsilon}(\mathbf{y}_j)}{\sum\limits_{\mathbf{y}_j \in V} \omega(\mathbf{y}_j)|J(\mathbf{y}_j)|\psi(||\mathbf{x}_i - \mathbf{y}_j||)} \tag{3.95}$$

In equation (3.95) $\omega(\mathbf{y}_j)$ represents the integration weight at the Gauss point $\mathbf{y}_j$ and $|J(\mathbf{y}_j)|$ the determinant of the Jacobian matrix of the isoparametric transformation at $\mathbf{y}_j$, determined in section 2.3.4.

It is noted that when relation (3.93) is adopted for the weight function, the sums in equation (3.95) do not need to be considered over all integration points of the model, but only over those points which are found inside a sphere or circle of centre at point $\mathbf{x}$ and radius $R$. Hence, a search algorithm should be used to identify all pairs of points within the radius R in the model, which adds to the computational effort required for the solution. However, the convergence rate is improved in comparison with the local damage model, due to the fact that the spatial distribution of strain is relatively smooth even after localization starts.

Finally, the remaining steps for the implementation of the non-local damage law are the same as described in sections 3.3.1 and 3.3.2.

# Chapter 4

# Contact mechanics

## 4.1 Introduction

*Unilateral contact* represents a model example of highly non-linear effects which can be treated by non-smooth tools. They have a variable, excitation-dependent mechanical behaviour, which can be described by an appropriate selection of classical models. The transition points between the various models make the problem non-smooth, indicating that it belongs to the field of *non-smooth mechanics*. Non-smooth analysis and optimization provide a powerful framework for the modelling and numerical treatment of mechanical problems with hard non-linearities, thus, is much more effective in comparison with the general purpose Newton-Raphson technique.

The links between non-smooth analysis, optimization and mechanics, which provide a more general theoretical framework for the formulation and study of non-smooth non-linearities, including the case of unilateral contact and the stick-slip motion in friction, are discussed in this text. The non-smoothness comes from the fact that one considers in a unique model all possible phases (states) of the non-linear joint, namely contact and separation areas and the free boundaries between them. The presented approach allows for automatic determination of the areas of contact and separation which are not known apriori.

The adoption of refined models, which are based on non-smooth mechanics, allows for the development of powerful solution algorithms. For instance, in the case of unilateral contact mechanics, one is able to formulate inequality constrained optimization problems or *complementarity problems*. Analogous formulations exist for elastoplasticity and no-tension models for masonry. Further material can be found, among others, in [3], [4], [9], [15], [30], [33], [60], [76], [103], [116], [131], [145], [149], [165], [166], [231] and the references given there.

The cost paid by the analyst for this unified formulation of an engineering problem with non-classical non-linearities of the *either-or* type is that, instead of linear or non-linear equations, one must solve more complicated linear or non-linear complementarity problems or set-valued inclusions. It must be noted that these problems appear in other branches of science and engineering

DOI: 10.1201/9781003017240-4

as well (mathematical programming and optimization, economics, traffic flow etc.) where reliable algorithms and useful theoretical results can be found.

Within multilevel schemes discussed in subsequent chapters, detailed non-smooth mechanics may be used for the analysis at the microstructural level, while the homogeneous upper level does not need to have this complexity.

---

## 4.2 Theoretical framework: NSO, NSA, VI, HI and CP

Elements of NSO (*non-smooth optimization*) and NSA (*non-smooth analysis*) are presented in this section. For non-smooth models the classical variational equations must be replaced by VI (*variational inequalities*), for *convex*, and HI (*hemivariational inequalities*), for *non-convex* problems [166]. Inequalities reflect the fact that involvement in systems with direction-dependent (unilateral) behaviour takes place. In some cases, further simplification is possible, leading to linear or non-linear CP (*complementarity problems*). These latter problems are well-known in several branches of science and engineering, from economics to transportation modelling, and have been studied from both theoretical and algorithmic point of view.

### 4.2.1 Smooth and non-smooth functions

Let $f$ be a real-valued finite function defined on the real line $R$. Function $f$ is called differentiable at $x \in R$ if its derivative $f'(x)$ at $x$ exists, which is defined by:

$$f'(x) = \lim_{\alpha \to 0} \frac{1}{\alpha}\left[f(x+\alpha) - f(x)\right] \tag{4.1}$$

If this limit exists for every point of some open set $S \in R$ the function $f$ is called differentiable on $S$. If the limit does not exist, the function $f$ is called non-differentiable or non-smooth at $x$.

Among the variety of applications of the derivative one recalls here the first-order approximation (linearization) of $f$ in the neighborhood of a point $x$:

$$f(x+\delta x) = f(x) + f'(x)\delta x + o_x(\delta x) \tag{4.2}$$

with

$$\frac{o_x(\delta x)}{\delta x} \longrightarrow 0 \text{ as } \delta x \to 0 \tag{4.3}$$

Moreover, $x^*$ is a minimum of the function $f$ if:

$$f'(x^*) = 0 \tag{4.4}$$

Relation (4.4) defines a *stationary point* of $f$, since it also holds true for a maximum and for a saddle point of $f$. In fact, higher-order derivatives are checked in order to specify the nature of the stationary point.

#### 4.2.1.1 One-sided differentials

Assume now that the limit (4.1) does not exist, but at the same time the following directional derivatives exist: the right-hand side derivative $f'_+(x)$ and the left-hand side derivative $f'_-(x)$ of $f$ at $x$. They are defined by:

$$f'_+(x) = \lim_{\alpha \downarrow 0} \frac{1}{\alpha}[f(x+\alpha) - f(x)], \; f'_-(x) = \lim_{\alpha \uparrow 0} \frac{1}{\alpha}[f(x+\alpha) - f(x)] \quad (4.5)$$

Here $\alpha \downarrow 0$ means that $\alpha \to 0$, by taking positive values $\alpha > 0$ and $\alpha \uparrow 0$ means that $\alpha \to 0$, with negative values $\alpha < 0$.

For a function $f$ to be differentiable at $x$, it is necessary and sufficient that $f'_+(x) = f'_-(x)$.

The *directional derivative* of a function $f$ at point $x$ and in the direction $x \in R$ is defined by:

$$f'(x, g) = \lim_{\alpha \downarrow 0} \frac{1}{\alpha}[f(x + \alpha g) - f(x)] \quad (4.6)$$

if this limit exists. This is a proper extension of the notion of the derivative. For example, it can be used to linearize a given function (equation (4.2)) along a direction $g$. In this case relation (4.2) holds along a given direction, a different value holds for the opposite direction, so that it provides the basis for a quasi-linearization of the function $f$.

From this definition, one may easily see that a necessary condition for a directionally differentiable function $f$ to attain a minimum at point $x^*$ is that:

$$f'(x^*, g) \geq 0 \qquad \forall g \in R \quad (4.7)$$

If strict inequality holds in (4.7) for every direction $g$ not equal to zero, the condition becomes also sufficient for $x^*$ to be a strict local minimum of $f$.

A point $x^*$ which satisfies relation (4.7) is called an inf-stationary point of $f$.

#### 4.2.1.2 Convex analysis

It is initially assumed a convex non-differentiable function $f(x)$. The set-valued *subdifferential* of this function, denoted by $\partial$ is written as:

$$w \in \partial f(x) \quad (4.8)$$

Relation (4.8) is by definition equivalent to the variational inequality:

$$w(x)\delta x \leq f(x + \delta x) - f(x) \quad \forall x, \delta x \in R^n \quad (4.9)$$

which defines for all elements of the subdifferential set (4.8) at a given point $x$, the subgradients $w(x)$, a lower linear approximation of the considered function, which holds for all small variations $\delta x$ around point $x$. In fact the name *sub*-gradient characterizes their main property. The directional derivative of function $f(x)$ at point $x$ in the direction $\delta x$ reads:

$$f^{'}(x, \delta x) = \max_{w \in \partial f(x)} \langle w, \delta x \rangle \tag{4.10}$$

In this case, $\langle ., . \rangle$ denotes the inner product of two vectors. For a piecewise smooth function which is defined as the maximum of two smooth functions:

$$f(x) = \max \{f_1(x), f_2(x)\} \tag{4.11}$$

the convex analysis subdifferential at the point of non-differentiability ($x_0$ such that $f_1(x_0) = f_2(x_0)$) is constructed by means of the convex hull of the classical gradients of the two smooth constituents, i.e., $\partial f(x_0) = co\{\nabla f_1(x_0), \nabla f_2(x_0)\}$.

A subdifferential relation which can be described by the convex analysis tools is the one describing the mechanical behaviour of a locking spring. For the mathematical modelling of the following non-linear, monotone constitutive law:

$$s = \begin{cases} s_2 + c_1 e & \text{for} \quad e \geq 0 \\ [-s_1, s_2] & \text{for} \quad e = 0 \\ -s_1 + c_2 e & \text{for} \quad e \leq 0 \end{cases} \tag{4.12}$$

between scalar stress $s$ and strain $e$, which are components of the corresponding tensors, one needs the introduction of a non-differentiable, convex potential (the so-called superpotential):

$$f(e) = \max \{f_1(e), f_2(e)\} = \max \left\{ s_2 + \frac{1}{2} c_1 e^2, -s_1 + \frac{1}{2} c_2 e^2 \right\} \tag{4.13}$$

At the point of non-differentiability, $e = 0$, the vertical branch of relation (4.12), i.e. the interval $[-s_1, s_2]$, is produced by the convex analysis subdifferential $\partial$ of the non-smooth function $f(e)$, which is constructed by means of the convex hull of the classical gradients of the two smooth constituents, i.e., $\partial f(e = 0) = co\{\nabla f_1(e), \nabla f_2(e)\}$. This is schematically shown in figure 4.1.

The normal cone to a set defined by: $C = \{x \in R^n | f(x) \leq 0\}$ at a point $x_0$ with $f(x_0) = 0$ is described by the relation:

$$N_C(x_0) \subset \{\lambda x^{\star} : x^{\star} \in R^n \lambda \geq 0, x^{\star} \in \partial f(x_0)\} \tag{4.14}$$

whenever $f$ is Lipschitzian on a neighbourhood of $x_0$ and $0 \notin \partial f(x_0)$.

For example, for the max-type function $f = \max\{\varphi_1, \ldots, \varphi_m\}$ one has:

$$\begin{aligned} N_C(x_0) =& \partial I_C(x_0) \\ =& \{z : z = \sum_{i=1}^{m} \lambda_i \text{grad}\, \varphi_i(x_0), \lambda_i \geq 0, \;\; \varphi_i(x_0) \leq 0, \;\; \lambda_i \varphi_i(x_0) = 0\} \end{aligned} \tag{4.15}$$

if $0 \notin \partial f(x_0)$. Here $I_C$ denotes the indicator function of convex set $C$ which is defined by $I_C(x) = 0$, if $x \in C$, $+\infty$ otherwise. The above relation constitutes the *Lagrange multiplier rule* for optimization problems subjected to convex inequality constraints $\varphi_i(x) \leq 0$, $i = 1, \ldots, m$. It is then considered the search for a local minimum problem of a continuously differentiable function $g : R^n \to R$ over $C = \{x \in R^n : \varphi_i(x) \leq 0 \ i = 1, \ldots, m\}$. A necessary condition is $0 \in \partial(g + I_C)(x)$ which implies that:

$$-\text{grad}\, g(x) \in \partial I_C(x) \tag{4.16}$$

which together with (4.15) leads to the Lagrange multiplier rule.

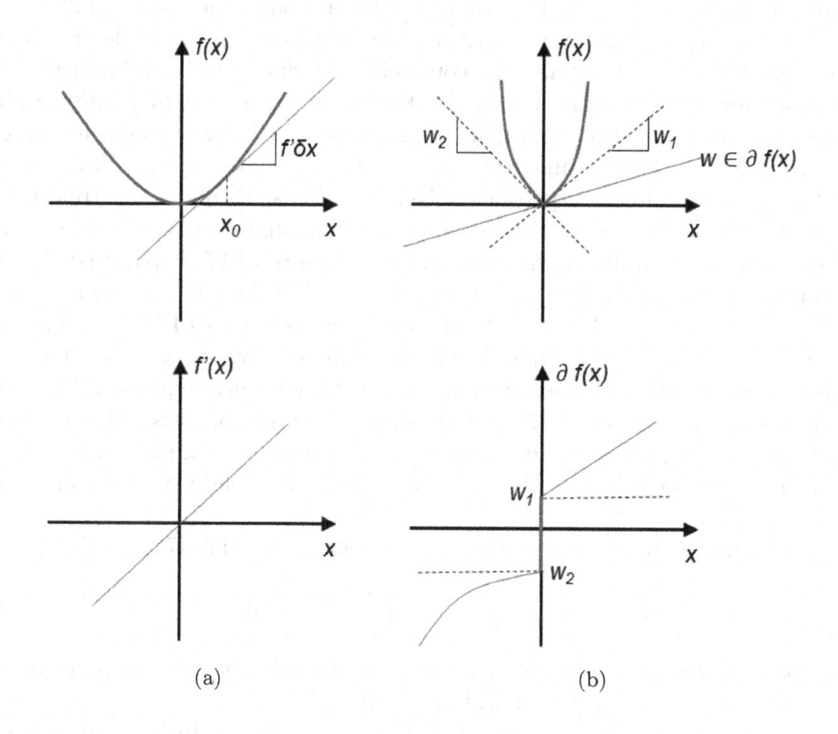


**FIGURE 4.1**: a) Classical potential energy and b) non-differentiable (non-smooth) potential energy adopted for unilateral contact-friction constitutive description.

### 4.2.2 Linear and non-linear complementarity problems

A *non-linear complementarity problem* has the following general form: find $x \in R^n$ such that to satisfy the following set of relations:

$$x \geq 0, \quad F(x) \geq 0, \quad x^T F(x) = 0 \tag{4.17}$$

where $F(x) : R^n \to R^n$ is a continuously differentiable function. In the case of a linear function $F(x)$, one gets the *linear complementarity problem* in the standard form. Note that, in general, the non-linear complementarity problem defined in (4.17) is equivalent to the variational inequality: find $x \in \mathrm{I\!R}^n_+$ such as to satisfy:

$$\langle F(x), x^* - x \rangle \geq 0 \quad \forall\, x^* \in \mathrm{I\!R}^n_+ \tag{4.18}$$

Furthermore, if the function $F(x)$ is the gradient of a continuously differentiable convex function $f(x)$, i.e., $F(x) = \partial f(x)$, then the above problems (4.17) and (4.18), are identified to be the solvability relations of an inequality constrained optimization problem with respect to the function $f(x)$. For a linear function $F(x) = Bx + c$, one gets a linear complementarity problem.

The non-classical character of the non-linearities which can be described by relation (4.17), and especially the combinatorial character introduced by the last complementarity relation, requires the use of non-standard solution techniques for the numerical treatment of problems which involve non-linearities of this kind. In order to integrate the non-linear relation (4.17) into a general purpose non-linear equations solver and, thus, to allow the treatment of inequality mechanics' applications by classical, smooth computational mechanics tools, a non-linear equation reformulation of (4.17) is required. Thus, a function $\phi(x, F(x)) : R^2 \to R^1$ is sought for which the following equivalence is true: $\phi(x, F(x)) = 0 \iff x, F(x)$ satisfy the relations (4.17). A function with this property is called non-linear complementarity, or NCP-function. By this way, an underdetermined system of linear or non-linear equations between variables $x$ and $F(x)$ of (4.17) is derived, based on the physics of the studied problem. Adding one NCP-function for each couple of complementary variables, enforces the relations of (4.17) and makes the whole system of equations solvable.

For example, the following Fischer-Burmeister function [74]:

$$\phi_{FB}(a, b) = \sqrt{a^2 + b^2} - (a + b) \tag{4.19}$$

can be used. More details on this direction of studying and solving complementarity problems can be found in [69], [104], [105].

It is worth-mentioning that equivalent smoothing techniques have been proposed in contact mechanics, in connection with finite element formulations. Details can be found in [17], [18], [43], [44], [110], [119], [247]. Unilateral contact problems in two-dimensional elastic structures with cracks have been solved with this method using boundary elements in statics [202] and dynamics [201].

It is also noted that this approach seems to be suitable for application of *physics informed neural networks* as a computing tool.

## 4.3 Interaction along boundaries and interfaces

### 4.3.1 Tie contact conditions

It is assumed that two regions are in *tie contact conditions* (*bilateral contact*) as shown in figure 4.2, with points along adjacent boundaries to have the same displacement. Due to this kinematic reaction a contact traction arises. Such a case may appear along interfaces between layers in composites. At each point $x$ of the common boundary one writes the coupling interaction equalities, which involve the compressive traction $S_N$ and the displacement $u$:

$$S^{I}_{N_i}(x) + S^{II}_{N_i}(x) = 0 \tag{4.20}$$

$$u^{I}_i(x) - u^{II}_i(x) = 0 \tag{4.21}$$

Here, superscripts $I$ and $II$ denote the regions $I$ and $II$, respectively, while subscript $i$ $(= 1, 2, 3)$ is defined according to the dimensionality of the problem.

Relation (4.21) can be enforced by adding a penalty term in the energy optimization formulation, which has the form of the indicator function $I_{U_{ad}}$ of the set of admissible displacements $U_{ad}$:

$$I_{U_{ad}} \quad \text{where} \quad U_{ad} = \{u \quad \text{s.t.} \quad u^{I}_i(x) - u^{II}_i(x) = 0\} \tag{4.22}$$

Relation (4.20) is automatically satisfied, noticing that the tractions $S_N$ are expressed by the Lagrange multipliers of the previous constraints.

For simplicity, the case of *node-to-node* coupling between regions in contact can be considered, in the context of small displacement or in geometrically non-linear finite element analysis. Technical modifications are needed in case different discretizations are used in the two neighbouring regions, which would require *node-to-surface* or *surface-to-surface* coupling techniques.

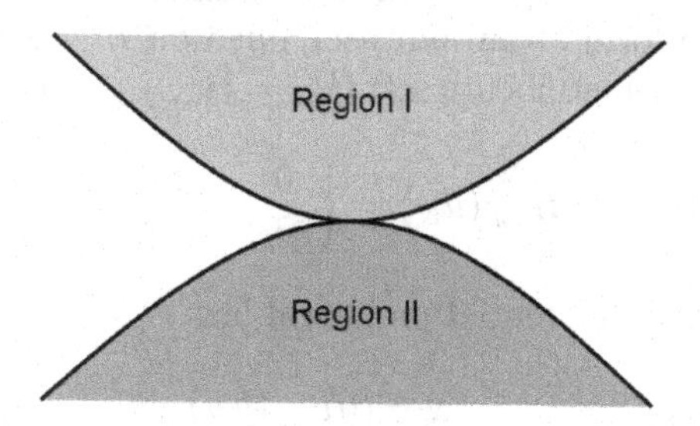


**FIGURE 4.2**: Regions in contact.

### 4.3.2 Unilateral contact

*Unilateral contact* arises, when points on the adjacent boundaries may or may not be in contact. The complete description of unilateral contact, which involves *non-penetration* inequalities, non-tensile contact stress inequalities and complementarity relations which exclude, pointwise, the one of the previous two cases, reads:

$$\mathbf{y}_N = \mathbf{d} - \mathbf{u}_N \geq \mathbf{0}, \quad -\mathbf{S}_N \geq \mathbf{0}, \quad \mathbf{y}_N^T\mathbf{S}_N = 0 \tag{4.23}$$

Here, $\mathbf{u}_N$ is the subvector of $\mathbf{u}$ with the displacements normal to the boundary and $\mathbf{S}_N$ the corresponding subvector of tractions $\mathbf{S}$, for the potential contact boundary. Normal boundary quantities are expressed in an appropriate local coordinate system such that the outward direction is positive. Moreover, $\mathbf{d}$ is the vector of initial gaps (e.g., initial distance from a rigid support or an initial crack opening).

Relations (4.23) constitute a linear complementarity problem (LCP). A frictionless contact mechanism is investigated here, indicating that zero tangential forces are considered with free, unrestricted tangential displacements along the contact boundary. The innovative aspect of formulating a LCP is that the arising system of equations is underdetermined, since both contact displacements and tractions (i.e., vectors $\mathbf{u}_N$ and $\mathbf{S}_N$) are involved. The complementarity condition of (4.23) makes the system solvable, since it enforces some of the elements of the two vectors to be equal to zero. From the mechanical point of view, instead of trying to find, by trial-and-error, the areas of contact and of separation, one has a more general model and allows a specialized algorithm find this information automatically.

The unilateral contact mechanism can also be seen as a multivalued, monotone contact law:

$$-S_N = \begin{cases} 0 & \text{for} \quad u_N \leq d \\ [0, +\infty] & \text{for} \quad u_N = d \end{cases} \tag{4.24}$$

The latter relation can be produced by subdifferentiating a potential function, which has the form of an indicator function $I_{U_{ad}}$ for the kinematically admissible relative displacements set: $U_{ad} = \{u_N \in R^1, u_N - d \leq 0\}$:

$$\phi_n(u_N) = I_{U_{ad}}(u_N) = \begin{cases} 0 & \text{for} \quad u_N \leq d \\ +\infty & \text{for} \quad u_N = d \end{cases} \tag{4.25}$$

Thus, one has the subdifferential unilateral law:

$$-S_N \in \partial I_{U_{ad}}(u_N) \tag{4.26}$$

In the framework of convex, non-smooth analysis, the unilateral contact relation (4.26) is connected to the following variational inequality:

$$-S_N\delta u \leq I_{U_{ad}}(u_N + \delta u) - I_{U_{ad}}(u_N) \quad \forall \delta u \in R^1 \tag{4.27}$$

or, taking into account the definition of the indicator function, equivalently, to:

$$-S_N \delta u \leq 0 \quad \forall \delta u \in U_{ad} \tag{4.28}$$

The above variational inequalities describe local non-linear effects which are expressed by means of monotone, possibly multivalued, subdifferential laws (cf. (4.25), (4.8)). They can be integrated along the whole non-linear boundary, introduced into the principle of virtual work of mechanics and they lead to variational formulations of structural analysis problems which have the form of variational inequalities (see, among others, [145] and [165] for more details).

The formulation of the unilateral contact interactions in the previous set of complementarity relations, or in the form of a set-valued subdifferential relation is best suited for use within the framework of energy methods in mechanics. This is the case where a structural analysis problem is seen as a minimization of the potential energy of the structure. For small displacement, small deformation linear elastostatics, this function is quadratic. The non-compatibility relations of unilateral contact introduce linear inequality restrictions in this optimization problem. The linear complementarity problem, which has been constructed here directly, arises then naturally from the Karush-Kuhn-Tucker optimality conditions of the inequality constrained quadratic programming problem.

### 4.3.3 Frictional stick-slip effects

It is now assumed a simplified friction law for a two-dimensional mechanical problem:

$$\begin{aligned} u_T = 0 \quad & \text{if} \quad && S_T \leq \mu \mid S_N \mid \\ u_T \geq 0 \quad & \text{for} \quad && -S_T = \mu S_N \\ u_T \leq 0 \quad & \text{for} \quad && -S_T = -\mu S_N \end{aligned} \tag{4.29}$$

with $u_T$ being the tangential displacement in the contact interface and $S_T$ the corresponding tangential (shear) traction. With $\mu$ is denoted the friction coefficient, used in the *Coulomb's friction law*, which is defined as:

$$\tau_T = \pm\mu \mid S_N \mid \tag{4.30}$$

where $\tau_T$ is the critical shear traction required to initiate sliding in the contact interface.

For a given contact traction $S_N$, one writes the subdifferential law:

$$S_T \in \partial\phi_T(u_T, S_N) \tag{4.31}$$

where the following convex and non-smooth, max-type subdifferential is considered (it is seen as a function of the first argument, parametrized by the second argument):

$$\phi(u_T, S_N) = \max\{\mu S_N, -\mu S_N\} \tag{4.32}$$

The connection with variational inequalities, cf. (4.27), is straightforward. Due to the implicit dependence on $S_N$, one gets implicit variational inequalities which are connected with quasi-variational inequalities.

It is noted that the law given in (4.29) can be rewritten in the primitive complementarity form used for the contact relation (4.23), by using appropriate *slack variables*, provided in equation (4.33) by $\lambda_1$, $\lambda_2$, $\tau_1$ and $\tau_2$:

$$\begin{aligned} u_T &= \lambda_1 - \lambda_2 \\ -S_T - \tau_2 &= -\mu S_N \\ -S_T + \tau_1 &= \mu S_N \\ \lambda_1 \geq 0,\ \tau_1 \geq 0,\ \lambda_1\tau_1 &= 0 \\ \lambda_2 \geq 0,\ \tau_2 \geq 0,\ \lambda_2\tau_2 &= 0 \end{aligned} \tag{4.33}$$

More complicated, monotone laws are easily formulated within the previous framework. Multi-dimensional laws can also be formulated by following the example of the plasticity or elastoplasticity with hardening. In particular, frictional laws for three-dimensional problems which involve the Coulomb friction cone, lead to non-linear complementarity problems, as it is discussed in the literature (see, for instance, [6]).

### 4.3.4 Adhesive contact

It is considered the following one-dimensional law which models *delamination effects* expressing perfect damage for crushing and cracking. For notational simplicity the same traction limit $t_1$ and delamination yield displacement $u_1$ are assumed. The non-monotone law with complete vertical (i.e., falling) branches reads:

$$-S_N = \begin{cases} 0 & \text{for} \quad u_N < -u_1 \\ [0, t_1] & \text{for} \quad u_N = -u_1 \\ c_1 = \frac{-t_1}{u_1} u_N & \text{for} \quad -u_1 < u_N < u_1 \\ [-t_1, 0] & \text{for} \quad u_N = u_1 \\ 0 & \text{for} \quad u_N > u_1 \end{cases} \tag{4.34}$$

This law can be derived by differentiating the following non-smooth and non-convex potential, which is expressed as a difference of convex functions (i.e., it is a so-called *d.c. function*):

$$\begin{aligned} \overline{\phi}_n(u_N) &= \min\left\{\frac{1}{2}c_1 {u_N}^2,\ \frac{1}{2}c_1 u_1^2\right\} \\ &= \frac{1}{2}c_1 u_N^2 - \left\{ \begin{array}{cc} 0 & \text{for } u_N < u_1 \\ s_1 u_N + \frac{1}{2}c_1(u_N - u_1)^2 & \text{for } u_N > u_1 \end{array} \right\} \\ &= \phi_1(u_N) - \phi_2(u_N) \end{aligned} \tag{4.35}$$

## 4.4 Boundary value problems in non-smooth mechanics

### 4.4.1 Variational Inequalities

Variational formulations are first considered, for a boundary value problem which is defined in a subset $\Omega$ of $R^n$, $n = 1, 2, \ldots, n$ with boundary $\Gamma$. Let $V$ be a real Hilbert space and $V'$ be its dual space. Let also $a(\cdot,\cdot)\colon V \times V \to R$ be a symmetric, continuous and coercive bilinear form and $(l,\cdot)$ be a continuous linear form on $V$. An abstract variational problem reads: find $u \in V$ such that:

$$a(u, v-u) = (l, v-u) \quad \forall v \in V \tag{4.36}$$

Let moreover $K$ be a closed convex subset of $V$ and assume that a solution of the boundary value problem within the set $K$ is sought. It can be shown that this solution is characterized by the following abstract variational inequality (of the G. Fichera type, according to [165]): find $u \in K \subset V$ such that:

$$a(u, v-u) \geq (l, v-u) \quad \forall v \in K \tag{4.37}$$

For a convex, proper functional $\Phi$ on $V$ one may define the more general variational inequality: find $u \in V$ such that:

$$\alpha(u, v) + \Phi(v) - \Phi(u) \geq (l, v-u) \quad \forall v \in V \tag{4.38}$$

It is obvious that (4.37) is a special case of (4.38), with $\Phi = I_K$, where the indicator function of the set $K$ is defined by $I_K(v) = \{0 \text{ if } v \in K, +\infty \text{ otherwise}\}$.

It is now assumed that on $\Gamma$ the general monotone multivalued boundary condition holds:

$$-S \in \partial j(u) \tag{4.39}$$

Here $j(u)$ is assumed to be a convex superpotential and $\partial$ denotes the subdifferential of convex analysis. Moreover, all (normal and tangential) contributions of boundary displacements $u$ and tractions $S$ are included in (4.39), which holds as a multi-dimensional boundary condition at each point of the boundary $\Gamma$. Relation (4.39) is, by definition of the subdifferential, equivalent to:

$$j(v) - j(u) \geq -S_i(v_i - u_i) \quad \forall v = \{v_i\} \in R^3 \tag{4.40}$$

By using (4.40) one gets the variational inequality: find $u \in V$ with $j(u) < \infty$, such that:

$$\alpha(u, v-u) + \int_\Gamma (j(v) - j(u))d\Gamma \geq \int_\Omega f_i(v_i - u_i)d\Omega \tag{4.41}$$

$\forall v \in V$ with $j(v) < \infty$.

### 4.4.2 Implicit variational inequalities

If it is assumed that the linear form $(l, \cdot)$ or the set $K$ in the previous relations depend on the solution $u$, then various types of implicit variational inequalities or quasi-variational inequalities can be obtained.

Let the set $K$ depend on the solution $u$. From (4.37), one gets the following quasi-variational inequality: find $u \in K(u) \subset V$ such that:

$$\alpha(u, v) \geq (l, v - u) \quad \forall v \in K(u) \tag{4.42}$$

Along the same lines, an implicit variational inequality is formulated using (4.38), as follows: find $u \in V$ such that:

$$\alpha(u, v) + \Phi(u, v) - \Phi(u, u) \geq (l, v - u) \quad \forall v \in V \tag{4.43}$$

Here the first argument in $\Phi(\cdot, \cdot)$ is tackled as a parameter.

### 4.4.3 The case of unilateral contact with friction

Let $V \in R^3$ be an open bounded subset occupied by a deformable body in its undeformed state. On the assumption of small displacement gradients, the principle of virtual work which has been defined in section 2.2.2.1, is expressed here using Einstein notation, as follows:

$$\begin{aligned} \int_V \sigma_{ij}\varepsilon_{ij}(v-u)dV &= \int_V \rho f_i(v_i - u_i)dV + \int_S t_i(v_i - u_i)dS \\ &+ \int_\Gamma S_N(v_N - u_N)d\Gamma + \int_\Gamma S_{T_i}(v_{T_i} - u_{T_i})d\Gamma \end{aligned} \tag{4.44}$$

The terms $f_i$, $t_i$, $S_N$ and $S_{T_i}$ represent the external body forces, the surface traction forces, as well as the normal and the shear tractions in the contact interface $\Gamma$. By comparing with the weak formulation (2.21) given in section 2.2.2.1, relation (4.44) shows that when unilateral contact with friction is added to the general description, the last two terms are introduced at the right-hand part of the principle of virtual work.

In addition, the abstract bilinear form $\alpha(\cdot, \cdot)$ in this case of linear elasticity reads:

$$\alpha(u, v) = \int_\Omega C_{ijhk}\varepsilon_{ij}(u)\varepsilon_{hk}(v)d\Omega \tag{4.45}$$

where $C_{ijhk}$ is the elasticity tensor which satisfies the well-known symmetry and ellipticity properties, with $i, j, h, k = 1, 2, 3$.

Similar to the descriptions presented in section 2.2.1, the underlying elastostatic equilibrium boundary value problem can also be expressed in the following, strong form:

$$\sigma_{ji,j} + \rho f_i = 0 \tag{4.46}$$

where inertial effects are neglected.

One recalls here the strain-displacement relation, within small displacement gradients:

$$\varepsilon_{ij} = \frac{1}{2}(u_{i,j} + u_{j,i}) \tag{4.47}$$

It can also be considered that the material follows a linearly elastic constitutive law:

$$\sigma_{ij} = C_{ijhk}\varepsilon_{hk}$$

Recall also here that on the assumption that classical support conditions hold on $\Gamma$ (i.e., say $u_N = 0$ and $u_{T_i} = 0, i = 1, 2, 3$) one gets the following variational equality: find $u \in V_0 = \{v \big| v \in V$, with $v_N = 0$, and $v_{T_i} = 0$ on $\Gamma\}$ such that:

$$\alpha(u, v) = \int_{\Omega} f_i v_i d\Omega \quad \forall v \in V_0 \tag{4.48}$$

### 4.4.4 Signorini-Coulomb unilateral frictional contact

Below are expressed the pointwise, frictionless, unilateral contact relations, which are also known as *Signorini conditions*:

$$-S_N \geq 0, \quad u_N - g \leq 0, \quad -S_N(u_N - g) = 0 \quad \text{in} \quad \Gamma \tag{4.49}$$

The inequality on the boundary normal tractions $S_N$ at the left-hand part of (4.49) corresponds to the mechanical restriction that no tensile tractions are permitted. Moreover, the normal boundary displacements $u_N$ are not permitted to be higher than a given initial distance $g$ as provided by the inequality at the middle of (4.49), which means that no penetration is allowed. Finally, the complementarity relation at the right-hand part of (4.49) expresses the physical fact that either contact is realized or a separation takes place.

A simplified, static version of the Coulomb's friction law connects the tangential (frictional or shear) traction vector $\mathbf{S}_T$ with the normal (contact) traction vector $\mathbf{S}_N$ by the relation:

$$\gamma = \mu||\mathbf{S}_N|| - ||\mathbf{S}_T|| \geq 0 \tag{4.50}$$

where $||*||$ denotes the norm in $R^3$ and $\mu$ is the friction coefficient. The friction mechanism is considered to work such, that if $||\mathbf{S}_T|| < \mu||\mathbf{S}_N||$ (i.e. $\gamma > 0$), the sliding $y_{T_i}$ along the interface must be equal to zero while if $||\mathbf{S}_T|| = \mu||\mathbf{S}_N||$ (i.e. $\gamma = 0$), then non-zero sliding appears in the opposite direction of $\mathbf{S}_T$. Explicitly, this statement is expressed as follows:

$$\begin{array}{lll} \text{If} \quad \gamma > 0 & \text{then} & y_T = 0 \\ \text{if} \quad \gamma = 0 & \text{then} & \text{there exists } \sigma > 0 \\ & \text{such that} & y_{T_i} = -\sigma S_{T_i} \end{array} \tag{4.51}$$

where $i = 1, 2, 3$ indicate the components of vector $\mathbf{S}_T$ with respect to a reference Cartesian coordinate system.

The contact law (4.49) can be written in the superpotential form:

$$-S_N \in \partial \mathbf{I}_{U_{ad}}(u_N) = \partial \Phi_N(u) = \mathcal{N}_{U_{ad}}(u_N) \tag{4.52}$$

Then, the following set of admissible displacements is introduced:

$$U_{ad} = \{u \in V \mid u_N - g \leq 0\} \tag{4.53}$$

and the notions of the convex analysis subdifferential and of the normal cone to a set have been used. The corresponding variational inequality reads:

$$-S_N(u_N)(v_N - u_N) \leq 0 \quad \forall v_N \in U_{ad} \tag{4.54}$$

For the friction law one writes, analogously:

$$-S_{T_i} \in \partial_{u_N} (\mu |S_N| |u_N|) = \partial \Phi_T(S_N, u_T) \tag{4.55}$$

where the involved potential is non-differentiable (due to the absolute value non-linearity of $|u_N|$) and depends on the normal contact force $S_N$. Thus, implicitly, the potential depends on the solution of the problem $u$, i.e. one may consider the parametrized potential $\Phi_T(u, u_T) = \Phi_T(S_N, u_T) = \mu |S_N| |u_N|$. The corresponding variational inequality can then be expressed as follows:

$$-S_{T_i}(u_T)(v_T - u_T) \leq \Phi_T(u, v_T) - \Phi_T(u, u_T) \quad \forall v_T \in U_{ad} \tag{4.56}$$

By combining relations (4.44), (4.46) and (4.52), the following implicit variational inequality is obtained: find $u \in U_{ad}$ such that:

$$\int_V \sigma_{ij}\varepsilon_{ij}(v-u)dV + \Phi_T(u, v_T) - \Phi_T(u, u_T) \geq \int_V f_i(v_i - u_i)dV \ \forall v \in U_{ad} \tag{4.57}$$

### 4.4.5 Computational algorithms

The complementarity problem description of unilateral effects provides a unified, general formulation which includes the either-or decisions of the corresponding non-linear mechanisms. Instead of trying to solve the problem by adjusting incomplete equations and adopting trial-and-error methods, one can use for the solution dedicated algorithms. These algorithms are iterative, due to the non-linearity of the problem. Nevertheless, the large experience on solving those problems in other scientific disciplines can be used in a beneficial way.

Linear complementarity problems can be solved by simplex-type methods. The most known (and robust) algorithm is the *Lemke's complementary pivoting algorithm* (see [153] and [204] for a mechanical interpretation of it), noticing that other techniques of non-smooth programming are also appropriate. In order to avoid using very specialized algorithms, which make the development of codes in general purpose software complicated, the use of special (smooth) *penalty functions* relying on the transformation of the complementarity problem into a set of non-linear equations, is also very promising. This is a special case of the smoothing technique, which has been tested in various finite element packages providing tools for contact mechanics simulations.

By considering that smoothing, in every case, hides the either-or information of the unilateral mechanism, one should be very careful with the use of this method. On the other hand, its use in frictionless contact problems for (at least) two-dimensional applications, can be recommended. Furthermore, for usage within computational homogenization calculations, the accuracy may be sufficient.

---

## 4.5 Discretized Problems

The solution of the previously outlined variational problems can be approximated numerically, after discretization, by using finite element or boundary element techniques. As an example, it is presented here the formulation of discretized hemivariational inequalities for non-linear material laws in a matricial form, which is more familiar to the matrix structural analysis community. More details can be found in [145], [166].

In particular, the static analysis problem is considered for an elastic structure with both classical, linearly elastic and degrading elements. Within the finite element / matrix structural analysis framework, the stress equilibrium equations read:

$$\overline{\mathbf{G}}\overline{\mathbf{s}} = \begin{bmatrix} \mathbf{G} & \mathbf{G}_N \end{bmatrix} \begin{bmatrix} \mathbf{s} \\ \mathbf{s}_N \end{bmatrix} = \mathbf{p} \tag{4.58}$$

where $\overline{\mathbf{G}}$ is the equilibrium matrix of the discretized structure which takes into account the stress contribution of the linear $\mathbf{s}$ and non-linear $\mathbf{s}_N$ elements and $\mathbf{p}$ is the loading vector.

The strain-nodal displacement compatibility equations take the form:

$$\overline{\mathbf{e}} = \begin{bmatrix} \mathbf{e} \\ \mathbf{e}_N \end{bmatrix} = \overline{\mathbf{G}}^T \mathbf{u} = \begin{bmatrix} \mathbf{G}^T \\ \mathbf{G}_N^T \end{bmatrix} \mathbf{u} \tag{4.59}$$

where $\overline{\mathbf{e}}$, $\mathbf{u}$ are the strain and displacement vectors, respectively.

The linear material constitutive law for the structure reads:

$$\mathbf{s} = \mathbf{K}_0(\mathbf{e} - \mathbf{e}_0) \tag{4.60}$$

where $\mathbf{K}_0$ is the stiffness matrix and $\mathbf{e}_0$ is the initial strain vector.

The non-linear material law is considered to be in the form:

$$\mathbf{s}_N \in \partial_{CL}\phi_N(\mathbf{e}_N) \tag{4.61}$$

where $\phi_N(\cdot)$ is a general non-convex superpotential and a suitable generalization of the convex analysis subdifferential is used, for example the Clarke subdifferential introduced in the theory of hemivariational inequalities [166], the summation of which over all non-linear elements gives the total strain energy contribution of them as:

$$\Phi_N(\mathbf{e}_N) = \sum_{i=1}^{q} \phi_N^{(i)}(\mathbf{e}_N) \tag{4.62}$$

Classical support boundary conditions complete the description of the problem. The principle of virtual work is then expressed as follows:

$$\mathbf{s}^T(\mathbf{e}^* - \mathbf{e}) + \mathbf{s}_N^T(\mathbf{e}_N^* - \mathbf{e}_N) = \mathbf{p}^T(\mathbf{u}^* - \mathbf{u}) \tag{4.63}$$

By substituting the elasticity law (4.60) and the strain-displacement equation (4.59) into the virtual work relation (4.63), it is obtained that:

$$\begin{aligned} &\mathbf{u}^T\mathbf{G}\mathbf{K}_0^T\mathbf{G}^T(\mathbf{u}^* - \mathbf{u}) - (\mathbf{p} + \mathbf{G}\mathbf{K}_0\mathbf{e}_0)^T(\mathbf{u}^* - \mathbf{u}) + \mathbf{s}_N^T(\mathbf{e}_N^* - \mathbf{e}_N) = 0 \\ &\forall \mathbf{u}^* \in V_{ad} \end{aligned} \tag{4.64}$$

where $\mathbf{K} = \mathbf{G}\mathbf{K}_0\mathbf{G}^T$ denotes the stiffness matrix of the structure, $\overline{\mathbf{p}} = \mathbf{p} + \mathbf{G}\mathbf{K}_0\mathbf{e}_0$ denotes the nodal equivalent loading vector and $V_{ad}$ includes all support boundary conditions of the structure.

Next, the following inequality is introduced for the elements obeying the non-linear material law (4.61):

$$\mathbf{s}_N^T(\mathbf{e}_N^* - \mathbf{e}_N) \leq \Phi_N'(\mathbf{e}_N^* - \mathbf{e}_N) \quad \forall \mathbf{e}_N^* \tag{4.65}$$

where $\Phi_N'(\mathbf{e}_N^* - \mathbf{e}_N)$ is the directional derivative of the potential $\Phi_N$. By replacing the inequality (4.65) into the virtual work equation (4.64), the following discretized hemivariational inequality is obtained: find kinematically admissible displacements $\mathbf{u} \in V_{ad}$ such that:

$$\mathbf{u}^T\mathbf{K}^T(\mathbf{u}^* - \mathbf{u}) - \overline{\mathbf{p}}^T(\mathbf{u}^* - \mathbf{u}) + \Phi_N'(\mathbf{u}_N^* - \mathbf{u}_N) \geq 0,\ \forall \mathbf{u}^* \in V_{ad} \tag{4.66}$$

Equivalently, a substationarity problem for the total potential energy can be written as follows: find $\mathbf{u} \in V_{ad}$ such that:

$$\Pi(\mathbf{u}) = \underset{\mathbf{v} \in V_{ad}}{\text{stat}} \{\Pi(\mathbf{v})\} \tag{4.67}$$

Here the potential energy reads $\Pi(\mathbf{v}) = \frac{1}{2}\mathbf{v}^T\mathbf{K}\mathbf{v} - \overline{\mathbf{p}}^T\mathbf{v} + \Phi_N(\mathbf{v})$, where the first two terms (quadratic potential) are well-known in the structural analysis community. The corresponding energy optimization problems or, in general, stationarity problems, are analogous to contact analysis formulations.

### 4.5.1 Optimization and contact analysis

In mechanics, a frictionless unilateral contact problem for a linear elastic structure in the framework of small displacement analysis, which is discretized by the finite element method, is formulated as the minimum of the potential energy function. Additional inequality constraints which arise from the unilateral contact (kinematic non-penetration) relations must be taken into account as well. Thus, the problem reads: find displacements $\mathbf{u} \in U_{ad}$ such that:

$$\mathbf{u} = \arg \min_{\mathbf{v} \in U_{ad}} \left\{ \Pi(\mathbf{v}) = \frac{1}{2}\mathbf{v}^T\mathbf{K}\mathbf{v} - \mathbf{p}^T\mathbf{v} \right\} \tag{4.68}$$

where $U_{ad} = \{\mathbf{u} \in \mathbb{R}^n : \mathbf{A}\mathbf{u} - \mathbf{b} \leq 0\}$ is the set of admissible displacements, $\mathbf{K}$ is the stiffness matrix of the discretized structure, $\mathbf{u}$ is the discrete nodal displacement vector, $\mathbf{p}$ is the loading vector and $\Pi(\mathbf{v})$ is the potential energy of the structure. Moreover, $n$ is the number of displacement degrees of freedom of the finite element model. By expressing the solvability relations of the linearly constrained quadratic programming minimization problem (4.68), the following mixed linear complementarity formulation arises:

$$\begin{aligned} &\mathbf{K}\mathbf{u} + \mathbf{A}^T\boldsymbol{\lambda} = \mathbf{p} \\ &\mathbf{A}\mathbf{u} - \mathbf{b} \leq 0, \quad \boldsymbol{\lambda} \geq 0, \quad \boldsymbol{\lambda}^T(\mathbf{A}\mathbf{u} - \mathbf{b}) = 0 \end{aligned} \tag{4.69}$$

Here, the vector of discrete contact forces $\boldsymbol{\lambda}$ (i.e., the Lagrange multipliers) appears.

Further elaboration of the problem (4.69) (static condensation of the displacement variables $u$ outside of the contact nodes) and the introduction of a set of non-negative slack variables $y$, leads to the following standard linear complementarity problem (LCP):

$$\begin{aligned} &\mathbf{y} = \mathbf{b} - \mathbf{A}\mathbf{u} = -\mathbf{A}\mathbf{K}^{-1}\mathbf{A}^T\boldsymbol{\lambda} + \mathbf{b} - \mathbf{A}\mathbf{K}^{-1}\mathbf{p} \\ &\mathbf{y} \geq 0, \quad \boldsymbol{\lambda} \geq 0, \quad \mathbf{y}^T\boldsymbol{\lambda} = 0 \end{aligned} \tag{4.70}$$

Here, the additional assumption that $\mathbf{K}$ is invertible, i.e., that no rigid body displacements or rotations arise, has been considered for simplicity (cf. [204]).

### 4.5.2 Existence of solution

Non-coercive problems arise in mechanics, when for example the existing classical boundary conditions are not sufficient to prevent (even infinitesimal) rigid body displacements and rotations of the structure. In classical, equality mechanics, conditions that guarantee the existence of a solution can be formulated for the considered boundary value problem. Nevertheless, uniqueness of some quantities may be lost. One may consider, as an example, a free elastic body subjected to self-equilibrated external forces. In this case, stresses and deformations of the elastic body can be determined, but its displacements are not uniquely determined due to the presence of rigid body displacements and rotations. In the presence of inequality constraints (e.g., unilateral contact), analogous relations have also been provided by G. Fichera [165].

It is interesting to observe that some inequality constraints may be activated and stabilize the body. Unilateral supports can be realized by fingers in a robotic grasping application: according to their placement, they may or may not stabilize a given object, depending on the applied loading. In another context, a stone wall or a vault constructed from stones without mortar is stable due to the interplay between self-weight and unilateral (no-tension) interfaces. Solvability may pose restrictions on the intensity and direction of any additional loading, coming for instance from a traffic loading on a stone bridge or from an earthquake. This strong relation between geometrical form and loading is well-known in structural analysis and design, especially for masonry (no-tension) or tensegrity (no-compression) structures.

Theoretically, one asks if a finite solution of the engineering problem exists for a given case. It is noted that the complementarity problem formulation of unilateral problems allows for a systematic investigation of this question. Solvability conditions for linear complementarity problems have been investigated in [153]. Analogous results exist if one considers the equivalent variational inequality or (if applicable) the energy minimization problem. In this respect the reader is referred to the classical work of [73] (see also [165]). Applications in the area of robotic hands have been given in [6], [7], [8], and [204]. In these cases the authors were able to study, in a systematic way, the (previously known) notions of form and force grasping within a robotic hand, using solvability conditions of the unilateral contact problem. Other applications on problems of higher dimensions have also been studied in [14], [71] and [154]. Rigid body motions for parts of a structure may appear during the analysis in a microscopic length scale, especially if open cracks and interfaces appear. Usually a displacement-driven solution can be used in order to resolve this issue.

# Chapter 5

# The Extended Finite Element Method

The numerical approach which is discussed in this chapter, focuses on the representation of damage in composite materials, using principles of fracture mechanics. Within this approach, discontinuities such as cracks, can be simulated by adopting an enriched version of the conventional finite element method, called Extended Finite Element Method, or XFEM. The main aspect of the XFEM method, is to allow for the representation of discontinuities without the need to align the geometry of each discontinuity with the finite element mesh. This is opposed to other discrete approaches used to simulate the interaction between constituents of composite materials, such as contact mechanics descriptions, where discontinuities like bimaterial interfaces should be aligned with the finite element mesh.

The general framework of linear fracture mechanics is first presented in this chapter. Then, the XFEM method within linear fracture mechanics as well as the extension of the method into non-linear problems involving cohesive discontinuities are developed.

## 5.1 Introduction to linear fracture mechanics

*Fracture mechanics* provides a description for the mechanical response of structures and materials, when discontinuities are present. As defined in section 3.3.4, a discontinuous displacement gradient (strain) is known as a weak discontinuity while a discontinuous displacement is known as a strong discontinuity. Weak discontinuities may appear at interfaces of composite materials while strong discontinuities are cracks which cross the material. The following descriptions are provided in the framework of *linear fracture mechanics* for which, the fracture process zone depicting the size of a crack is much smaller than the characteristic dimensions of the structure, as given in sections 3.3.4 and 3.3.5.

DOI: 10.1201/9781003017240-5

### 5.1.1 The stress intensity factor

The sketch of figure 5.1 shows a crack which has been developed in a body. The opposite boundaries at the opening of the crack are the crack faces and at the end of the crack there is the *crack tip*, or crack front. In figure 5.2 are shown the three widely used modes of failure for problems with cracks. Namely, *failure mode I* involves opening of the crack faces, *mode II* depicts sliding and *mode III* includes transverse sliding.

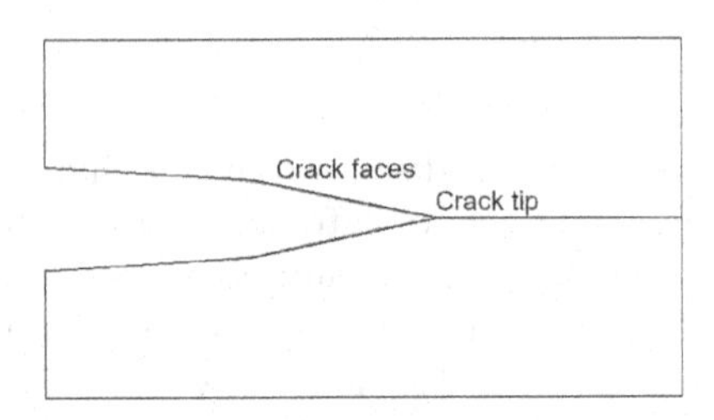


**FIGURE 5.1**: Sketch of a body with a crack.

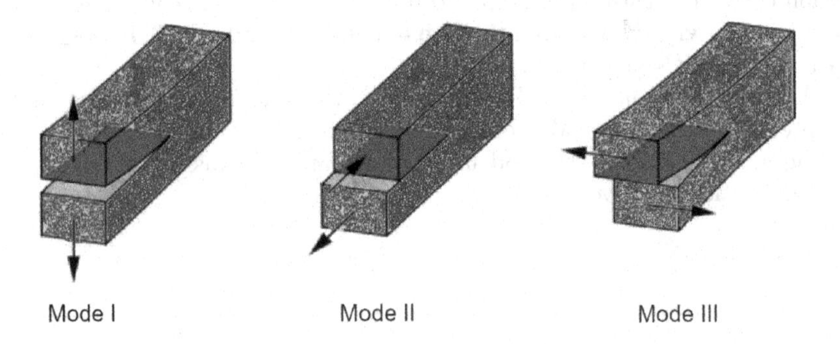


**FIGURE 5.2**: Basic failure modes.

A significant task within fracture mechanics, is to provide a description for the crack tip field, thus, the stress and displacement field, in the proximity of the crack tip. A polar coordinate system, consisting of the polar coordinates $\theta$, $r$ is defined in the crack tip as shown in figure 5.3. The stress and displacement fields are then investigated in the vicinity of the tip, as defined by the circle which includes the tip.

For the description of the crack tip field, a factor called *stress intensity factor*, $K$, is introduced. This factor can be considered as a strength measure for the crack tip field and once it is known, the crack tip field can be determined. Expressions depicting the stress and displacement fields near the crack tip, in respect to the crack intensity factor, can be found in [83]. For

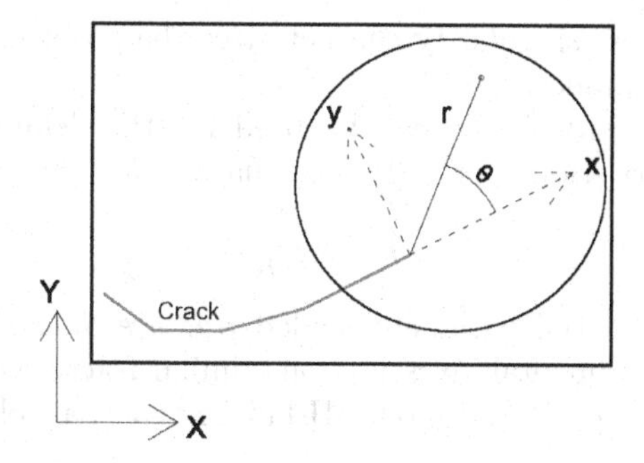


**FIGURE 5.3**: Polar coordinates and the near crack tip field.

two-dimensional plane stress and plane strain problems, these fields near the crack tip are provided by relations (5.1) for mode I failure [83]:

$$\begin{bmatrix}\sigma_x\\ \sigma_y\\ \tau_{xy}\end{bmatrix} = \frac{K_I}{\sqrt{2\pi r}}\cos(\theta/2)\begin{bmatrix}1-\sin(\theta/2)\sin(3\theta/2)\\ 1+\sin(\theta/2)\sin(3\theta/2)\\ \sin(\theta/2)\cos(3\theta/2)\end{bmatrix}$$
$$\begin{bmatrix}u\\ v\end{bmatrix} = \frac{K_I}{2G}\sqrt{\frac{r}{2\pi}}(\kappa-\cos(\theta))\begin{bmatrix}\cos(\theta/2)\\ \sin(\theta/2)\end{bmatrix} \tag{5.1}$$

and by relations (5.2) for mode II failure type [83]:

$$\begin{bmatrix}\sigma_x\\ \sigma_y\\ \tau_{xy}\end{bmatrix} = \frac{K_{II}}{\sqrt{2\pi r}}\begin{bmatrix}-\sin(\theta/2)[2+\cos(\theta/2)\cos(3\theta/2)]\\ \sin(\theta/2)\cos(\theta/2)\cos(3\theta/2)\\ \cos(\theta/2)[1-\sin(\theta/2)\sin(3\theta/2)]\end{bmatrix}$$
$$\begin{bmatrix}u\\ v\end{bmatrix} = \frac{K_{II}}{2G}\sqrt{\frac{r}{2\pi}}\begin{bmatrix}\sin(\theta/2)[\kappa+2+\cos(\theta)]\\ \cos(\theta/2)[\kappa-2+\cos(\theta)]\end{bmatrix} \tag{5.2}$$

where, for plane strain conditions:

$$\kappa = 3-4\nu, \quad \sigma_z = \nu(\sigma_x+\sigma_y) \tag{5.3}$$

and for plane stress conditions:

$$\kappa = \frac{3-\nu}{1+\nu}, \quad \sigma_z = 0 \tag{5.4}$$

In the above equations, $G$ is the Shear modulus and $\nu$ is Poisson's ratio. It can be noticed from equations (5.1) and (5.2), that for $r \to 0$ the stress field increases infinitely, resulting in a *singular* stress field, with a singularity of type $r^{-\frac{1}{2}}$. This stress singularity characterizes the problems which belong to linear fracture mechanics.

The stress intensity factor can be used for the definition of a fracture criterion. Such a criterion is provided in equation (5.5) for pure mode I failure type:

$$K_I = K_{IC} \tag{5.5}$$

where $K_{IC}$ is a material parameter called *fracture toughness*, which can be determined by experimental investigation. Similar fracture criteria can be formulated for pure mode II and mode III failure types, as follows:

$$K_{II} = K_{IIC} \tag{5.6}$$

$$K_{III} = K_{IIIC} \tag{5.7}$$

For a mixed mode type, failure is a combination of modes I, II and III and a generalized criterion of the following form can be adopted:

$$f(K_I, K_{II}, K_{III}) = 0 \tag{5.8}$$

### 5.1.2 Energy balance and the energy release rate

Next, the energy balance is discussed for problems which involve propagation of cracks. According to the conservation of energy law, the rate of change of the work of external forces $\dot{\Omega}$ is equal to the summation of the rate of change of the internal elastic (strain) energy $\dot{U}$ and the rate of change of the work $\dot{\Gamma}$ which is consumed for the crack propagation, considering that no plastic (deformation) energy and kinetic energy arise:

$$\dot{\Omega} = \dot{U} + \dot{\Gamma} \tag{5.9}$$

Assuming that a body with a crack is subject to external forces, the crack may propagate by the surface (or length in two-dimensional analysis) $\Delta A$. By introducing the crack surface growth rate $\dot{A}$ and by using the following relation:

$$\frac{\partial}{\partial t} = \frac{\partial}{\partial A}\frac{\partial A}{\partial t} = \frac{\partial}{\partial A}\dot{A} \tag{5.10}$$

equation (5.9) can be rewritten as follows:

$$\frac{\partial \Omega}{\partial A} = \frac{\partial U}{\partial A} + \frac{\partial \Gamma}{\partial A} \tag{5.11}$$

Then, as provided in equation (2.28) of section 2.2.2.2, the potential energy $\Pi$ of a structure is defined as follows:

$$\Pi = U - \Omega \tag{5.12}$$

By substituting (5.12) into (5.11), it is obtained that:

$$-\frac{\partial \Pi}{\partial A} = \frac{\partial \Gamma}{\partial A} \tag{5.13}$$

Equation (5.13) indicates that the reduction of the potential energy is equal to the energy dissipation which is required for the crack growth. In addition, from equation (5.13) the *energy release rate* $\mathcal{G}$ expressing the released energy during an infinitesimal crack advance, is defined as:

$$\mathcal{G} = -\frac{\partial \Pi}{\partial A} \tag{5.14}$$

### 5.1.3 J-integral and the interaction integral

Another concept which is used for the numerical implementation of fracture mechanics is introduced by the *J-integral.* For a homogeneous body with a crack as shown in figure 5.4, it is assumed that the stress state in the crack tip field is expressed by the tensor $\sigma_{ij}$, the displacement by the vector $u_i$ and the elastic, strain energy by $U$. By ignoring the volume forces and considering infinitesimal strains, the J-integral is defined as follows:

$$J = \int_{\partial V} (U\delta_{jk} - \sigma_{ij}\frac{\partial u_i}{\partial x_k})n_j dV \tag{5.15}$$

where $\delta_{jk}$ is Kronecker-delta and $n_j$ the outward unit normal to the surface $\partial V$.

Next, the J-integral concept is further elaborated for a cracked body, by considering a present state (1) denoted by $(u_i^{(1)}, \varepsilon_{ij}^{(1)}, \sigma_{ij}^{(1)})$ and an auxiliary state (2) denoted by $(u_i^{(2)}, \varepsilon_{ij}^{(2)}, \sigma_{ij}^{(2)})$. The J-integral can then be provided for the sum of the two states, after a proper manipulation of relation (5.15) [107]:

$$J^{(1+2)} = \int_{\partial V} \left(\frac{1}{2}(\sigma_{ij}^{(1)} + \sigma_{ij}^{(2)})(\varepsilon_{ij}^{(1)} + \varepsilon_{ij}^{(2)})\delta_{jk} - (\sigma_{ij}^{(1)} + \sigma_{ij}^{(2)})\frac{\partial}{\partial x_k}(u_i^{(1)} + u_i^{(2)})\right) n_j dV \tag{5.16}$$

By expanding relation (5.16), the following equation arises:

$$J^{(1+2)} = J^{(1)} + J^{(2)} + I^{(1,2)} \tag{5.17}$$

The term $I^{(1,2)}$ is called *interaction integral* for the states (1), (2) and is provided by:

$$I^{(1,2)} = \int_{\partial V} \left(U^{(1,2)}\delta_{jk} - \sigma_{ij}^{(1)}\frac{\partial u_i^{(2)}}{\partial x_k} - \sigma_{ij}^{(2)}\frac{\partial u_i^{(1)}}{\partial x_k}\right) n_j dV \tag{5.18}$$

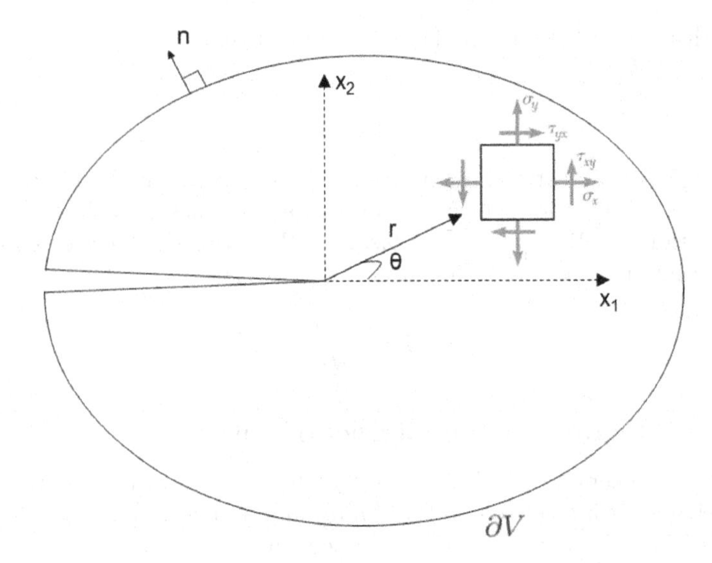


**FIGURE 5.4**: Schematic representation of a cracked body for the definition of the J-integral.

where the strain energy $U^{(1,2)}$ is defined as follows:

$$U^{(1,2)} = \sigma_{ij}^{(1)} \varepsilon_{ij}^{(2)} = \sigma_{ij}^{(2)} \varepsilon_{ij}^{(1)} \tag{5.19}$$

As it will be shown here, the interaction integral can be used for the calculation of the stress intensity factors within numerical applications, in the framework for instance, of the Extended Finite Element Method. Hence, in the following lines, it is presented how the stress intensity factors can be correlated to the J-integral and the interaction integral. It has been found that the J-integral is equal to the variation of the potential energy for an infinitesimal crack propagation and according to (5.14), it is also equal to the energy release rate $\mathcal{G}$ [107]. For a mixed mode I and mode II failure types of an elastic body, the energy release rate can be defined using the corresponding stress intensity factors $K_I$ and $K_{II}$ as follows [83, 107]:

$$J = \mathcal{G} = \frac{1}{E'}(K_I^2 + K_{II}^2) \tag{5.20}$$

noticing that for plane stress problems $E' = E$ and for plane strain problems $E' = E/(1-\nu^2)$. For the combined states (1) and (2), the J-integral provided in equation ((5.20)) can be expressed as follows:

$$J^{(1+2)} = J^{(1)} + J^{(2)} + \frac{2}{E'}\left(K_I^{(1)} K_I^{(2)} + K_{II}^{(2)} K_{II}^{(1)}\right) \tag{5.21}$$

By conducting the comparison between equations (5.17) and (5.21), it is derived that the interaction integral $I^{(1,2)}$ is given by:

$$I^{(1,2)} = \frac{2}{E'}\left(K_I^{(1)}K_I^{(2)} + K_{II}^{(2)}K_{II}^{(1)}\right) \tag{5.22}$$

Then, relations which provide the stress intensity factors can be obtained, by adopting the following assumptions for the auxiliary state (2). By assuming that the auxiliary state (2) corresponds to pure mode I failure, indicating that for this case $K_I^{(2)} = 1$ and $K_{II}^{(2)} = 0$, equation (5.22) leads to the following expression for the stress intensity factor of mode I, $K_I^{(1)}$, of the real state:

$$K_I^{(1)} = \frac{E'}{2} I_{\text{mode}I}^{(1)} \tag{5.23}$$

By assuming that the auxiliary state (2) corresponds to pure mode II failure, then $K_{II}^{(2)} = 1$ and $K_I^{(2)} = 0$, indicating that equation (5.22) leads to the following expression for the stress intensity factor of mode II, $K_{II}^{(1)}$, of the real state:

$$K_{II}^{(1)} = \frac{E'}{2} I_{\text{mode}II}^{(1)} \tag{5.24}$$

Thus, for both cases the stress intensity factors can be derived using the corresponding interaction integral. Towards the implementation of fracture mechanics within the Extended Finite Element Method, an alternative relation has been proposed for the calculation of the J-integral, relying on the transformation of the J-integral defined in (5.15) to an area integral [121], according to the following expression:

$$J = \int_S (\sigma_{ij}\frac{\partial u_i}{\partial x_k} - U\delta_{jk})\frac{\partial q}{\partial x_j} dS \tag{5.25}$$

with $q$ denoting a weighting factor over the domain of integration, such that it is equal to 1 in the crack tip and equal to zero on an outer prescribed contour. From this new definition of the J-integral, the interaction integral can be derived in numerical applications by the following relation:

$$I^{(1,2)} = \int_S \left(-U^{(1,2)}\delta_{jk} + \sigma_{ij}^{(1)}\frac{\partial u_i^{(2)}}{\partial x_k} + \sigma_{ij}^{(2)}\frac{\partial u_i^{(1)}}{\partial x_k}\right)\frac{\partial q}{\partial x_j} dS \tag{5.26}$$

Within a numerical (finite element) code, the stress and displacement terms which appear in relation (5.26) are practically calculated in the integration points of the elements in the prescribed area $S$, using numerical integration. After the interaction integral provided in (5.26) is calculated, the stress intesity factors $K_I$ and $K_{II}$ can be determined using relations (5.23) and (5.24), respectively.

## 5.2 The Extended Finite Element Method (XFEM) within linear fracture mechanics

When problems involving discontinuities, such as interfaces or cracks, are about to be studied within the finite element method, the discretization should properly be built, such that it conforms to the geometry of each discontinuity. Thus, the mesh in this case is properly developed and the finite element edges/faces are adjacent to the faces of the discontinuities. For several applications, this approach is appropriate and can provide quite accurate results.

However, when the geometry of the discontinuity is expected to change during analysis, which would be the case for problems with evolving discontinuities such as cracks, then the application of the finite element method becomes cumbersome. The reason for this, is that the mesh should properly be updated during the simulation, in order to conform to the geometry of the discontinuity, when propagation takes place. Thus, advanced re-meshing techniques must be adopted in this case.

To overcome the need for re-meshing, an alternative approach has been introduced in [139] which is called *partition of unity* finite element method. The core concept of this method is to exploit the partition of unity property, which characterizes the shape functions of the finite element method as given in section 2.3.1, in order to provide an *enrichment* of the finite element approximation. This enrichment is implemented by using proper *enrichment functions*, which are chosen for every particular type of discontinuity. The necessity for introducing these enrichment functions is derived from the fact that the traditional polynomial (shape) functions, which are used to implement the finite element approximation, are not able to provide an accurate approximation when discontinuities arbitrarily oriented in the finite element mesh appear.

Hence, the fundamental aspect of this category of methods, is to provide a tool for the simulation of evolving discontinuities, without the need to align the geometry of each discontinuity with the mesh. The *Extended Finite Element Method* or *XFEM*, which was first presented in [22], [147], belongs to the same category since it uses the partition of unity property in order to introduce an enrichment to the finite element approximation. This enrichment allows the simulation of one or more cracks arbitrarily oriented in the finite element mesh, crossing for instance some elements as shown in figure 5.5.

### 5.2.1 Enrichment for weak discontinuities

A weak discontinuity, which introduces a discontinuous displacement gradient, arises for instance when an interface between different materials appears. Such a case is depicted in figure 5.6, for the matrix-fibre system of

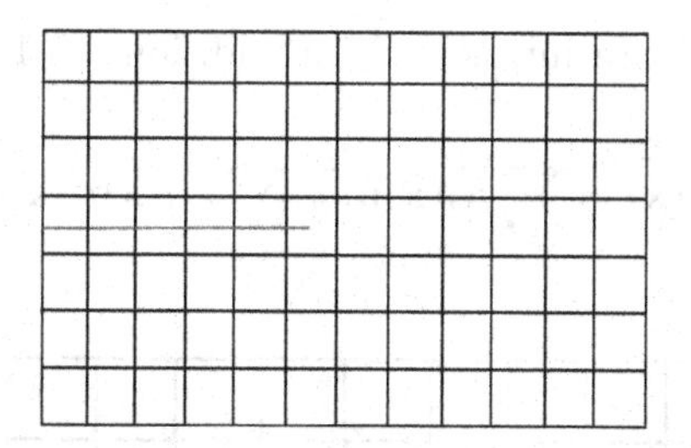

**FIGURE 5.5**: Geometry of a crack within the Extended Finite Element Method.

a two-dimensional composite geometry. In particular, the matrix domain is represented by $\Omega_A$, the fibre by $\Omega_B$ and their interface by $\Gamma_d$.

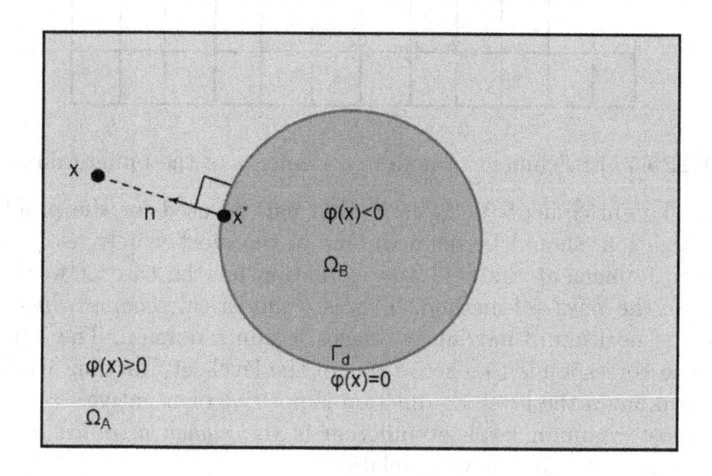


**FIGURE 5.6**: Two-dimensional, two-phase fibre reinforced composite geometry.

The enrichment of the displacement field will be applied to nodes of the elements which are crossed by the discontinuity $\Gamma_d$. These nodes are shown in a circle, in figure 5.7. As provided by the first component of the right-hand part of equation (5.27), the approximation of the displacement field $\mathbf{u}$ is implemented for all the elements of the model (assigned a node set $I$), using the standard shape functions $N_i$ which have been defined in section 2.3, within the traditional finite element method. In addition, in order to include in the approximation the bimaterial interface, all the nodes of the elements which are crossed by the interface (assigned a node set $J$) are enriched by additional enrichment functions $\psi$, as given by the second component of the right-hand part of equation (5.27). For these enrichment functions, new nodal degrees of freedom $\mathbf{a}$ are introduced, in addition to the conventional nodal displacements $\mathbf{u}$.

$$\mathbf{u}(\mathbf{x}) = \sum_{i \in I} N_i(\mathbf{x})\mathbf{u}_i + \sum_{j \in J} N_j(\mathbf{x})\psi(\mathbf{x})\mathbf{a}_j \tag{5.27}$$

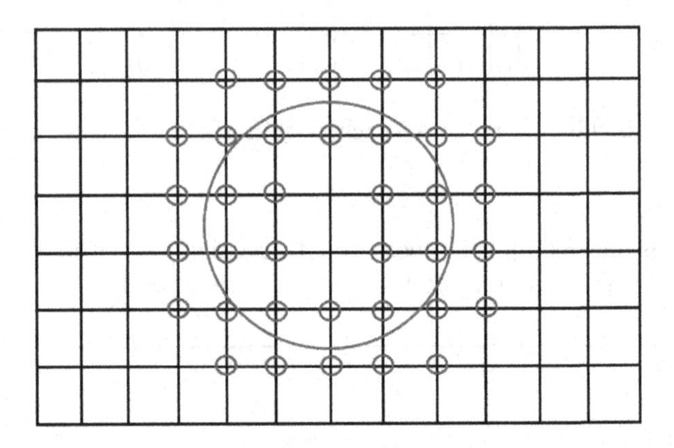

**FIGURE 5.7**: Enrichment of nodes on elements of the bimaterial interface.

Next, the enrichment functions $\psi$ that can be used for the problem under investigation, should be defined. One of the most widely used methods, adopted to implement the field approximation for the case of weak discontinuities, is the *level-set* method. This is a numerical scheme which is used to track the position of moving interfaces within a domain. The interface in this scheme corresponds to a zero value of the level-set function, while in the remaining domain the level-set function is positive or negative.

The most common level-set function is the *signed distance function* $\phi$, which is defined by the following relation:

$$\phi(\mathbf{x}) = \|\mathbf{x} - \mathbf{x}^*\| \, \mathrm{sign}(\mathbf{n} \cdot (\mathbf{x} - \mathbf{x}^*)) \tag{5.28}$$

where the Euclidean norm defined in section 1.2.1.1 between the position vector $\mathbf{x}$ of a point on the domain and its closest point projection $\mathbf{x}^*$ on the discontinuity $\Gamma_d$ is used, noticing that $\mathbf{n}$ is the unit normal vector to the interface at point $\mathbf{x}^*$, as shown in figure 5.6. According to this definition, the inner product between vectors $\mathbf{n}$ and $\mathbf{x} - \mathbf{x}^*$ will determine the sign of the distance function, since the cosine of their angle is used for this, as shown in section 1.2.1.1. Thus, a different sign will appear for points which belong to the domain $\Omega_A$ and $\Omega_B$. In addition, for every point on the interface, $\mathbf{x} = \mathbf{x}^*$ indicating that the distance function is zero in this case. Based on these descriptions, the distance function is expressed by the following relation:

$$\phi(\mathbf{x}) \begin{cases} > 0 & \text{if } \mathbf{x} \in \Omega_A \\ = 0 & \text{if } \mathbf{x} \in \Gamma_d \\ < 0 & \text{if } \mathbf{x} \in \Omega_B \end{cases} \tag{5.29}$$

Another quite common selection for the enrichment of the displacement field when weak discontinuities are investigated, is the absolute value of the signed distance function, $|\phi(\mathbf{x})|$, known as *ramp function*:

$$|\phi(\mathbf{x})| = \begin{cases} -\phi(\mathbf{x}) & \text{if } \phi(\mathbf{x}) < 0 \\ +\phi(\mathbf{x}) & \text{if } \phi(\mathbf{x}) \geq 0 \end{cases} \tag{5.30}$$

By introducing this ramp enrichment function in relation (5.27), it is obtained that:

$$\mathbf{u}(\mathbf{x}) = \sum_{i \in I} N_i(\mathbf{x})\mathbf{u}_i + \sum_{j \in J} N_j(\mathbf{x})|\phi(\mathbf{x})|\mathbf{a}_j \tag{5.31}$$

A slightly different expression is often adopted for the enrichment, as shown below:

$$\mathbf{u}(\mathbf{x}) = \sum_{i \in I} N_i(\mathbf{x})\mathbf{u}_i + \sum_{j \in J} N_j(\mathbf{x})(|\phi(\mathbf{x})| - |\phi(\mathbf{x}_j)|)\mathbf{a}_j \tag{5.32}$$

The gradient of the displacement field can then be provided as follows:

$$\begin{aligned} \nabla\mathbf{u}(\mathbf{x}) = &\sum_{i \in I} \nabla N_i(\mathbf{x})\mathbf{u}_i + \sum_{j \in J} \nabla N_j(\mathbf{x})(|\phi(\mathbf{x})| - |\phi(\mathbf{x}_j)|)\mathbf{a}_j + \\ &\sum_{j \in J} N_j(\mathbf{x})\nabla(|\phi(\mathbf{x})| - |\phi(\mathbf{x}_j)|)\mathbf{a}_j \end{aligned} \tag{5.33}$$

where:

$$\nabla(|\phi(\mathbf{x})| - |\phi(\mathbf{x}_j)|) = \text{sign}(\phi(\mathbf{x}))\nabla\phi(\mathbf{x}) = \text{sign}(\phi(\mathbf{x}))\mathbf{n} \tag{5.34}$$

Since the $\text{sign}(\phi(\mathbf{x}))$ is different for points in the domains $\Omega_A$ and $\Omega_B$ surrounding the interface $\Gamma_d$, the term $\nabla(|\phi(\mathbf{x})| - |\phi(\mathbf{x}_j)|)$ in relation (5.34) which appears also in (5.33), causes the discontinuity or *jump*, in the gradient of the displacement.

By substituting equation (5.34) into (5.33) is obtained that:

$$\begin{aligned} \nabla\mathbf{u}(\mathbf{x}) = &\sum_{i \in I} \nabla N_i(\mathbf{x})\mathbf{u}_i + \sum_{j \in J} \nabla N_j(\mathbf{x})(|\phi(\mathbf{x})| - |\phi(\mathbf{x}_j)|)\mathbf{a}_j + \\ &\sum_{j \in J} N_j(\mathbf{x})\text{sign}(\phi(\mathbf{x}))\mathbf{n}\mathbf{a}_j \end{aligned} \tag{5.35}$$

The jump of the displacement gradient is then defined as:

$$||\nabla\mathbf{u}(\mathbf{x})|| = \nabla\mathbf{u}(\mathbf{x}^+) - \nabla\mathbf{u}(\mathbf{x}^-) \tag{5.36}$$

noticing that $\mathbf{x}^+$ and $\mathbf{x}^-$ are position vectors of points adjacent to the bimaterial interface, belonging to domains $\Omega_A$ and $\Omega_B$, respectively. Eventually, the jump of the displacement gradient is provided by replacing (5.35) to (5.36):

$$||\nabla\mathbf{u}(\mathbf{x})|| =$$

$$\sum_{i\in I}\nabla N_i(\mathbf{x}^+)\mathbf{u}_i + \sum_{j\in J}\nabla N_j(\mathbf{x}^+)(|\phi(\mathbf{x}^+)| - |\phi(\mathbf{x}_j)|)\mathbf{a}_j + \sum_{j\in J} N_j(\mathbf{x}^+)\text{sign}(\phi(\mathbf{x}^+))\mathbf{n}\mathbf{a}_j$$

$$-\sum_{i\in I}\nabla N_i(\mathbf{x}^-)\mathbf{u}_i - \sum_{j\in J}\nabla N_j(\mathbf{x}^-)(|\phi(\mathbf{x}^-)| - |\phi(\mathbf{x}_j)|)\mathbf{a}_j - \sum_{j\in J} N_j(\mathbf{x}^-)\text{sign}(\phi(\mathbf{x}^-))\mathbf{n}\mathbf{a}_j \tag{5.37}$$

Since $\mathbf{x}^+$ and $\mathbf{x}^-$ are position vectors of points adjacent to the bimaterial interface, the corresponding shape functions are equal and thus, $N(\mathbf{x}^+) = N(\mathbf{x}^-)$. Using this statement, relation (5.37) can further be simplified as follows:

$$\begin{aligned} ||\nabla\mathbf{u}(\mathbf{x})|| &= \sum_{j\in J} N_j(\mathbf{x})[\text{sign}(\phi(\mathbf{x}^+)) - \text{sign}(\phi(\mathbf{x}^-))]\mathbf{n}\mathbf{a}_j \\ &= 2\sum_{j\in J} N_j(\mathbf{x})\mathbf{n}\mathbf{a}_j \end{aligned} \tag{5.38}$$

According to relation (5.38), the jump of the displacement gradient in the direction normal to the bimaterial interface, is provided by:

$$||\nabla\mathbf{u}(\mathbf{x})\mathbf{n}|| = 2\sum_{j\in J} N_j(\mathbf{x})\mathbf{a}_j \tag{5.39}$$

### 5.2.2 Enrichment for strong discontinuities

When strong discontinuities such as cracks are investigated within the XFEM method, a different enrichment scheme is applied to the nodes of elements which are split by the discontinuity and to the nodes of the crack tip elements. As given in figure 5.8, nodes of split elements are depicted with a circle while crack tip nodes are shown in a square.

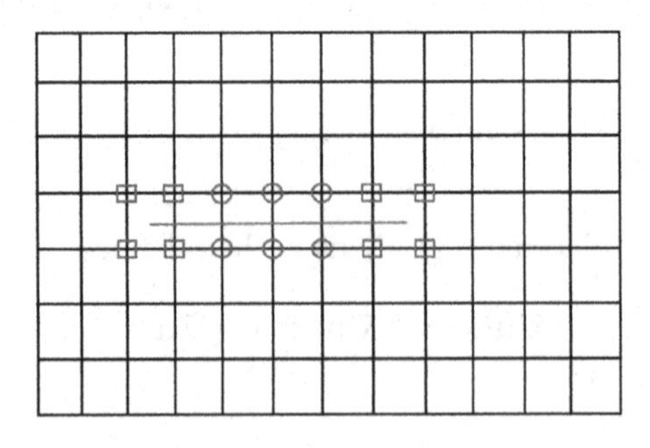

**FIGURE 5.8**: Split (in circle) and tip (in square) element nodes in a strong discontinuity.

Since strong discontinuities introduce a jump in the displacement field, a widely used enrichment function which is adopted for nodes of split elements to capture this jump, is the *Heaviside function*:

$$H(\mathbf{x}) = \begin{cases} +1 & \text{if } \phi(\mathbf{x}) > 0 \\ -1 & \text{if } \phi(\mathbf{x}) < 0 \end{cases} \tag{5.40}$$

where $\phi(\mathbf{x})$ is the signed distance function provided in relation (5.28).

The enrichment on the crack tip will depend on the type of the problem which is under investigation. If the problem belongs to linear fracture mechanics, which is the case for cracks developed in brittle materials, then the crack tip field is singular, depicting infinite stresses, as described in section 5.1.1. To capture this singular behaviour in the crack tip, the following four crack tip *asymptotic* or *branch enrichment functions* have been defined in a polar coordinate system at the crack tip [147]:

$$\psi(r,\theta) = \left\{\sqrt{r}\sin(\theta/2), \sqrt{r}\cos(\theta/2), \sqrt{r}\sin(\theta/2)\sin(\theta), \sqrt{r}\cos(\theta/2)\sin(\theta)\right\} \tag{5.41}$$

To study strong discontinuities of composite materials depicting an orthotropic response within linear fracture mechanics, the following modified crack tip enrichment functions have been proposed in [12]:

$$\psi(r,\theta) =$$

$$\left\{\sqrt{r}\cos(\theta_1/2)\sqrt{g_1(\theta)}, \sqrt{r}\cos(\theta_2/2)\sqrt{g_2(\theta)}, \sqrt{r}\sin(\theta_1/2)\sqrt{g_1(\theta)}, \sqrt{r}\sin(\theta_2/2)\sqrt{g_2(\theta)}\right\} \tag{5.42}$$

where the functions $g_i$, $i = 1, 2$ can be provided, for different classes of composites defined in [12], by:

$$\begin{aligned} g_i(\theta) &= \left(\cos^2(\theta) + \frac{1}{p_i^2}\sin^2(\theta)\right)^{1/2} \\ \theta_i &= \arctan\left(\frac{1}{p_i}\tan(\theta)\right) \end{aligned} \tag{5.43}$$

or:

$$\begin{aligned} g_i(\theta) &= \left[\cos^2(\theta) + l^2\sin^2(\theta) + (-1)^i l^2\sin(2\theta)\right]^{1/2} \\ \theta_i &= \arctan\left(\frac{\gamma_2 l^2\sin(\theta)}{\cos(\theta) + (-1)^i\gamma_1 l^2\sin(\theta)}\right) \end{aligned} \tag{5.44}$$

noticing that the parameters $p_i$, $l$ and $\gamma_i$ are related to material constants.

The approximation of the displacement field can now be developed for a strong discontinuity with two crack tips. The enrichment in this case, is considered for the split nodes as well as for the two crack tip nodes, as shown below:

$$\mathbf{u}(\mathbf{x}) = \sum_{i\in I} N_i(\mathbf{x})\mathbf{u}_i + \sum_{j\in J} N_j(\mathbf{x})H(\mathbf{x})\mathbf{a}_j + \sum_{k\in K_1} N_k(\mathbf{x}) \sum_{q=1}^{n} (\psi_q^{(1)}(\mathbf{x}))\mathbf{b}_k^{q(1)}$$
$$+ \sum_{k\in K_2} N_k(\mathbf{x}) \sum_{q=1}^{n} (\psi_q^{(2)}(\mathbf{x}))\mathbf{b}_k^{q(2)} \tag{5.45}$$

It is noted that the same crack tip enrichment functions $\psi$ for both tips 1 and 2 are used, in the local-polar coordinate system of each crack tip. In addition, $J$ is the set of the split element nodes while $K_1$ and $K_2$ denote the set of the enriched nodes at each of the two crack tips. In the third and fourth components of the right-hand side, $n$ represents the number of the asymptotic functions which are adopted to provide the enrichment on the crack tip. For a linear fracture mechanics discontinuity $n = 4$ while for a cohesive crack $n = 1$.

Similar to the weak discontinuity enrichment, a correction of the relation (5.45) is considered:

$$\mathbf{u}(\mathbf{x}) = \sum_{i\in I} N_i(\mathbf{x})\mathbf{u}_i + \sum_{j\in J} N_j(\mathbf{x})(H(\mathbf{x}) - H(\mathbf{x}_j))\mathbf{a}_j$$
$$+ \sum_{k\in K_1} N_k(\mathbf{x}) \sum_{q=1}^{n} (\psi_q^{(1)}(\mathbf{x}) - \psi_q^{(1)}(\mathbf{x}_k))\mathbf{b}_k^{q(1)} + \sum_{k\in K_2} N_k(\mathbf{x}) \sum_{q=1}^{n} (\psi_q^{(2)}(\mathbf{x}) - \psi_q^{(2)}(\mathbf{x}_k))\mathbf{b}_k^{q(2)} \tag{5.46}$$

Relation (5.46) can be rewritten in the following, simplified form:

$$\mathbf{u}(\mathbf{x}) = \mathbf{N}(\mathbf{x})\mathbf{u} + \mathbf{N}^{split}(\mathbf{x})\mathbf{a} + \mathbf{N}^{tip(1)}(\mathbf{x})\mathbf{b}^{(1)} + \mathbf{N}^{tip(2)}(\mathbf{x})\mathbf{b}^{(2)} \tag{5.47}$$

where $\mathbf{N}$ is the matrix with the shape functions defined in the traditional finite element method as shown in section 2.3.2, $\mathbf{N}^{split}$ is the matrix of shape functions for split element nodes and $\mathbf{N}^{tip(1)}$, $\mathbf{N}^{tip(2)}$ the ones for tip element nodes.

As it is discussed in section 5.3, the displacement jump will also be used for the discretization of the governing equations, in case a cohesive crack is simulated with the XFEM method. Hence, the displacement jump is defined as follows:

$$||\mathbf{u}(\mathbf{x})|| = \mathbf{u}(\mathbf{x}^+) - \mathbf{u}(\mathbf{x}^-) \tag{5.48}$$

where $\mathbf{u}(\mathbf{x}^+)$, $\mathbf{u}(\mathbf{x}^-)$ are the displacements at the two faces $\Gamma^+_{coh}$, $\Gamma^-_{coh}$ of the cohesive discontinuity, as depicted in figure 5.12 of section 5.3.

By substituting the approximation of the displacement field from equation (5.46) into equation (5.48), the following expression for the displacement jump is obtained:

$$\begin{aligned} ||\mathbf{u}(\mathbf{x})|| =& \sum_{j\in J} N_j(\mathbf{x})(H(\mathbf{x}^+) - H(\mathbf{x}^-))\mathbf{a}_j \\ &+ \sum_{k\in K_1} N_k(\mathbf{x}) \sum_{q=1}^{n} (\psi_q^{(1)}(\mathbf{x}^+) - \psi_q^{(1)}(\mathbf{x}^-))\mathbf{b}_k^{q(1)} \\ &+ \sum_{k\in K_2} N_k(\mathbf{x}) \sum_{q=1}^{n} (\psi_q^{(2)}(\mathbf{x}^+) - \psi_q^{(2)}(\mathbf{x}^-))\mathbf{b}_k^{q(2)} \end{aligned} \tag{5.49}$$

where it has been considered that $N(\mathbf{x}^+) = N(\mathbf{x}^-)$. By applying the simplified format of equation (5.47), relation (5.49) can be rewritten as follows:

$$||\mathbf{u}(\mathbf{x})|| = \overline{\mathbf{N}}^{split}(\mathbf{x})\mathbf{a} + \overline{\mathbf{N}}^{tip(1)}(\mathbf{x})\mathbf{b}^{(1)} + \overline{\mathbf{N}}^{tip(2)}(\mathbf{x})\mathbf{b}^{(2)} \tag{5.50}$$

### 5.2.3 Discretization of the governing equations within linear fracture mechanics

For the discretization of the governing equations within linear fracture mechanics, the weak formulation which is derived from the principle of virtual work will be used. For a body with a strong discontinuity, for instance a crack with a single crack tip, the governing equation presented in section 2.2.2.1 of chapter 2, is used:

$$\int_V \delta\boldsymbol{\varepsilon}^T \boldsymbol{\sigma} dV = \int_V \rho \delta\mathbf{u}^T \mathbf{f} dV + \int_S \delta\mathbf{u}^T \mathbf{t} dS \tag{5.51}$$

The relation between strains and displacements for small displacement gradients is provided by:

$$\boldsymbol{\varepsilon} = \boldsymbol{\partial}\mathbf{u} \tag{5.52}$$

where matrix $\boldsymbol{\partial}$ has been defined in equation (2.13) of section 2.2.2.1. By substituting the approximation of the displacement field given in (5.47) into relation (5.52), it is obtained that:

$$\begin{aligned} \boldsymbol{\varepsilon} =& \boldsymbol{\partial}[\mathbf{N}\mathbf{u} + \mathbf{N}^{split}\mathbf{a} + \mathbf{N}^{tip}\mathbf{b}] \\ =& \boldsymbol{\partial}\mathbf{N}\mathbf{u} + \boldsymbol{\partial}\mathbf{N}^{split}\mathbf{a} + \boldsymbol{\partial}\mathbf{N}^{tip}\mathbf{b} \end{aligned} \tag{5.53}$$

Equation (5.53) represents the strain-nodal displacement relation which includes in this case standard as well as enriched approximations. In addition, as presented in section 2.3.2, the strain-displacement matrix $B$ is defined for classical finite elements as follows:

$$\mathbf{B} = \boldsymbol{\partial}\mathbf{N} \tag{5.54}$$

Following a similar concept, the strain-displacement matrices $\mathbf{B}^{split}$ and $\mathbf{B}^{tip}$ are defined below:

$$\mathbf{B}^{split} = \boldsymbol{\partial}\mathbf{N}^{split}, \quad \mathbf{B}^{tip} = \boldsymbol{\partial}\mathbf{N}^{tip} \tag{5.55}$$

for the enriched split and tip element nodes, respectively. Then, relations (5.54) and (5.55) are substituted into (5.53), resulting in the following equation for the strain within the XFEM method:

$$\boldsymbol{\varepsilon} = \mathbf{Bu} + \mathbf{B}^{split}\mathbf{a} + \mathbf{B}^{tip}\mathbf{b} \tag{5.56}$$

The virtual strain $\delta\boldsymbol{\varepsilon}$ can now be determined using equation (5.56) and substituted in the weak formulation (5.51):

$$\begin{aligned}\int_V \delta(\mathbf{Bu} + \mathbf{B}^{split}\mathbf{a} + \mathbf{B}^{tip}\mathbf{b})^T \boldsymbol{\sigma} dV &= \int_V \rho\delta(\mathbf{Nu} + \mathbf{N}^{split}\mathbf{a} + \mathbf{N}^{tip}\mathbf{b})^T \mathbf{f} dV \\ &+ \int_S \delta(\mathbf{Nu} + \mathbf{N}^{split}\mathbf{a} + \mathbf{N}^{tip}\mathbf{b})^T \mathbf{t} dS\end{aligned} \tag{5.57}$$

noticing that the approximation of the displacement field provided by (5.47) has also been substituted in the right hand-part of (5.57). After a simple manipulation, equation (5.57) becomes:

$$\begin{aligned}&\int_V \delta\mathbf{u}^T\mathbf{B}^T\boldsymbol{\sigma} dV + \int_V \delta\mathbf{a}^T(\mathbf{B}^{split})^T\boldsymbol{\sigma} dV + \int_V \delta\mathbf{b}^T(\mathbf{B}^{tip})^T\boldsymbol{\sigma} dV = \\ &\int_V \rho\delta\mathbf{u}^T\mathbf{N}^T\mathbf{f} dV + \int_V \rho\delta\mathbf{a}^T(\mathbf{N}^{split})^T\mathbf{f} dV + \int_V \rho\delta\mathbf{b}^T(\mathbf{N}^{tip})^T\mathbf{f} dV \\ &+ \int_S \delta\mathbf{u}^T\mathbf{N}^T\mathbf{t} dS + \int_S \delta\mathbf{a}^T(\mathbf{N}^{split})^T\mathbf{t} dS + \int_S \delta\mathbf{b}^T(\mathbf{N}^{tip})^T\mathbf{t} dS\end{aligned} \tag{5.58}$$

or:

$$\begin{aligned}&\delta\mathbf{u}^T\left(\int_V \mathbf{B}^T\boldsymbol{\sigma} dV - \int_V \rho\mathbf{N}^T\mathbf{f} dV - \int_S \mathbf{N}^T\mathbf{t} dS\right) \\ &+\delta\mathbf{a}^T\left(\int_V (\mathbf{B}^{split})^T\boldsymbol{\sigma} dV - \int_V \rho(\mathbf{N}^{split})^T\mathbf{f} dV - \int_S (\mathbf{N}^{split})^T\mathbf{t} dS\right) \\ &+\delta\mathbf{b}^T\left(\int_V (\mathbf{B}^{tip})^T\boldsymbol{\sigma} dV - \int_V \rho(\mathbf{N}^{tip})^T\mathbf{f} dV - \int_S (\mathbf{N}^{tip})^T\mathbf{t} dS\right) = 0\end{aligned} \tag{5.59}$$

Due to the fact that virtual displacements $\delta\mathbf{u}$, $\delta\mathbf{a}$ and $\delta\mathbf{b}$ receive independent, arbitrary values, relation (5.59) leads to the following system of equations:

$$\delta\mathbf{u}^T\left(\int_V \mathbf{B}^T\boldsymbol{\sigma} dV - \int_V \rho\mathbf{N}^T\mathbf{f} dV - \int_S \mathbf{N}^T\mathbf{t} dS\right) = 0 \tag{5.60a}$$

$$\delta\mathbf{a}^T\left(\int_V (\mathbf{B}^{split})^T\boldsymbol{\sigma}dV - \int_V \rho(\mathbf{N}^{split})^T\mathbf{f}dV - \int_S (\mathbf{N}^{split})^T\mathbf{t}dS\right) = 0 \quad (5.60b)$$

$$\delta\mathbf{b}^T\left(\int_V (\mathbf{B}^{tip})^T\boldsymbol{\sigma}dV - \int_V \rho(\mathbf{N}^{tip})^T\mathbf{f}dV - \int_S (\mathbf{N}^{tip})^T\mathbf{t}dS\right) = 0 \quad (5.60c)$$

which leads to:

$$\int_V \mathbf{B}^T\boldsymbol{\sigma}dV - \int_V \rho\mathbf{N}^T\mathbf{f}dV - \int_S \mathbf{N}^T\mathbf{t}dS = 0 \quad (5.61a)$$

$$\int_V (\mathbf{B}^{split})^T\boldsymbol{\sigma}dV - \int_V \rho(\mathbf{N}^{split})^T\mathbf{f}dV - \int_S (\mathbf{N}^{split})^T\mathbf{t}dS = 0 \quad (5.61b)$$

$$\int_V (\mathbf{B}^{tip})^T\boldsymbol{\sigma}dV - \int_V \rho(\mathbf{N}^{tip})^T\mathbf{f}dV - \int_S (\mathbf{N}^{tip})^T\mathbf{t}dS = 0 \quad (5.61c)$$

Within linear elasticity, the constitutive stress-strain response is provided below:

$$\boldsymbol{\sigma} = \mathbf{C}\boldsymbol{\varepsilon} \Rightarrow \boldsymbol{\sigma} = \mathbf{C}(\mathbf{B}\mathbf{u} + \mathbf{B}^{split}\mathbf{a} + \mathbf{B}^{tip}\mathbf{b}) \quad (5.62)$$

where the strain provided by relation (5.56) has been used in (5.62). The final system of equations is derived, by substituting (5.62) into equations (5.61):

$$\begin{aligned}&\int_V \mathbf{B}^T\mathbf{C}\mathbf{B}dV\mathbf{u} + \int_V \mathbf{B}^T\mathbf{C}\mathbf{B}^{split}dV\mathbf{a} + \int_V \mathbf{B}^T\mathbf{C}\mathbf{B}^{tip}dV\mathbf{b} = \\ &\int_V \rho\mathbf{N}^T\mathbf{f}dV + \int_S \mathbf{N}^T\mathbf{t}dS\end{aligned} \quad (5.63a)$$

$$\begin{aligned}&\int_V (\mathbf{B}^{split})^T\mathbf{C}\mathbf{B}dV\mathbf{u} + \int_V (\mathbf{B}^{split})^T\mathbf{C}\mathbf{B}^{split}dV\mathbf{a} + \int_V (\mathbf{B}^{split})^T\mathbf{C}\mathbf{B}^{tip}dV\mathbf{b} = \\ &\int_V \rho(\mathbf{N}^{split})^T\mathbf{f}dV + \int_S (\mathbf{N}^{split})^T\mathbf{t}dS\end{aligned} \quad (5.63b)$$

$$\begin{aligned}&\int_V (\mathbf{B}^{tip})^T\mathbf{C}\mathbf{B}dV\mathbf{u} + \int_V (\mathbf{B}^{tip})^T\mathbf{C}\mathbf{B}^{split}dV\mathbf{a} + \int_V (\mathbf{B}^{tip})^T\mathbf{C}\mathbf{B}^{tip}dV\mathbf{b} = \\ &\int_V \rho(\mathbf{N}^{tip})^T\mathbf{f}dV + \int_S (\mathbf{N}^{tip})^T\mathbf{t}dS\end{aligned} \quad (5.63c)$$

The system of equations (5.63) can be expressed in the following form:

$$\begin{bmatrix}\mathbf{K}_{uu} & \mathbf{K}_{ua} & \mathbf{K}_{ub}\\ \mathbf{K}_{au} & \mathbf{K}_{aa} & \mathbf{K}_{ab}\\ \mathbf{K}_{bu} & \mathbf{K}_{ba} & \mathbf{K}_{bb}\end{bmatrix}\begin{bmatrix}\mathbf{u}\\ \mathbf{a}\\ \mathbf{b}\end{bmatrix} = \begin{bmatrix}\mathbf{F}_u\\ \mathbf{F}_a\\ \mathbf{F}_b\end{bmatrix} \quad (5.64)$$

where the indices $u$, $a$ and $b$ denote degrees of freedom for standard, split and tip element nodes. In addition, the stiffness terms $\mathbf{K}$ and the external force terms $\mathbf{F}$ which appear in relation (5.64) are provided below:

$$
\begin{aligned}
&\mathbf{K}_{uu}=\int_V \mathbf{B}^T\mathbf{C}\mathbf{B}dV, \quad \mathbf{K}_{ua}=\int_V \mathbf{B}^T\mathbf{C}\mathbf{B}^{split}dV, \quad \mathbf{K}_{ub}=\int_V \mathbf{B}^T\mathbf{C}\mathbf{B}^{tip}dV \\
&\mathbf{K}_{au}=\int_V (\mathbf{B}^{split})^T\mathbf{C}\mathbf{B}dV, \quad \mathbf{K}_{aa}=\int_V (\mathbf{B}^{split})^T\mathbf{C}\mathbf{B}^{split}dV \\
&\mathbf{K}_{ab}=\int_V (\mathbf{B}^{split})^T\mathbf{C}\mathbf{B}^{tip}dV, \quad \mathbf{K}_{bu}=\int_V (\mathbf{B}^{tip})^T\mathbf{C}\mathbf{B}dV \\
&\mathbf{K}_{ba}=\int_V (\mathbf{B}^{tip})^T\mathbf{C}\mathbf{B}^{split}dV, \quad \mathbf{K}_{bb}=\int_V (\mathbf{B}^{tip})^T\mathbf{C}\mathbf{B}^{tip}dV
\end{aligned}
\tag{5.65}
$$

$$
\begin{aligned}
\mathbf{F}_u &= \int_V \rho\mathbf{N}^T\mathbf{f}dV + \int_S \mathbf{N}^T\mathbf{t}dS \\
\mathbf{F}_a &= \int_V \rho(\mathbf{N}^{split})^T\mathbf{f}dV + \int_S (\mathbf{N}^{split})^T\mathbf{t}dS \\
\mathbf{F}_b &= \int_V \rho(\mathbf{N}^{tip})^T\mathbf{f}dV + \int_S (\mathbf{N}^{tip})^T\mathbf{t}dS
\end{aligned}
\tag{5.66}
$$

### 5.2.4 Crack propagation criteria

A significant aspect related to the implementation of the XFEM method within linear fracture mechanics, is to provide criteria for crack propagation. These criteria involve a description for when a crack propagates as well as for the direction of the crack propagation. In case a pure mode I failure type is assumed, a crack is expected to propagate along the already chosen direction at the crack tip and for a straight crack, along the extension of its direction. However, for a mixed mode I and mode II failure type, the direction of the crack propagation will be such, that forms an angle in respect to the tangent at the crack tip. For this case, a criterion should be adopted in order to provide the value of this angle.

One of the crack propagation criteria relies on the concept of the *maximum circumferential tensile stress*. According to this concept, the crack propagates in a radial direction at the crack tip, on a plane which is perpendicular to the direction of the maximum, principal tensile stress and thus, on a plane where the shear stress is zero. In addition, crack propagation takes place when the maximum principal tensile stress reaches the tensile strength of the material.

An expression providing the circumferential stress $\sigma_{\theta\theta}$ and the shear stress $\sigma_{r\theta}$ can be derived, by transforming equations (5.1) and (5.2) into polar coordinates:

$$\begin{bmatrix} \sigma_{\theta\theta} \\ \sigma_{r\theta} \end{bmatrix} =$$

$$\frac{K_I}{4\sqrt{2\pi r}} \begin{bmatrix} 3\cos(\theta/2) + \cos(3\theta/2) \\ \sin(\theta/2) + \sin(3\theta/2) \end{bmatrix} + \frac{K_{II}}{4\sqrt{2\pi r}} \begin{bmatrix} -3\sin(\theta/2) - 3\sin(3\theta/2) \\ \cos(\theta/2) + 3\cos(3\theta/2) \end{bmatrix} \tag{5.67}$$

where the stress intensity factors $K_I$, $K_{II}$ for failure modes I and II respectively, are defined in section 5.1.1. By assigning a zero value for the shear stress $\sigma_{r\theta}$ in equations (5.67), it is obtained the critical angle of crack propagation $\theta_c$, defined as the angle between the existing crack and the crack growth direction:

$$\theta_c = 2\text{arctan}\frac{1}{4}\left(\frac{K_I}{K_{II}} \pm \sqrt{\left(\frac{K_I}{K_{II}}\right)^2 + 8}\right) \tag{5.68}$$

which is also written as:

$$\theta_c = 2\text{arctan}\left(\frac{-2K_{II}/K_I}{1 + \sqrt{1 + 8(K_{II}/K_I)^2}}\right) \tag{5.69}$$

It is noted that the stress integrity factors which are needed for the calculation of the crack propagation angle in equations (5.68) and (5.69), can be determined using the interaction integrals as presented in section 5.1.3.

Another crack propagation concept relies on the minimum strain energy density criterion [107]. According to this criterion, the crack propagates in the direction of the minimum strength of the material, in the direction that the strain energy density becomes minimum.

A third criterion is based on the concept that the crack propagates in the direction of the maximum energy release rate. Crack propagation occurs in this case, when the energy release rate becomes equal to a critical value. By assuming that an infinitesimal crack kink with an angle $\theta$ is developed in respect to an existing crack, it can be proved that the energy release rate $\mathcal{G}(\theta)$ is provided by the following expression [107]:

$$\mathcal{G}(\theta) = \frac{1}{4E'}g^2(\theta)[(1+3\cos^2\theta)K_I^2 + 8\sin\theta\cos\theta K_I K_{II} + (9-5\cos^2\theta)K_{II}^2] \tag{5.70}$$

where $g(\theta)$ is provided by:

$$g(\theta) = \left(\frac{4}{3+\cos^2\theta}\right)\left(\frac{1-\theta/\pi}{1+\theta/\pi}\right)^{\theta/2\pi} \tag{5.71}$$

To obtain the angle of the propagation, the energy release rate $\mathcal{G}(\theta)$ in equation (5.70) is maximized, by demanding its first derivative in respect to the angle $\theta$ to be zero and its second derivative to be negative.

### 5.2.5 Crack tip enrichment approaches and numerical integration

Two general approaches are used to implement the growth of a crack within an XFEM code, when linear fracture is considered. The first approach relies on the enrichment of the nodes of the element which contains the crack tip, using the corresponding crack tip enrichment functions presented in section 5.2.2. The enriched crack tip nodes at an initial and subsequent position of the crack, are depicted in a square, in figure 5.9. This type of enrichment is adopted for the calculation of the stress and displacement fields in the tip element, which are then used for the implementation of the crack propagation criteria as described in section 5.2.4.

This scheme may result in a slow convergence of the XFEM solution, especially in case the size of the crack tip element is significantly small. To improve the rate of convergence another approach has been introduced, according to which the crack tip enrichment is not applied to nodes of the single element which contains the tip, but to nodes of a group of elements, surrounding the crack tip. As shown in figure 5.10 these elements are found within the dashed circle, the radius of which will be constant for every crack propagation step and determined as a fraction of the characteristic length of the structure. According to this approach, the crack tip enrichment is applied to the nodes depicted in a square, noticing that there are also nodes inside the circle, which should be enriched with both the split (Heaviside) and crack tip enrichment functions, given within square-circle symbols in figure 5.10.

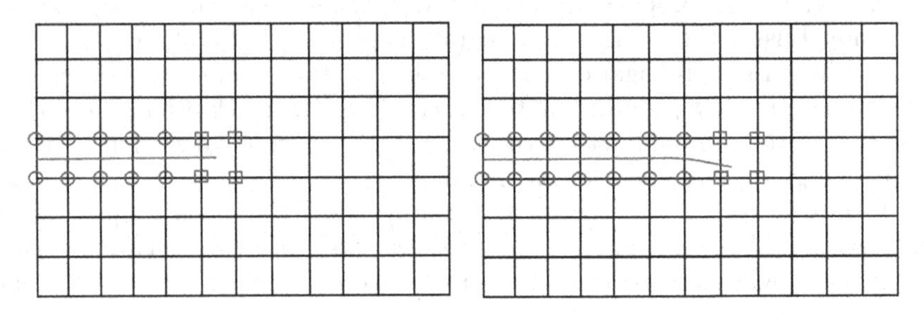

**FIGURE 5.9**: Crack growth using enrichment on the nodes of a single crack tip element.

A last concept of practical importance towards the implementation of the XFEM method refers to numerical integration, which is needed for the calculation of the integrals used in the governing equations (5.63). The critical point for a proper numerical integration, is to take the discontinuities into account since if not, inaccurate results may arise. This is done by dividing the domain between each discontinuity and every element which is cut by the discontinuity, into several sub-domains of rectangular or triangular shape, in both sides of the discontinuity. The Gauss integration scheme presented in

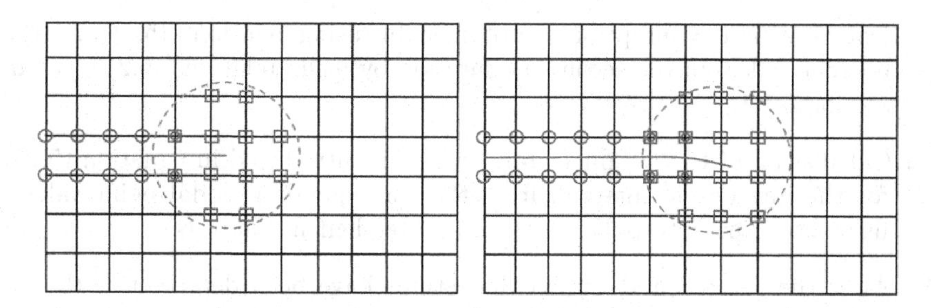

**FIGURE 5.10**: Crack growth using enrichment on the nodes of a group of elements surrounding the crack tip.

section 2.3.6 for continuous elements can then be applied to each of these sub-domains. For the numerical integration at the tip of the discontinuity, the crack tip element is also divided into sub-domains following the previous concept. In general, a higher number of sub-domains results in a more accurate numerical integration but it increases the computational cost.

The concept of numerical integration using Gauss quadrature is also adopted for the calculation of the interaction integral provided by equation (5.26), which is needed for the determination of the stress intensity factors $K_I$, $K_{II}$. A circle with centre on the crack tip and a specified radius is considered, crossing some elements surrounding the crack tip as shown in figure 5.11. The interaction integral is calculated for the elements which are crossed by the circle, using numerical integration.

In particular, as it is described in section 5.1.3, for the calculation of the interaction integral the present, real state (1) for the studied problem denoted by $(u_i^{(1)}, \varepsilon_{ij}^{(1)}, \sigma_{ij}^{(1)})$ and an auxiliary state (2) denoted by $(u_i^{(2)}, \varepsilon_{ij}^{(2)}, \sigma_{ij}^{(2)})$, are considered. The following steps are then followed for the determination of the interaction integral using numerical integration and the calculation of the stress intensity factors within an XFEM code:

**1.**: Calculation of the displacements after the governing equations (5.63) are solved and determination of the present state for the displacement, strain and stress for every Gauss points of the elements which are crossed by the circle at the crack tip.

**2.**: Transformation of the present state quantities from the global to the local coordinate system which is defined at the crack tip as shown in figure 5.3.

**3.**: Calculation of the auxiliar state for the displacement, strain and stress. This is done by assuming a) that the auxiliary state corresponds to pure mode I failure, indicating that for this case $K_I^{(2)} = 1$ and $K_{II}^{(2)} = 0$ and b) that the auxiliary state corresponds to pure mode II failure, then $K_{II}^{(2)} = 1$ and $K_I^{(2)} = 0$. For a two-dimensional problem, the auxiliar state derived from the first assumption is obtained by using relations (5.1) and the one derived from the second assumption by using relations (5.2) given in section 5.1.1.

**4.:** Calculation of the weight factor $q$ which is introduced in equation (5.25) for the interaction integral, from the corresponding nodal point values, using standard interpolation within finite element analysis.

**5.:** After the present and the auxiliar states have been determined, the interaction integral provided by relation (5.26) is calculated using Gauss integration, as follows:

$$I^{(1,2)} = \sum_{r=1}^{N} \left[ \left( -U^{(1,2)}\delta_{jk} + \sigma_{ij}^{(1)} \frac{\partial u_i^{(2)}}{\partial x_k} + \sigma_{ij}^{(2)} \frac{\partial u_i^{(1)}}{\partial x_k} \right) \frac{\partial q}{\partial x_j} \right]_r \alpha_r \det(\mathbf{J}) \tag{5.72}$$

where $\alpha_r$ is the weight factor used in Gauss quadrature as discussed in section 2.3.6, $\det(\mathbf{J})$ is the determinant of the Jacobian matrix defined in section 2.3.4 and $N$ is the number of Gauss points per element. It is noticed that two values of the interaction integral are calculated, one using the present state and pure failure mode I assumption for the auxiliar state and the other using the present state and pure failure mode II assumption for the auxiliar state.

**6.:** Determination of the stress intensity factors $K_I^{(1)}$, $K_{II}^{(1)}$ using relation (5.23) for pure mode I and relation (5.24) for pure mode II assumption for the auxiliar state, respectively. In each of these equations, the corresponding interaction integral calculated in the previous step, is used.

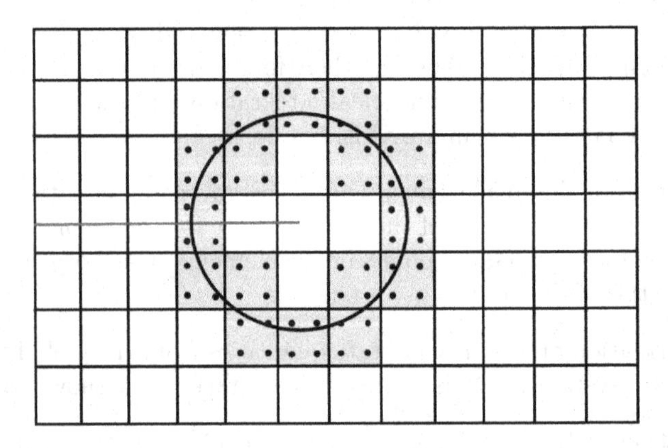

**FIGURE 5.11:** Elements surrounding the crack tip and Gauss integration points used for the numerical integration of the interaction integral.

## 5.3 The XFEM method for cohesive cracks

### 5.3.1 Definition and weak formulation

Contrary to problems which are related to linear fracture mechanics, for which the size of the fracture process zone is small in comparison with the characteristic dimensions of the structure, there is another class of problems where the size of the fracture process zone is not negligible and some cohesive forces are developed in the faces of the crack. As shown in figure 5.12, the point which separates the traction-free area and the fracture process zone is the *real* or *physical* crack tip, while the point which separates the fracture process zone and the uncracked material defines the *fictitious* or *mathematical* crack tip. Within cohesive models, the stress in the fictitious tip is finite and generally equal to the tensile strength of the material, indicating that the concept of singular stresses, which was met in linear fracture mechanics, is not yet applicable.

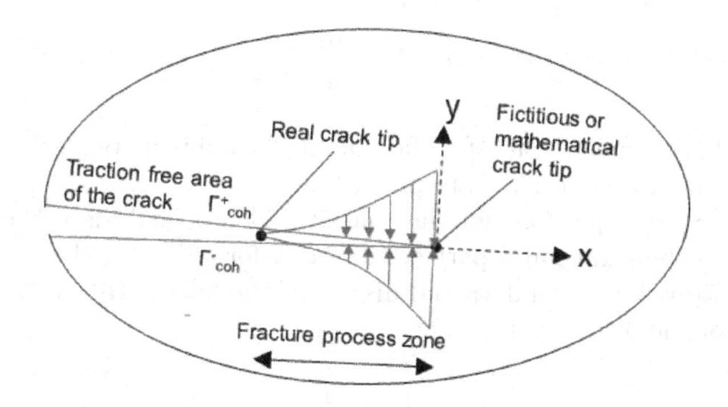


**FIGURE 5.12**: Cohesive crack.

In section 3.3.5 is presented the approach of cohesive zone models, which can be used to capture the response of cohesive cracks within the finite element method. According to this approach, cohesive zone elements are introduced between continuum elements to depict the mentioned response. As shown in figure 3.6 of section 3.3.5 damage growth takes place with opening of the cohesive elements, indicating that the direction of crack propagation is determined by the position of these elements on the finite element mesh.

An alternative approach which enables the simulation of arbitrarily oriented cohesive cracks developed within the finite elements, is provided by the XFEM method. Thus, there is no need in this case, to consider known crack paths or adopt any re-meshing technique. For the representation of the main steps of cohesive XFEM analysis, a *traction-separation* law is introduced in the fracture process zone as discussed in section 3.3.5. This law is often non-linear as shown by the diagrams of figure 3.7 in section 3.3.5 and may include a softening branch, which characterizes the behaviour of quasi-brittle materials like concrete. The area under the traction-separation law is the fracture energy, which provides the energy dissipation taking place when separation occurs.

The traction vector $\mathbf{t}$ and the separation or jump vector $||\mathbf{u}||$ of this cohesive law have been defined in section 3.3.5 for a two-dimensional problem and are also provided here for completeness:

$$\mathbf{t} = \begin{bmatrix} t_t \\ t_n \end{bmatrix} \tag{5.73}$$

$$||\mathbf{u}|| = \begin{bmatrix} ||u||_t \\ ||u||_n \end{bmatrix} \tag{5.74}$$

where $t_t$, $t_n$ are the traction components in the tangential and normal direction of the discontinuity and $||u||_t$, $||u||_n$ the corresponding displacement jump components. The traction-separation law can be formulated using the potential $\phi$ of the cohesive discontinuity, according to the following relation:

$$\mathbf{t} = \frac{\partial \phi(||\mathbf{u}||)}{\partial ||\mathbf{u}||} \tag{5.75}$$

In section 3.3.5 an expression for the potential $\phi$ and the related formulation of the traction-separation law are presented.

The governing equations for the cohesive XFEM analysis will be derived, by adding in the right-hand part of the weak formulation, the work created by the tractions $\mathbf{t}^+$, $\mathbf{t}^-$ and virtual displacements $\delta\mathbf{u}^+$, $\delta\mathbf{u}^-$ at the two sides $\Gamma^+_{coh}$, $\Gamma^-_{coh}$ of the discontinuity shown in figure 5.12:

$$\int_V \delta\boldsymbol{\varepsilon}^T \boldsymbol{\sigma} dV = \int_V \rho \delta\mathbf{u}^T \mathbf{f} dV + \int_S \delta\mathbf{u}^T \bar{\mathbf{t}} dS + \int_{\Gamma^+_{coh}} \delta(\mathbf{u}^+)^T \mathbf{t}^+ d\Gamma + \int_{\Gamma^-_{coh}} \delta(\mathbf{u}^-)^T \mathbf{t}^- d\Gamma \tag{5.76}$$

By considering that equal tractions of opposite direction appear at the two faces of the discontinuity, it is obtained that:

$$\mathbf{t} = \mathbf{t}^- = -\mathbf{t}^+ \tag{5.77}$$

In addition, equation (5.48) leads to the following relation which defines the virtual displacement jump:

$$\delta||\mathbf{u}|| = \delta\mathbf{u}^+ - \delta\mathbf{u}^- \tag{5.78}$$

By substituting relations (5.77), (5.78) in the last two terms of the right-hand part of equation (5.76), the governing equation becomes:

$$\int_V \delta\boldsymbol{\varepsilon}^T \boldsymbol{\sigma} dV + \int_{\Gamma_{coh}} \delta(||\mathbf{u}||)^T \mathbf{t} d\Gamma = \int_V \rho \delta\mathbf{u}^T \mathbf{f} dV + \int_S \delta\mathbf{u}^T \bar{\mathbf{t}} dS \tag{5.79}$$

It is noticed that the same governing equation is adopted for the case of cohesive zone models in section 3.3.5. However, as it will be shown in the following lines, discretization of the governing equation (5.79) within the XFEM method will differentiate the description.

### 5.3.2 Discretization and linearization of the governing equation

To proceed with discretization of the governing equation, the enrichment of the split and tip element nodes should be defined. As it has been mentioned at section 5.3.1, when cohesive, strong discontinuities arise, the stress field in the crack tip is not singular, indicating that finite stresses are developed and proper crack tip enrichment functions should be considered within the XFEM method. In [146] the following crack tip enrichment is proposed, for the case of cohesive XFEM discontinuities:

$$\psi(r, \theta) = r^k \sin(\theta/2) \tag{5.80}$$

where $k$ is equal to 1, or 1.5 or 2.

To capture the jump in the split element nodes, the Heaviside enrichment function introduced in relation (5.40) of section 5.2.2 is used.

These enrichment functions are introduced in relations (5.46), (5.47) in order to provide the approximation of the displacement field for cohesive crack growth. Similarly, they are included in relations (5.49) and (5.50) in order to determine the relevant displacement jump at the two faces of the crack and in equation (5.56) to provide the approximation of the strain field.

Before the substitution of these approximation fields in the governing equation, it should be noted that the traction and the displacement jump vectors are provided in equations (5.73), (5.74) in a local coordinate system with axes along the direction and perpendicular to the discontinuity. Therefore, the transformation of the displacement jump from the local to the global coordinate system is required, according to the following relation:

$$||\mathbf{u}(\mathbf{x})|| = \mathbf{u}(\mathbf{x}^+)_{local} - \mathbf{u}(\mathbf{x}^-)_{local} = \mathbf{T}\mathbf{u}(\mathbf{x}^+) - \mathbf{T}\mathbf{u}(\mathbf{x}^-) \tag{5.81}$$

where $\mathbf{T}$ is the transformation matrix which is provided in equation (3.63) of section 3.3.6. By adopting this transformation, the approximation of the displacement jump given in equation (5.50) is modified as follows:

$$||\mathbf{u}(\mathbf{x})|| = \mathbf{T}\overline{\mathbf{N}}^{split}(\mathbf{x})\mathbf{a} + \mathbf{T}\overline{\mathbf{N}}^{tip}(\mathbf{x})\mathbf{b} \tag{5.82}$$

Then, the approximations of displacement $\mathbf{u}$, strain $\boldsymbol{\varepsilon}$ and displacement jump $||\mathbf{u}||$ are substituted from relations (5.47), (5.56) and (5.82), into the governing equation (5.79):

$$\int_V \delta(\mathbf{Bu}+\mathbf{B}^{split}\mathbf{a}+\mathbf{B}^{tip}\mathbf{b})^T\boldsymbol{\sigma}dV+\int_{\Gamma_{coh}}\delta(\mathbf{T}\overline{\mathbf{N}}^{split}\mathbf{a}+\mathbf{T}\overline{\mathbf{N}}^{tip}\mathbf{b})^T\mathbf{t}d\Gamma=$$
$$\int_V \rho\delta(\mathbf{Nu}+\mathbf{N}^{split}\mathbf{a}+\mathbf{N}^{tip}\mathbf{b})^T\mathbf{f}dV+\int_S\delta(\mathbf{Nu}+\mathbf{N}^{split}\mathbf{a}+\mathbf{N}^{tip}\mathbf{b})^T\bar{\mathbf{t}}dS \tag{5.83}$$

Equation (5.83) is further developed into:

$$\begin{aligned}&\int_V\delta\mathbf{u}^T\mathbf{B}^T\boldsymbol{\sigma}dV+\int_V\delta\mathbf{a}^T(\mathbf{B}^{split})^T\boldsymbol{\sigma}dV+\int_V\delta\mathbf{b}^T(\mathbf{B}^{tip})^T\boldsymbol{\sigma}dV\\&+\int_{\Gamma_{coh}}\delta\mathbf{a}^T(\overline{\mathbf{N}}^{split})^T\mathbf{T}^T\mathbf{t}d\Gamma+\int_{\Gamma_{coh}}\delta\mathbf{b}^T(\overline{\mathbf{N}}^{tip})^T\mathbf{T}^T\mathbf{t}d\Gamma=\\&\int_V\rho\delta\mathbf{u}^T\mathbf{N}^T\mathbf{f}dV+\int_V\rho\delta\mathbf{a}^T(\mathbf{N}^{split})^T\mathbf{f}dV+\int_V\rho\delta\mathbf{b}^T(\mathbf{N}^{tip})^T\mathbf{f}dV\\&+\int_S\delta\mathbf{u}^T\mathbf{N}^T\bar{\mathbf{t}}dS+\int_S\delta\mathbf{a}^T(\mathbf{N}^{split})^T\bar{\mathbf{t}}dS+\int_S\delta\mathbf{b}^T(\mathbf{N}^{tip})^T\bar{\mathbf{t}}dS\end{aligned} \tag{5.84}$$

or:

$$\begin{aligned}&\delta\mathbf{u}^T\Big(\int_V\mathbf{B}^T\boldsymbol{\sigma}dV-\int_V\rho\mathbf{N}^T\mathbf{f}dV-\int_S\mathbf{N}^T\bar{\mathbf{t}}dS\Big)\\&+\delta\mathbf{a}^T\Big(\int_V(\mathbf{B}^{split})^T\boldsymbol{\sigma}dV+\int_{\Gamma_{coh}}(\overline{\mathbf{N}}^{split})^T\mathbf{T}^T\mathbf{t}d\Gamma-\int_V\rho(\mathbf{N}^{split})^T\mathbf{f}dV-\int_S(\mathbf{N}^{split})^T\bar{\mathbf{t}}dS\Big)\\&+\delta\mathbf{b}^T\Big(\int_V(\mathbf{B}^{tip})^T\boldsymbol{\sigma}dV+\int_{\Gamma_{coh}}(\overline{\mathbf{N}}^{tip})^T\mathbf{T}^T\mathbf{t}d\Gamma-\int_V\rho(\mathbf{N}^{tip})^T\mathbf{f}dV-\int_S(\mathbf{N}^{tip})^T\bar{\mathbf{t}}dS\Big)=0\end{aligned} \tag{5.85}$$

which leads to the following system of equations:

$$\int_V\mathbf{B}^T\boldsymbol{\sigma}dV-\int_V\rho\mathbf{N}^T\mathbf{f}dV-\int_S\mathbf{N}^T\bar{\mathbf{t}}dS=0 \tag{5.86a}$$

$$\int_V(\mathbf{B}^{split})^T\boldsymbol{\sigma}dV+\int_{\Gamma_{coh}}(\overline{\mathbf{N}}^{split})^T\mathbf{T}^T\mathbf{t}d\Gamma-\int_V\rho(\mathbf{N}^{split})^T\mathbf{f}dV-\int_S(\mathbf{N}^{split})^T\bar{\mathbf{t}}dS=0 \tag{5.86b}$$

$$\int_V(\mathbf{B}^{tip})^T\boldsymbol{\sigma}dV+\int_{\Gamma_{coh}}(\overline{\mathbf{N}}^{tip})^T\mathbf{T}^T\mathbf{t}d\Gamma-\int_V\rho(\mathbf{N}^{tip})^T\mathbf{f}dV-\int_S(\mathbf{N}^{tip})^T\bar{\mathbf{t}}dS=0 \tag{5.86c}$$

The system (5.86) is generally non-linear, due to the non-linear traction-separation law which is adopted for the discontinuities. Therefore, linearization of these equations should be adopted within the Newton-Raphson incremental-iterative process. This is implemented by the expansion in Taylor series of the displacement jump and traction at iteration $i+1$, as follows:

$$||\mathbf{u}||^{i+1} = ||\mathbf{u}||^{i} + \Delta||\mathbf{u}|| \tag{5.87}$$

$$\mathbf{t}^{i+1} = \mathbf{t}^{i} + \Delta\mathbf{t} \Rightarrow \mathbf{t}^{i+1} = \mathbf{t}^{i} + \frac{\partial \mathbf{t}}{\partial ||\mathbf{u}||^{i}} \Delta||\mathbf{u}|| \tag{5.88}$$

noticing that the Newton-Raphson scheme provided by equations (2.81) and (2.82) in section 2.4.1 is used. Then, $\Delta||\mathbf{u}||$ is determined using relation (5.82), and substituted into equations (5.87), (5.88):

$$||\mathbf{u}||^{i+1} = ||\mathbf{u}||^{i} + \mathbf{T}\overline{\mathbf{N}}^{split}\Delta\mathbf{a} + \mathbf{T}\overline{\mathbf{N}}^{tip}\Delta\mathbf{b} \tag{5.89}$$

$$\begin{aligned} \mathbf{t}^{i+1} &= \mathbf{t}^{i} + \frac{\partial \mathbf{t}}{\partial ||\mathbf{u}||^{i}} (\mathbf{T}\overline{\mathbf{N}}^{split}\Delta\mathbf{a} + \mathbf{T}\overline{\mathbf{N}}^{tip}\Delta\mathbf{b}) \\ &= \mathbf{t}^{i} + \mathbf{D}_{coh} (\mathbf{T}\overline{\mathbf{N}}^{split}\Delta\mathbf{a} + \mathbf{T}\overline{\mathbf{N}}^{tip}\Delta\mathbf{b}) \end{aligned} \tag{5.90}$$

where $\mathbf{D}_{coh}$ is the cohesive tangent stiffness which is derived for the chosen constitutive traction-separation law. As presented in section 3.3.6 this tangent stiffness may be formulated either for coupled or uncoupled traction-separation response.

In a more general description, material non-linearity may also be considered for the bulk elements away of the discontinuity, indicating that linearization of the corresponding stress-strain law should be applied:

$$\boldsymbol{\sigma}^{i+1} = \boldsymbol{\sigma}^{i} + \Delta\boldsymbol{\sigma} \Rightarrow \boldsymbol{\sigma}^{i+1} = \boldsymbol{\sigma}^{i} + \frac{\partial \boldsymbol{\sigma}}{\partial \boldsymbol{\varepsilon}^{i}} \Delta\boldsymbol{\varepsilon} \tag{5.91}$$

which leads to the following equation, after substituting the approximation of the strain provided in relation (5.56):

$$\boldsymbol{\sigma}^{i+1} = \boldsymbol{\sigma}^{i} + \mathbf{D}(\mathbf{B}\Delta\mathbf{u} + \mathbf{B}^{split}\Delta\mathbf{a} + \mathbf{B}^{tip}\Delta\mathbf{b}) \tag{5.92}$$

noticing that $\mathbf{D}$ is the tangent tensor expressing the constitutive response of the bulk elements. Linearization can now be completed, by replacing the linearized terms in the governing equations (5.86). Substituting the stress $\boldsymbol{\sigma}$ from equation (5.92) into the governing equation (5.86a), results in:

$$\int_V \mathbf{B}^T[\boldsymbol{\sigma}^{i} + \mathbf{D}(\mathbf{B}\Delta\mathbf{u} + \mathbf{B}^{split}\Delta\mathbf{a} + \mathbf{B}^{tip}\Delta\mathbf{b})]dV - \int_V \rho\mathbf{N}^T\mathbf{f}dV - \int_S \mathbf{N}^T\bar{\mathbf{t}}dS = 0 \tag{5.93}$$

which leads to the final form:

$$\begin{aligned} &\int_V \mathbf{B}^T\mathbf{D}\mathbf{B}dV\Delta\mathbf{u} + \int_V \mathbf{B}^T\mathbf{D}\mathbf{B}^{split}dV\Delta\mathbf{a} + \int_V \mathbf{B}^T\mathbf{D}\mathbf{B}^{tip}dV\Delta\mathbf{b} = \\ &\int_V \rho\mathbf{N}^T\mathbf{f}dV + \int_S \mathbf{N}^T\bar{\mathbf{t}}dS - \int_V \mathbf{B}^T\boldsymbol{\sigma}^{i}dV \end{aligned} \tag{5.94}$$

This equation can be expressed in the following form, which represents the equilibrium within the Newton-Raphson process:

$$\mathbf{K}_{uu}\Delta\mathbf{u} + \mathbf{K}_{ua}\Delta\mathbf{a} + \mathbf{K}_{ub}\Delta\mathbf{b} = \mathbf{F}_u^{ext} - \mathbf{F}_u^{int} \tag{5.95}$$

where the tangent stiffness terms as well as the external and the internal force vectors are provided by:

$$\begin{aligned}
&\mathbf{K}_{uu} = \int_V \mathbf{B}^T\mathbf{D}\mathbf{B}dV,\ \mathbf{K}_{ua} = \int_V \mathbf{B}^T\mathbf{D}\mathbf{B}^{split}dV,\ \mathbf{K}_{ub} = \int_V \mathbf{B}^T\mathbf{D}\mathbf{B}^{tip}dV\\
&\mathbf{F}_u^{ext} = \int_V \rho\mathbf{N}^T\mathbf{f}dV + \int_S \mathbf{N}^T\bar{\mathbf{t}}dS\\
&\mathbf{F}_u^{int} = \int_V \mathbf{B}^T\boldsymbol{\sigma}^i dV
\end{aligned} \tag{5.96}$$

The same concept is followed, for the linearization of the second and third governing equations (5.86b) and (5.86c), respectively. In particular, linearization of the second equation takes place after substituting the stress $\boldsymbol{\sigma}$ from (5.92) and the traction vector $\mathbf{t}$ from (5.90) into the governing equation (5.86b):

$$\begin{aligned}
&\int_V(\mathbf{B}^{split})^T\mathbf{D}\mathbf{B}dV\Delta\mathbf{u} + \int_V(\mathbf{B}^{split})^T\mathbf{D}\mathbf{B}^{split}dV\Delta\mathbf{a} + \int_V(\mathbf{B}^{split})^T\mathbf{D}\mathbf{B}^{tip}dV\Delta\mathbf{b}\\
&+\int_{\Gamma_{coh}}(\overline{\mathbf{N}}^{split})^T\mathbf{T}^T\mathbf{D}_{coh}\mathbf{T}\overline{\mathbf{N}}^{split}d\Gamma\Delta\mathbf{a} + \int_{\Gamma_{coh}}(\overline{\mathbf{N}}^{split})^T\mathbf{T}^T\mathbf{D}_{coh}\mathbf{T}\overline{\mathbf{N}}^{tip}d\Gamma\Delta\mathbf{b} =\\
&\int_V \rho(\mathbf{N}^{split})^T\mathbf{f}dV + \int_S(\mathbf{N}^{split})^T\bar{\mathbf{t}}dS - \left(\int_V(\mathbf{B}^{split})^T\boldsymbol{\sigma}^i dV + \int_{\Gamma_{coh}}(\overline{\mathbf{N}}^{split})^T\mathbf{T}^T\mathbf{t}^i d\Gamma\right)
\end{aligned} \tag{5.97}$$

It is noted that the component $\mathbf{T}^T\mathbf{D}_{coh}\mathbf{T}$ expresses the cohesive tangent tensor in the global coordinate system. Equation (5.97) can be rewritten in the following, brief form:

$$\mathbf{K}_{au}\Delta\mathbf{u} + \mathbf{K}_{aa}\Delta\mathbf{a} + \mathbf{K}_{ab}\Delta\mathbf{b} = \mathbf{F}_a^{ext} - \mathbf{F}_a^{int} \tag{5.98}$$

where:

$$\begin{aligned}
&\mathbf{K}_{au} = \int_V(\mathbf{B}^{split})^T\mathbf{D}\mathbf{B}dV\\
&\mathbf{K}_{aa} = \int_V(\mathbf{B}^{split})^T\mathbf{D}\mathbf{B}^{split}dV + \int_{\Gamma_{coh}}(\overline{\mathbf{N}}^{split})^T\mathbf{T}^T\mathbf{D}_{coh}\mathbf{T}\overline{\mathbf{N}}^{split}d\Gamma\\
&\mathbf{K}_{ab} = \int_V(\mathbf{B}^{split})^T\mathbf{D}\mathbf{B}^{tip}dV + \int_{\Gamma_{coh}}(\overline{\mathbf{N}}^{split})^T\mathbf{T}^T\mathbf{D}_{coh}\mathbf{T}\overline{\mathbf{N}}^{tip}d\Gamma\\
&\mathbf{F}_a^{ext} = \int_V \rho(\mathbf{N}^{split})^T\mathbf{f}dV + \int_S(\mathbf{N}^{split})^T\bar{\mathbf{t}}dS\\
&\mathbf{F}_a^{int} = \int_V(\mathbf{B}^{split})^T\boldsymbol{\sigma}^i dV + \int_{\Gamma_{coh}}(\overline{\mathbf{N}}^{split})^T\mathbf{T}^T\mathbf{t}^i d\Gamma
\end{aligned} \tag{5.99}$$

Finally, linearization of the third governing equation takes place by substituting the stress $\boldsymbol{\sigma}$ from (5.92) and the traction vector $\mathbf{t}$ from (5.90) into (5.86c):

$$\int_V (\mathbf{B}^{tip})^T \mathbf{D}\mathbf{B} dV \Delta\mathbf{u} + \int_V (\mathbf{B}^{tip})^T \mathbf{D}\mathbf{B}^{split} dV \Delta\mathbf{a} + \int_V (\mathbf{B}^{tip})^T \mathbf{D}\mathbf{B}^{tip} dV \Delta\mathbf{b}$$
$$+ \int_{\Gamma_{coh}} (\overline{\mathbf{N}}^{tip})^T \mathbf{T}^T \mathbf{D}_{coh} \mathbf{T} \overline{\mathbf{N}}^{split} d\Gamma \Delta\mathbf{a} + \int_{\Gamma_{coh}} (\overline{\mathbf{N}}^{tip})^T \mathbf{T}^T \mathbf{D}_{coh} \mathbf{T} \overline{\mathbf{N}}^{tip} d\Gamma \Delta\mathbf{b} =$$
$$\int_V \rho (\mathbf{N}^{tip})^T \mathbf{f} dV + \int_S (\mathbf{N}^{tip})^T \bar{\mathbf{t}} dS - \left( \int_V (\mathbf{B}^{tip})^T \boldsymbol{\sigma}^i dV + \int_{\Gamma_{coh}} (\overline{\mathbf{N}}^{tip})^T \mathbf{T}^T \mathbf{t}^i d\Gamma \right) \tag{5.100}$$

which is rewritten as:

$$\mathbf{K}_{bu} \Delta\mathbf{u} + \mathbf{K}_{ba} \Delta\mathbf{a} + \mathbf{K}_{bb} \Delta\mathbf{b} = \mathbf{F}_b^{ext} - \mathbf{F}_b^{int} \tag{5.101}$$

where:

$$\begin{aligned}
\mathbf{K}_{bu} &= \int_V (\mathbf{B}^{tip})^T \mathbf{D}\mathbf{B} dV \\
\mathbf{K}_{ba} &= \int_V (\mathbf{B}^{tip})^T \mathbf{D}\mathbf{B}^{split} dV + \int_{\Gamma_{coh}} (\overline{\mathbf{N}}^{tip})^T \mathbf{T}^T \mathbf{D}_{coh} \mathbf{T} \overline{\mathbf{N}}^{split} d\Gamma \\
\mathbf{K}_{bb} &= \int_V (\mathbf{B}^{tip})^T \mathbf{D}\mathbf{B}^{tip} dV + \int_{\Gamma_{coh}} (\overline{\mathbf{N}}^{tip})^T \mathbf{T}^T \mathbf{D}_{coh} \mathbf{T} \overline{\mathbf{N}}^{tip} d\Gamma \\
\mathbf{F}_b^{ext} &= \int_V \rho (\mathbf{N}^{tip})^T \mathbf{f} dV + \int_S (\mathbf{N}^{tip})^T \bar{\mathbf{t}} dS \\
\mathbf{F}_b^{int} &= \int_V (\mathbf{B}^{tip})^T \boldsymbol{\sigma}^i dV + \int_{\Gamma_{coh}} (\overline{\mathbf{N}}^{tip})^T \mathbf{T}^T \mathbf{t}^i d\Gamma
\end{aligned} \tag{5.102}$$

The system of the linearized equations can eventually be formulated as follows:

$$\begin{bmatrix} \mathbf{K}_{uu} & \mathbf{K}_{ua} & \mathbf{K}_{ub} \\ \mathbf{K}_{au} & \mathbf{K}_{aa} & \mathbf{K}_{ab} \\ \mathbf{K}_{bu} & \mathbf{K}_{ba} & \mathbf{K}_{bb} \end{bmatrix} \begin{bmatrix} \Delta\mathbf{u} \\ \Delta\mathbf{a} \\ \Delta\mathbf{b} \end{bmatrix} = \begin{bmatrix} \mathbf{F}_u^{ext} - \mathbf{F}_u^{int} \\ \mathbf{F}_a^{ext} - \mathbf{F}_a^{int} \\ \mathbf{F}_b^{ext} - \mathbf{F}_b^{int} \end{bmatrix} \tag{5.103}$$

### 5.3.3 Crack propagation and numerical integration for cohesive cracks

Similar to linear XFEM analysis, crack propagation criteria should be defined for a cohesive crack growth. A criterion for the crack initiation and propagation may be formulated using the stress field in the crack tip. The maximum principal tensile stresses in the Gauss points of the element ahead of the crack tip are calculated in this case and the crack propagates once the maximum of these principal stresses reaches the strength of the material.

The direction of the crack propagation is perpendicular to the direction of the maximum principal stress. A disadvantage of this approach, which is used to impose criteria for crack propagation, is that it uses the stresses at the crack tip field, which is a process with questionable accuracy. To improve the accuracy, non-local calculation of (average) stresses can be adopted at the crack tip [107].

An alternative approach, which bypasses the need of calculating the stresses at the crack tip field, introduces the concept of vanishing stress intensity factors at the crack tip as a criterion for crack propagation in the cohesive zone. In [146], it is suggested a zero value for the stress intensity factor for mode I failure, thus, $K_I = 0$, indicating crack growth of the cohesive zone under mode I loading condition. A zero value for the stress intensity factor at the crack tip, implies that the stresses at the crack tip are finite. An equivalent criterion is that crack propagation takes place when the component of the stress at the crack tip, perpendicular to crack direction, becomes equal to the tensile strength of the material [107].

The direction of the crack growth can be calculated within this approach, using principles of linear fracture mechanics, which are presented in the context of the XFEM method in section 5.2.4. In particular, the direction of the crack growth is derived using the concept of the maximum circumferential tensile stress, according to which the crack propagates in a radial direction at the crack tip, on a plane which is perpendicular to the direction of the maximum, principal tensile stress. The shear stress in this plane is zero, a condition which leads to relation (5.68) of section 5.2.4 providing the crack propagation direction. The stress integrity factors used in this equation, can be calculated from the interaction integrals as presented in section 5.1.3.

Towards the implementation of cohesive cracks within the XFEM method, numerical integration is adopted for the calculation of the integrals in the governing equations. The concept of numerical integration which is presented analytically in section 5.2.5 for linear fracture problems, is also adopted for cohesive cracks. Briefly, the domain between each discontinuity and the element which is cut by it, is divided into several sub-domains (triangular or rectangular) in both sides of the discontinuity and Gauss integration is considered for each sub-domain. A similar approach is followed at the crack tip.

A point which differentiates the descriptions of cohesive crack growth as compared to linear fracture analysis, is that the governing equations include cohesive term integrals, indicating that numerical integration should be applied to these terms too. For a two-dimensional problem, the numerical integration over the cohesive zone is implemented by representing the zone as a sequence of linear segments. Each of these segments is treated as an one-dimensional element and numerical integration is performed at chosen Gauss points located on the segment.

# Chapter 6

# Homogenization

## 6.1 Introduction to the concept of homogenization

As presented in chapter 3, continuum mechanics descriptions assigned in the macroscopic length scale of composite materials, cannot capture the interaction between the constituent materials and the effect of this on the structural response. To overcome the phenomenological nature of this class of constitutive descriptions, alternative methods have been developed, aiming in investigating the response of the constituent materials in a microscopic length scale and its impact on the macroscopic, structural scale.

*Homogenization* is one of the widely used methods adopted to derive the *overall* or *effective* material properties of a heterogeneous structure, which can then be used by an equivalent homogeneous one. To derive the effective properties, all the constituent materials as well as their structural response are explicitly considered, highlighting a significant aspect of homogenization. This usually takes place in a microscopic or mesoscopic length scale and after the effective material properties are calculated, they are used in the macroscopic length scale by the equivalent homogeneous structure. In this sense, information related to the structural response is communicated between multiple length scales and for this reason the term *multi-scale* is quite often adopted in this class of problems. It is noted that the lower length scales are usually called *fine* scales since they depict all the details of the microstructural formulation, while the upper, macroscopic (structural) length scale is called *coarse* scale.

Although several homogenization approaches have been proposed in literature, someone may identify three major categories, namely the *mathematical/analytical homogenization*, the *numerical homogenization* and the *computational homogenization*. The first two will briefly be presented in this chapter while the third in chapter 7 of the book.

Mathematical/analytical homogenization describes the response of heterogeneous structures by adopting analytical relations, obtained for instance from an *asymptotic expansion* approximation of the displacement field. This method can provide accurate prediction of the effective material properties of heterogeneous structures with low computational cost, but its implementation becomes difficult when complex geometries or non-linear constitutive laws are considered.

DOI: 10.1201/9781003017240-6

Within numerical homogenization, finite element analysis computations on a representative sample of the material with all its constituents in lower length scales are conducted, to provide necessary data and fit the parameters of a macroscopic constitutive description used to depict the structural response in the coarse scale. Since the microscopic simulations are considered a priori or *offline*, in a representative sample of the material, the computational cost of applying the method is manageable, making it attractive for practical problems involving large structural systems. On the other hand, this a priori selection of the constitutive description in the macroscopic scale restricts the possibility of capturing evolving microstructures which appear for non-linear problems.

In computational homogenization, the macroscopic constitutive description is determined during simulation (*on the fly*), by solving several microstructural problems, for every integration point and load step of the macroscopic structure. This formulation provides the opportunity to depict evolving microstructures but increases the computational cost.

Quite often in composite materials, a repetitive geometric pattern which is known as *periodicity*, is recognized. A periodic microstructural description is found for instance in fibre reinforced composite materials, where fibre reinforcements are placed on a polymer matrix. Periodic patterns are also found in building materials, such as masonry. A masonry wall consists of stone blocks and the mortar which is used to connect the blocks.

Homogenization takes advantage of these periodic microstructural patterns by considering a representative sample of the material, called *Representative Volume Element* or *RVE*. The RVE includes all the constituent materials, e.g. fibres and the matrix or masonry blocks and the mortar. Homogenization methods are then applied to the RVE in order to derive the effective material properties which will be adopted in the structural scale. It is noted that quite often in analytical or numerical homogenization, the term *unit cell* is also used to express this representative microstructural sample. An example of a periodic microstructure is schematically presented in figure 6.1.

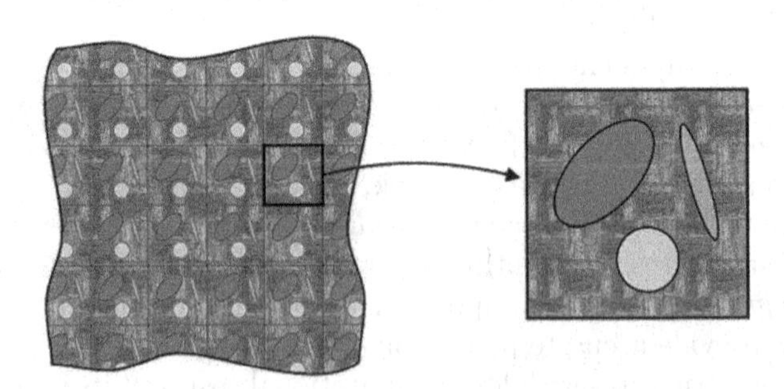

**FIGURE 6.1**: Periodic microstructure and the corresponding RVE.

In figure 6.2 is shown a masonry wall which consists of masonry blocks and the mortar joints, placed in a periodic pattern. Within homogenization, someone needs to select an appropriate RVE which represents this pattern, and derive its effective material properties by considering the mortar and the masonry blocks. Focusing on the RVE makes the simulation of all constituents and their response manageable, since otherwise, simulation of all constituents directly on the structural scale would lead to a quite large computational cost. Direct macroscopic modelling of all materials could become more difficult or even impossible, for larger structures and when complicated constitutive descriptions are involved, like opening-sliding between the constituents. Hence, the need for using a homogenization method is highlighted.

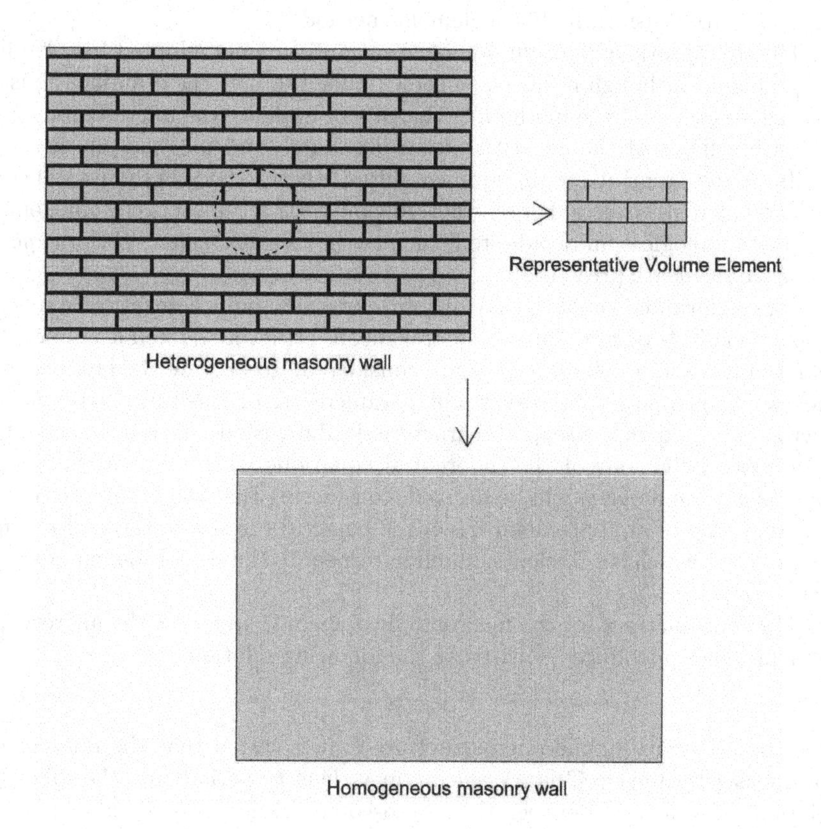


**FIGURE 6.2**: Homogenization scheme for a masonry wall.

## 6.2 Asymptotic expansion homogenization

Mathematical theory of homogenization has been studied extensively in 1970s for composite heterogeneous structures [27], [188]. One of the most significant methods of mathematical homogenization relies on the asymptotic behaviour of heterogeneous, fine scale unit cells with a periodic or semi-periodic microstructure. The method is known as *asymptotic expansion homogenization* and it can be formulated using the strong form (differential governing equations), which provides a robust mathematical description. It can also be formulated in terms of the weak form, which is appropriate for its numerical implementation using the finite element method.

The asymptotic expansion homogenization is mainly adopted to solve linear problems, although it has also been extended to non-linear applications. A distinctive aspect of the method, is that it can evaluate the microscopic stress and strain fields of the constituents, using macro and microscopic displacements. A requirement for the applicability of the method, is the existence of periodic microstructures. For non-periodic microstructures, other homogenization methodologies can be adopted, such as the *G-convergence*, *H-convergence* and *Γ-convergence* [51], [152].

For the formulation of the asymptotic expansion homogenization, a macroscopic domain $\Omega$ of a composite heterogeneous structure as well as a periodic, representative unit cell domain Y are considered. To provide the link between the two length scales, a very small parameter $\epsilon$, of the ratio between the microscopic and macroscopic characteristic dimensions, is introduced. The asymptotic behaviour of the differential equations which govern the initial, *boundary value problem* will be derived, considering that the parameter $\epsilon$ goes to zero. Quite often, parameter $\epsilon$ is taken equal to the size of the periodic microstructure which is sufficiently small in respect to the size of the macroscopic structure [213].

The coordinates $\mathbf{x}$ for the macroscopic domain $\Omega$ and $\mathbf{y}$ for the microscopic domain Y are introduced, satisfying the following relation:

$$\mathbf{y} = \mathbf{x}/\epsilon \tag{6.1}$$

Due to periodicity in the microstructure Y, it is stated that the microstructural elasticity tensor $\mathbf{C}$ is *Y-periodic* in $\mathbf{y}$. Due to periodicity, the following relation is formulated between the macroscopic elastic tensor $\mathbf{C}^{\epsilon}$ at $\mathbf{x}$, and the microscopic elastic tensor $\mathbf{C}$:

$$C^{\epsilon}_{ijkl}(\mathbf{x}) = C_{ijkl}(\mathbf{x}/\epsilon) = C_{ijkl}(\mathbf{y}) \tag{6.2}$$

where Einstein convention has been used. A linear elastic equilibrium problem can then be formulated for the macroscopic scale, according to the following equations:

$$\frac{\partial \sigma^{\epsilon}_{ij}}{\partial x_j} + f_i = 0 \quad in \quad \Omega \tag{6.3a}$$

$$\varepsilon_{ij}^{\epsilon} = \frac{1}{2}\left(\frac{\partial u_i^{\epsilon}}{\partial x_j^{\epsilon}} + \frac{\partial u_j^{\epsilon}}{\partial x_i^{\epsilon}}\right) \quad in \quad \Omega \tag{6.3b}$$

$$\sigma_{ij}^{\epsilon} = C_{ijkl}^{\epsilon}\varepsilon_{kl}^{\epsilon} \quad in \quad \Omega \tag{6.3c}$$

$$u_i^{\epsilon} = \bar{u}_i \quad in \quad \Gamma_u \tag{6.3d}$$

$$\sigma_{ij}^{\epsilon} n_j = \bar{t}_i \quad in \quad \Gamma_t \tag{6.3e}$$

The differential equation (6.3a) represents the equilibrium in strong form as presented in section 2.2.1, equation (6.3b) provides the strain-displacement relation given in section 1.2.4 in the framework of small displacement gradients and equation (6.3c) is the constitutive stress-strain law given in section 1.2.6. Relations (6.3d) and (6.3e) depict the displacement (*Dirichlet's*) and traction (*Neumann's*) boundary conditions, respectively, on boundaries $\Gamma_u$ and $\Gamma_t$ of the macroscale domain $\Omega$. With $\sigma_{ij}^{\epsilon}$, $\varepsilon_{ij}^{\epsilon}$ are denoted the stress and strain tensors, $f_i$, $u_i^{\epsilon}$ are the force per unit volume and displacement vectors, respectively and $n_j$ is the outward unit normal vector to the boundary $\Gamma_t$ of the macroscale domain $\Omega$. Vectors $\bar{u}_i$ and $\bar{t}_i$ represent prescribed displacements and tractions on the boundaries of $\Omega$.

Goal of the asymptotic expansion homogenization, is the determination of the solution of the displacement field $\mathbf{u}^{\epsilon}$. First, the asymptotic expansion of $\mathbf{u}^{\epsilon}$ is formulated according to the following equation:

$$u_i^{\epsilon}(\mathbf{x}) = u_i^{(0)}(\mathbf{x}) + \epsilon u_i^{(1)}(\mathbf{x},\mathbf{y}) + \epsilon^2 u_i^{(2)}(\mathbf{x},\mathbf{y}) + \dots \tag{6.4}$$

where the functions $u_i^{(r)}(\mathbf{x},\mathbf{y})$ are known as *correctors of order $r$* of the displacement field, noticing also that they are Y-periodic in the variable $\mathbf{y}$. The term $u_i^{(0)}(\mathbf{x})$ depends only on the macroscopic coordinates $\mathbf{x}$, representing the homogenized macroscale material [173], [218]. This term is accompanied by the parameter $\epsilon^0 = 1$, and thus, $\epsilon^0$ will correspond to the macroscale domain in the following formulations. In addition, equation (6.4) consists of the smooth function of the macroscale displacement field $u_i^{(0)}(\mathbf{x})$, plus some oscillating terms. These periodic oscillations, appear also for the stress and strain fields and they are attributed to the periodicity of the microstuctural heterogeneities. In figure 6.3 is schematically shown the first-order approximation of the asymptotic expansion of the displacement field. The oscillations of the displacement field due to the periodic microstructure are also depicted in this figure.

Before presenting the next steps of the formulation, the chain rule of differentiation is provided by the equation below, where relation (6.1) has also been considered:

$$\frac{\partial}{\partial x_i^{\epsilon}} = \frac{\partial}{\partial x_i} + \frac{1}{\epsilon}\frac{\partial}{\partial y_i} \tag{6.5}$$

Next, the asymptotic expansion of the displacement field given in relation (6.4) is substituted in the strain-displacement equation (6.3b). In addition, by using

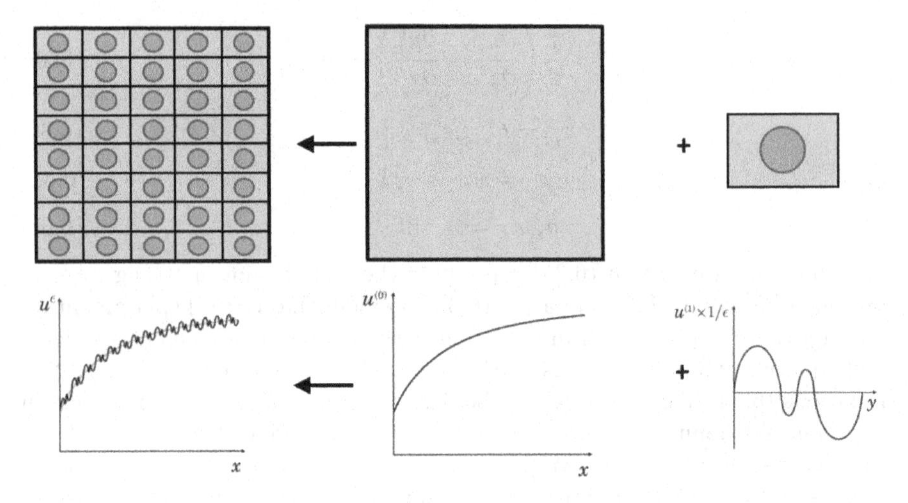

**FIGURE 6.3**: First-order approximation of the asymptotic expansion of the displacement field.

relation (6.5) and properly adjusting the indices, the following expression for the strain field is formulated:

$$\varepsilon_{ij}^{\epsilon} = \epsilon^{-1}\varepsilon_{ij}^{(0)} + \epsilon^{0}\varepsilon_{ij}^{(1)} + \epsilon^{1}\varepsilon_{ij}^{(2)} + \dots \tag{6.6}$$

where:

$$\varepsilon_{ij}^{(0)} = \frac{1}{2}\left(\frac{\partial u_i^{(0)}}{\partial y_j} + \frac{\partial u_j^{(0)}}{\partial y_i}\right) \tag{6.7}$$

$$\varepsilon_{ij}^{(r)} = \frac{1}{2}\left(\frac{\partial u_i^{(r-1)}}{\partial x_j} + \frac{\partial u_j^{(r-1)}}{\partial x_i} + \frac{\partial u_i^{(r)}}{\partial y_j} + \frac{\partial u_j^{(r)}}{\partial y_i}\right) \tag{6.8}$$

Then, by substituting relation (6.6) into the constitutive equation (6.3c) and also considering relations (6.1) and (6.2), the following equation is derived, for the stress field:

$$\sigma_{ij}^{\epsilon} = \epsilon^{-1}\sigma_{ij}^{(0)} + \epsilon^{0}\sigma_{ij}^{(1)} + \epsilon^{1}\sigma_{ij}^{(2)} + \dots \tag{6.9}$$

where using relations (6.7) and (6.8) is obtained that:

$$\sigma_{ij}^{(0)} = \frac{1}{2}C_{ijkl}(\mathbf{y})\left(\frac{\partial u_k^{(0)}}{\partial y_l} + \frac{\partial u_l^{(0)}}{\partial y_k}\right) \tag{6.10}$$

$$\sigma_{ij}^{(r)} = \frac{1}{2}C_{ijkl}(\mathbf{y})\left(\frac{\partial u_k^{(r-1)}}{\partial x_l} + \frac{\partial u_l^{(r-1)}}{\partial x_k} + \frac{\partial u_k^{(r)}}{\partial y_l} + \frac{\partial u_l^{(r)}}{\partial y_k}\right) \tag{6.11}$$

Next, the stress field provided by equation (6.9) is substituted in the differential equilibrium equation (6.3a) and the chain rule (6.5) is used, resulting in:

$$\epsilon^{-2}\frac{\partial \sigma_{ij}^{(0)}}{\partial y_j}+\epsilon^{-1}\left(\frac{\partial \sigma_{ij}^{(0)}}{\partial x_j}+\frac{\partial \sigma_{ij}^{(1)}}{\partial y_j}\right)+\epsilon^{0}\left(\frac{\partial \sigma_{ij}^{(1)}}{\partial x_j}+\frac{\partial \sigma_{ij}^{(2)}}{\partial y_j}+f_i\right)+\dots=0 \quad (6.12)$$

Since relation (6.12) should be valid for any value of the parameter $\epsilon$, where $\epsilon \to 0$, it is concluded that all the terms of equation (6.12), which stand as coefficients of the powers of $\epsilon$, must be zero. Thus:

$$\epsilon^{-2} \to \frac{\partial \sigma_{ij}^{(0)}}{\partial y_j}=0 \quad (6.13a)$$

$$\epsilon^{-1} \to \frac{\partial \sigma_{ij}^{(0)}}{\partial x_j}+\frac{\partial \sigma_{ij}^{(1)}}{\partial y_j}=0 \quad (6.13b)$$

$$\epsilon^{0} \to \frac{\partial \sigma_{ij}^{(1)}}{\partial x_j}+\frac{\partial \sigma_{ij}^{(2)}}{\partial y_j}+f_i=0 \quad (6.13c)$$

$$\epsilon^{r} \to \frac{\partial \sigma_{ij}^{(r+1)}}{\partial x_j}+\frac{\partial \sigma_{ij}^{(r+2)}}{\partial y_j}=0 \quad (6.13d)$$

noticing that for higher power orders of $\epsilon$, thus, for higher-orders of correction, more equations would be formulated. In this group of relations, equation (6.13c) corresponds to order of correction 0 ($\epsilon^0$) and thus, it characterizes the macroscale domain, while the remaining terms with different orders of correction stand for the microscale domain.

The boundary conditions given by relations (6.3d) and (6.3e) can also be expressed with respect to the asymptotic expansion terms. In particular, the asymptotic expansion displacement field provided by relation (6.4) is substituted in the displacement boundary condition (6.3d), leading to the following equation:

$$\epsilon^{0}u_i^{(0)}+\epsilon^{1}u_i^{(1)}+\epsilon^{2}u_i^{(2)}+\dots=\bar{u}_i \quad in \quad \Gamma_u \quad (6.14)$$

Similarly, the asymptotic expansion stress field given in relation (6.9) is substituted in the traction boundary condition (6.3e), resulting in:

$$\left(\epsilon^{-1}\sigma_{ij}^{(0)}+\epsilon^{0}\sigma_{ij}^{(1)}+\epsilon^{1}\sigma_{ij}^{(2)}+\dots\right)n_j=\bar{t}_i \quad in \quad \Gamma_t \quad (6.15)$$

By linking the proper order of correction to the above boundary conditions, noticing that they correspond to the macroscale order $\epsilon^0$, the following equations are derived:

$$\epsilon^{0} \to u_i^{(0)}=\bar{u}_i \quad in \quad \Gamma_u \quad (6.16a)$$

$$\epsilon^{r} \to u_i^{(r)}=0 \quad in \quad \Gamma_u \quad (6.16b)$$

$$\epsilon^{-1} \rightarrow \sigma_{ij}^{(0)} n_j = 0 \quad in \quad \Gamma_t \tag{6.16c}$$

$$\epsilon^{0} \rightarrow \sigma_{ij}^{(1)} n_j = \bar{t}_i \quad in \quad \Gamma_t \tag{6.16d}$$

$$\epsilon^{r} \rightarrow \sigma_{ij}^{(r+1)} n_j = 0 \quad in \quad \Gamma_t \tag{6.16e}$$

The next step of the formulation, is to provide the path for the solution of the differential equations (6.13), satisfying the boundary conditions (6.16). As presented in the next sections, this will be implemented by developing further the micro and macroscale descriptions.

### 6.2.1 Microscale description

For the description of the microscale problem in the framework of asymptotic expansion homogenization, the differential equations (6.13a) and (6.13b) are properly developed, resulting in the following relation [173], [188]:

$$\frac{\partial}{\partial y_j}\left[C_{ijkl}(\mathbf{y})\left(\frac{\partial u_k^{(0)}}{\partial x_l} + \frac{\partial u_k^{(1)}}{\partial y_l}\right)\right] = 0 \tag{6.17}$$

Relation (6.17) is the governing microscale equation and solution of the microstructural displacement can be derived from (6.17), by assuming that the macroscale displacement field $u_k^{(0)}$ is known (noticing that the path for the solution of $u_k^{(0)}$ will be presented in the next section). Thus, the solution of (6.17) representing the microstructural displacement, is provided by:

$$u_i^{(1)}(\mathbf{x},\mathbf{y}) = -\chi_i^{kl}(\mathbf{y})\frac{\partial u_k^{(0)}}{\partial x_l}(\mathbf{x}) + \bar{u}_i^{(1)}(\mathbf{x}) \tag{6.18}$$

where $\chi_i^{kl}$ are the Y-periodic components of the characteristic displacement field tensor $\boldsymbol{\chi}$ and $\bar{u}_i^{(1)}(\mathbf{x})$ are integration constants in $\mathbf{y}$. Towards determining the characteristic displacement field tensor, equation (6.11) is written using the symmetry of the elasticity tensor and $r = 1$, as follows:

$$\sigma_{ij}^{(1)} = C_{ijkl}(\mathbf{y})\left(\frac{\partial u_k^{(0)}}{\partial x_l} + \frac{\partial u_k^{(1)}}{\partial y_l}\right) \tag{6.19}$$

By replacing relation (6.18) into (6.19), it is obtained that:

$$\sigma_{ij}^{(1)} = \hat{\sigma}_{ij}^{mn}\frac{\partial u_m^{(0)}}{\partial x_n} \tag{6.20}$$

where:

$$\hat{\sigma}_{ij}^{mn} = C_{ijkl}(\mathbf{y})\left(I_{kl}^{mn} - \frac{\partial \chi_k^{mn}}{\partial y_l}\right), \quad I_{kl}^{mn} = \delta_{km}\delta_{ln} \tag{6.21}$$

are Y-periodic functions, with $\delta_{ij}$ denoting Kronecker-delta. It can be proved that the differential microscale equilibrium equation can be written as:

$$\frac{\partial \hat{\sigma}_{ij}^{mn}}{\partial y_j} = 0 \tag{6.22}$$

Substitution of equation (6.21) into (6.22) leads to the following partial differential equation that can be used to provide the solution of the characteristic displacement field $\chi$:

$$\frac{\partial}{\partial y_j}\left[C_{ijkl}(\mathbf{y})\left(I_{kl}^{mn} - \frac{\partial \chi_k^{mn}}{\partial y_l}\right)\right] = 0 \tag{6.23}$$

Alternatively, the characteristic displacement field tensor components are determined as the solutions $\chi_i^{kl} \in \tilde{V}_Y$ of the auxiliary variational problem:

$$\int_Y C_{ijkl} \frac{\partial \chi_k^{mn}}{\partial y_l} \frac{\partial v_i}{\partial y_j} dY = \int_Y C_{ijmn} \frac{\partial v_i}{\partial y_j} dY, \quad \forall v_i \in \tilde{V}_Y \tag{6.24}$$

with $\tilde{V}_Y$ being the set of Y-periodic continuous and regular functions with zero average value in Y. It is noted that equations (6.17), (6.18), (6.23) and (6.24) define the *unit cell* problem.

### 6.2.2 Macroscale description

After the calculation of the characteristic displacement field tensor, the macroscale problem is defined and solved, in order to provide the solution for the macroscopic displacement field. First, the average value is defined, for a Y-periodic function $\phi(\mathbf{x}, \mathbf{y})$ in Y:

$$\langle \phi \rangle_Y = \frac{1}{|Y|} \int_Y \phi(\mathbf{x}, \mathbf{y}) dY \tag{6.25}$$

By applying the average operator in Y, on the macroscale differential equilibrium equations (6.13c) and using the divergence theorem, the following homogenized macroscale equilibrium equation is obtained:

$$\frac{\partial \langle \sigma_{ij}^{(1)} \rangle_Y}{\partial x_j} + f_i = 0 \tag{6.26}$$

By substituting (6.21) into (6.20) and applying the average operator on the resulting expression, it is derived that:

$$\langle \sigma_{ij}^{(1)} \rangle_Y = C_{ijmn}^h \frac{\partial u_m^{(0)}}{\partial \chi_n} \tag{6.27}$$

where

$$C_{ijmn}^h = \frac{1}{|Y|} \int_Y C_{ijkl}(\mathbf{y}) \left[I_{kl}^{mn} - \frac{\partial \chi_k^{mn}}{\partial y_l}\right] dY \tag{6.28}$$

Equation (6.27) stands for the homogenized macroscopic constitutive relation and (6.28) provides the homogenized elasticity tensor $\mathbf{C}^h$. Therefore, relations (6.26), (6.27) and (6.28), as well as boundary conditions (6.16a) and (6.16d), define the macroscopic problem. The solution of this problem will provide the macrostructural displacement field $\mathbf{u}^{(0)}$.

After the solution of this problem, the asymptotic expansion displacement, stress, strain as well as the microstructural stress and strain can eventually be determined. To summarize the main steps of asymptotic expansion homogenization, the following points are highlighted:

**1:** Calculation of characteristic displacement field tensor $\chi$ using equation (6.23) or (6.24).

**2:** Calculation of the homogenized elasticity tensor $\mathbf{C}^h$ and solution of the macroscale displacement field $\mathbf{u}^{(0)}$ using equations (6.26), (6.27), (6.28), as well as boundary conditions (6.16a) and (6.16d).

**3:** Calculation of the microstructural displacement $\mathbf{u}^{(1)}$ using equation (6.18).

**4:** Determination of the asymptotic expansion displacement $\mathbf{u}^{\epsilon}$, according to relation (6.4), using a first-order approximation ($r = 1$). The assumption to ignore higher-order terms is satisfactory for a wide range of problems, provided that the *separation of scales principle* is valid and thus, $\epsilon \ll 1$. If, however, higher terms of the asymptotic expansion expression are needed, a similar path to the solution as the one presented in this chapter should be followed.

**5:** Calculation of the stress and strain fields, $\boldsymbol{\sigma}^{\epsilon}$ and $\boldsymbol{\varepsilon}^{\epsilon}$ using relations (6.9) and (6.6). Due to consideration of first-order expansion, these quantities are approximated by $\tilde{\boldsymbol{\sigma}}^{\epsilon}$ and $\tilde{\boldsymbol{\varepsilon}}^{\epsilon}$. The microstructural stress field $\boldsymbol{\sigma}^{(1)}$ of relation (6.9) is determined using equations (6.20) and (6.21). The microstructural strain field $\boldsymbol{\varepsilon}^{(1)}$ of (6.6) is calculated, by substituting the micro-displacement field given in (6.18) into (6.8), for $r = 1$. The implementation of these statements results in the following relations:

$$\sigma_{ij}^{\epsilon}(\mathbf{x}) \approx \tilde{\sigma}_{ij}^{\epsilon}(\mathbf{x}) = \epsilon^0 \sigma_{ij}^{(1)}(\mathbf{x}, \mathbf{y}), \quad \varepsilon_{ij}^{\epsilon}(\mathbf{x}) \approx \tilde{\varepsilon}_{ij}^{\epsilon}(\mathbf{x}) = \epsilon^0 \varepsilon_{ij}^{(1)}(\mathbf{x}, \mathbf{y}) \tag{6.29}$$

$$\sigma_{ij}^{(1)}(\mathbf{x}, \mathbf{y}) = C_{ijkl}(\mathbf{y}) \left( I_{kl}^{mn} - \frac{\partial \chi_k^{mn}}{\partial y_l} \right) \frac{\partial u_m^{(0)}}{\partial \chi_n} \tag{6.30}$$

$$\varepsilon_{ij}^{(1)}(\mathbf{x}, \mathbf{y}) = T_{ij}^{kl} \left( I_{kl}^{mn} - \frac{\partial \chi_k^{mn}}{\partial y_l} \right) \frac{\partial u_m^{(0)}}{\partial \chi_n}, \quad T_{ij}^{kl} = \frac{1}{2}(\delta_{ik}\delta_{jl} + \delta_{il}\delta_{jk}) \tag{6.31}$$

where conditions $\sigma_{ij}^{(0)} = 0$ and $\epsilon_{ij}^{(0)} = 0$ have been taken into account [173], [188].

### 6.2.3 Implementation within finite element analysis

For the implementation of asymptotic expansion homogenization within finite element analysis, it is required to provide the discretization of the continuum mechanics formulations, which have been presented in the previous sections. Thus, discretization of the microscale equation (6.24), results in the following relation, using vector notation:

$$\int_{Y^e} \mathbf{B}^T \mathbf{C} \mathbf{B} dY \boldsymbol{\chi} = \int_{Y^e} \mathbf{B}^T \mathbf{C} dY \tag{6.32}$$

where integration takes place in respect to elements of the unit cell ($Y^e$). In addition, known relations presented in section 2.3 have been used for the discretization of the displacement field, using the element nodal displacement vector $\mathbf{u}_e$, the matrix of shape functions $\mathbf{N}$ and the strain-displacement matrix $\mathbf{B}$:

$$\mathbf{u} = \mathbf{N}\mathbf{u}_e \tag{6.33}$$

$$\boldsymbol{\varepsilon} = \mathbf{B}\mathbf{u}_e \quad \text{where} \quad \mathbf{B} = \boldsymbol{\partial}\mathbf{N} \tag{6.34}$$

$$\boldsymbol{\sigma} = \mathbf{C}\boldsymbol{\varepsilon} \Rightarrow \boldsymbol{\sigma} = \mathbf{C}\mathbf{B}\mathbf{u}_e \tag{6.35}$$

The characteristic displacement field tensor $\boldsymbol{\chi}$ of relation (6.32), which is called *corrector*, is a matrix. The right-hand part of (6.32) is also a matrix, with six columns, each of which representing an external load vector. Hence, six systems of equations need to be solved, in order to provide the solution of the microstructural problem. The solution of each system, will provide the result for every column of matrix $\boldsymbol{\chi}$.

Another step towards the implementation of the method, is the application of *periodic boundary conditions* to the representative unit cell. These boundary conditions are used to apply periodicity in opposite boundaries of the unit cell. For a hexahedral unit cell with boundary faces denoted by $y_1 \in [0, y_1^0]$, $y_2 \in [0, y_2^0]$ and $y_3 \in [0, y_3^0]$, where $y_1^0$, $y_2^0$, $y_3^0$ indicate the length for each of the edges of the unit cell in a coordinate system with origin at point $(0, 0, 0)$, the periodic boundary conditions are provided below:

$$\chi_i^{jk}(0, y_2, y_3) = \chi_i^{jk}(y_1^0, y_2, y_3) \tag{6.36a}$$

$$\chi_i^{jk}(y_1, 0, y_3) = \chi_i^{jk}(y_1, y_2^0, y_3) \tag{6.36b}$$

$$\chi_i^{jk}(y_1, y_2, 0) = \chi_i^{jk}(y_1, y_2, y_3^0) \tag{6.36c}$$

To avoid rigid body motions, displacement constraints need to be imposed to a point of the unit cell model, for instance a corner point. After the unit cell problem is solved, the homogenized elasticity tensor is calculated, by applying equation (6.28) using vector notation, as expressed below:

$$\mathbf{C}^h = \sum_{k=1}^{n_e} \frac{Y^k}{Y} \mathbf{C}^k (\mathbf{I} - \mathbf{B}^k \boldsymbol{\chi}^k) \tag{6.37}$$

where $n_e$ is the number of finite elements of the unit cell, $Y^k$ is the volume of element $k$ and Y is the volume of the unit cell. Finally, the microscale strain and stress are calculated using relations (6.31) and (6.30), as follows:

$$\boldsymbol{\varepsilon}^{(1)}(\mathbf{x},\mathbf{y}) = (\mathbf{I} - \mathbf{B}^e\boldsymbol{\chi}^e)\boldsymbol{\varepsilon}^{(0)}(\mathbf{x},\mathbf{y}) \tag{6.38}$$

$$\boldsymbol{\sigma}^{(1)}(\mathbf{x},\mathbf{y}) = \mathbf{C}^e(\mathbf{I} - \mathbf{B}^e\boldsymbol{\chi}^e)\boldsymbol{\varepsilon}^{(0)}(\mathbf{x},\mathbf{y}) \tag{6.39}$$

More details related to the implementation of asymptotic homogenization in the context of finite element analysis can be found in a number of older or more recent publications. Some of them provide a general description of the numerical implementation, like for instance the work presented in [161], while others emphasize in applications using commercial finite element packages, such as Abaqus [45], [49].

### 6.2.4 A brief on mathematical homogenization for non-linear problems

Several research efforts have been conducted focusing on the simulation of the non-linear response of composite materials, by adopting principles of mathematical homogenization. The detailed description of these methods is beyond the scope of this book. Hence, a brief literature review is only presented here.

In [212] the asymptotic homogenization for linear elasticity problems is extended to non-linear ones, using a rate formulation of the updated Lagrangian scheme. A variational formulation is proposed in [213], within a two-scale modelling scheme, for the simulation of heterogeneous, periodic structures. In [52] asymptotic expansion homogenization principles are adopted for the description of damage evolution, stiffness degradation, softening behaviour, size effects and fracture, in heterogeneous structures. Microscopic energy analysis has been used in this work, to describe the damage evolution conditions. In [240] non-linear asymptotic homogenization is used to derive the effective material properties of layered thermoelectric composites, converting heat into electricity.

More recently, an asymptotic homogenization approach is applied in [177] to the equations describing the dynamic response of heterogeneous materials with evolving micro-structures, depicting an hyperelastic behaviour. A second-order reduced asymptotic homogenization scheme is developed in [75], to capture the non-linear response of heterogeneous materials with large periodic microstructure. In [193] a reduced-order asymptotic expansion homogenization technique is proposed, for the simulation of the inelastic, damaged behaviour of heterogeneous materials.

## 6.3 Numerical homogenization

Within numerical homogenization, the mechanical response of a Representative Volume Element is determined and its average (effective) material properties are calculated. Then, a macroscopic constitutive law is chosen a priori, to describe the corresponding homogeneous structure in the coarse scale. The key concept of this approach, is that the material parameters of the macroscopic constitutive law, are defined using the effective material properties calculated from the microstructural simulation.

This parameter fitting process, involves the upward (micro to macro) transfer of information only once, contrary to computational homogenization, where the effective micro properties are sent to macro scale repetitively, for every load step of the overall coarse scale model. This point constitutes an advantage of numerical homogenization, since once it is formulated for a composite heterogeneous material, it can be implemented with relatively low computation cost. Another advantage of this method is that it can be applied even by commercial finite element analysis packages, using standard constitutive laws, after introducing subroutines providing the effective material response. Thus, the method can be used in big structural problems, involving for instance three-dimensional masonry buildings or complex patterns of fibre reinforced composite laminates. On the other hand, the a priori selection of the macroscopic constitutive law does not allow for the representation of evolving microstructures, which stands as a restriction of this approach.

The average macroscopic strain tensor $<\boldsymbol{\varepsilon}>_{V_m}$ and stress tensor $<\boldsymbol{\sigma}>_{V_m}$ can generally be defined by simulations on the RVE of volume $V_m$, using the following, *averaging relations*:

$$<\boldsymbol{\varepsilon}>_{V_m}=\frac{1}{V_m}\int_{V_m}\boldsymbol{\varepsilon}^m dV_m, \quad <\boldsymbol{\sigma}>_{V_m}=\frac{1}{V_m}\int_{V_m}\boldsymbol{\sigma}^m dV_m \tag{6.40}$$

The average stress versus strain relation can then be expressed as follows, for linear elastic analysis:

$$<\boldsymbol{\sigma}>_{V_m}=\mathbf{C}<\boldsymbol{\varepsilon}>_{V_m} \tag{6.41}$$

where the effective elasticity tensor $\mathbf{C}$ consists of 36 unknown components for the general case of anisotropic materials, while each of the tensors $<\boldsymbol{\varepsilon}>$ and $<\boldsymbol{\sigma}>$ has six unknowns, as presented in section 1.2.6 of chapter 1. These components can easily be calculated, by applying six linearly independent strain loadings on the RVE and calculating the average stress tensor which is obtained from each loading. A system of 36 equations with 36 unknows is then built and solved, to determine all the components of the effective elasticity tensor. Detailed descriptions for applying strain loading to the RVE are presented in chapter 7.

However, when non-linearities are considered, the formulation of the macro-constitutive law using the effective microscopic description is a complex process. Several models have been proposed in literature to describe the non-linear homogenized response of composite structures, using numerical homogenization. These models focus on different materials, such as masonry and fibre reinforced composites. Depending on the type of the material under investigation, a proper constitutive description is chosen for both the microstructure and for the macroscopic structure.

In the micro, fine scale, a non-linear material law is considered for some or all of the constituent materials. In the context of fibre reinforced composites, the non-linear law is applied to the matrix and/or to the surrounding reinforcing fibres of the structure. The fine scale of masonry structures consists of stone blocks and the mortar joints, considered as elements of zero or non-zero thickness. The material law can be assigned to the mortar, to the stone blocks or even to both materials.

Then, finite element analysis simulations are conducted in the mircostructure. The most common boundary conditions which are adopted for these fine scale simulations, are *periodic boundary conditions*. These boundary conditions require periodic displacements and antiperiodic tractions, in the opposite boundaries of the RVE. More details about periodic boundary conditions and their implementation within finite element analysis of the fine scale are given in chapter 7. After the microstructural simulations are completed, the effective constitutive stress-strain law is derived. Next, this effective description is properly considered by the macroscopic constitutive law, which has been defined a priori. This is the most sophisticated step of numerical homogenization, since it requires the fitting of the effective material properties calculated in the microstructure, to every component of the macroscale constitutive law.

For the coarse and the fine scale description are adopted models which include elastoplastic, elasto-visco-plastic and continuum damage laws. Often, the averaging process derived from the microstructure takes place individually for the linear, elastic material properties as well as for the non-linear ones [221]. In [96] a three-dimensional homogenization-based continuum damage model is adopted in the coarse scale, to simulate the damage on fibre reinforced composites, due to matrix-fibre debonding. A fourth-order damage tensor is defined in terms of macroscopic strain components and calibrated by microstructural simulations, in which a cohezive zone model is adopted to capture debonding. The commercial software package Abaqus is used to implement the homogeneous macro constitutive law, by introducing Fortran user subroutines (UMAT, UEL).

In [191] a discrete macroscopic model is used to study the non-linear response of masonry structures, using numerical homogenization. A damage plasticity continuum law is adopted to depict failure in the interfaces between stones in the coarse scale, calibrated using average stress-strain and moment-curvature diagrams, derived from fine scale simulations on masonry unit cells. In [124] a homogenized viscoelastic-viscoplastic model is proposed to study the non-linear response of fibre reinforced polymers.

It is noted that more models can be found in literature, presenting some differences in respect to the general aspects of the method of numerical homogenization. Some of these models may refer to efforts focusing on establishing the link between the micro and the macro response, without considering any macroscopic constitutive law [126]. Instead, the macro description is achieved *a posteriori*, by determining the average stress and tangent stiffness from microstructural calculations. Other approaches provide a homogenization description towards calculating macro-parameters from unit cell simulations, without formulating a macro constitutive law [10]. An alternative procedure towards numerical homogenization for masonry structures is presented in [187], relying on the *Transformation Field Analysis*, which is based on the technique of superposition of the effects of micro simulations.

# Chapter 7

# *Multi-scale analysis for composite materials*

## 7.1 Introduction to multi-scale computational homogenization

The term *multi-scale* refers to models which make use of more than one length scales, in order to provide the constitutive response of composite materials. As presented in chapter 6, the concept of introducing multiple length scales is elaborated in different homogenization methods, aiming in depicting the effective material properties of heterogeneous structures. Thus, the terms multi-scale and homogenization are often used concurrently, since they both define a framework of study in multiple length scales, from nano or micro, to meso or macro scale, which is adopted for the determination of the effective material properties.

Among the several multi-scale approaches which have been developed, some broad categories can be recognized, based on the concept adopted for exchanging information between the length scales. According to a widely used method, data is exchanged between micro and macro scales in a two-way information flow, thus, from the macro to the micro and vice versa. This two-way information flow takes place between two different finite element models, each representing a different length scale. This method is known as *multi-scale computational homogenization* or $FE^2$ and is presented analytically in the following sections.

Alternatively, both length scales may be included in one, unique macroscopic model, where a coarse and a fine mesh are used to simulate the macro and the micro scale, respectively. This approach is called *concurrent*, but the same terminology is often used in literature, to express also the multi-scale computational homogenization.

The computational homogenization scheme which is presented in this chapter, offers an alternative solution to the evaluation of the non-linear response of composite materials, since it can be used to simulate evolving microstructures. In addition, the implementation of the method is relatively simple, though it involves some complexity in manipulating the microstructural system of equilibrium equations, which is needed for the determination of the

DOI: 10.1201/9781003017240-7

effective stress-strain response. On the other hand, the computational cost of the method is significantly high while its application within commercial software is not always an easy task.

Comparing to other homogenization schemes presented in chapter 6, computational homogenization presents some advantages and disadvantages. In particular, analytical homogenization becomes quite complex when non-linear phenomena are involved but once it is formulated, it can accurately simulate the mechanical response with a low computational effort. Numerical homogenization does not allow for the simulation of evolving microstructures and presents an inherent complexity in fitting the macroscopic response using the effective microscopic description. However, its application within commercial software is relatively simple.

For the implementation of multi-scale computational homogenization, a classical continuum mechanics formulation is initially adopted, in the framework of non-linear finite element analysis. According to the main concept of the method, the original, heterogeneous body is substituted by an equivalent, homogeneous one, the material properties of which are derived by considering and solving a proper microstructure, which corresponds to every Gauss point of the macroscopic structure. Thus, for this homogeneous macroscopic structure no material properties are assigned. Instead, its constitutive description will be built numerically, using the microstructural simulations.

The chosen microstructure needs to be representative, in the sense that it fully characterizes the microstructural pattern with all constituents. The macroscopic structure can then be conceptually derived by considering a repetition of the microstructural pattern. This implies one of the main assumptions of the method, also defined in section 6.1, called *periodicity*. In particular, the computational homogenization scheme can be applied when global or local periodicity appears. The first suggests that the whole macroscopic structure consists of a spatially repeated microstructure, while the second that different macroscopic regions may correspond to different microstructural patterns. Due to the representative nature of the microstructure, it is usually called *Representative Volume Element* (RVE).

Since emphasis is given to non-linear problems, the Newton-Raphson incremental-iterative process can be adopted to provide the solution to the coarse scale simulation, for every type of non-linearity applied to the RVE, such as geometric, material or boundary non-linearity. Within every load increment of the Newton-Raphson process, the macroscopic strain $\boldsymbol{\varepsilon}^M$ is calculated at every Gauss point and imposed as displacement loading to the corresponding RVE. The RVE model is solved and the effective stress $\boldsymbol{\sigma}^M$ and consistent tangent stiffness $\mathbf{D}^M$ are calculated and sent back to the macroscopic simulation. The effective stress-strain response is eventually determined numerically, for every macroscopic increment-iteration. This approach is schematically shown in figure 7.1 for the case of a fibre reinforced composite material.

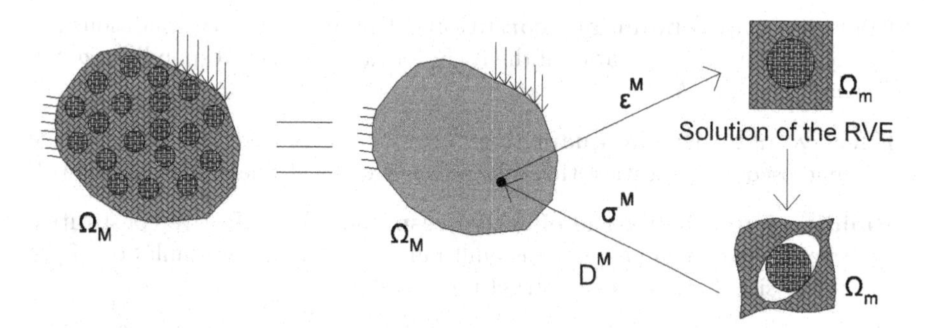


**FIGURE 7.1**: Schematic representation of multi-scale computational homogenization for a fibre reinforced composite material.

The presented numerical scheme is also known as *first-order computational homogenization* [72], [143], [144], [194]. Although it is robust and provides a solution to the difficult problem of the evaluation of the non-linear response and failure of heterogeneous materials, there are some restrictions which should be highlighted. According to a core assumption, the method works well when the principle of *separation of scales* is valid. This principle states that the size of the microscopic structure is much smaller than the characteristic length of the macroscopic structure or the size over which the macroscopic loading varies in space. According to this definition, the RVE will be sufficiently small, such that the macroscopic deformation field can be considered as uniform over it. This infinitesimal nature of the RVE also imposes that an increase of its size will not influence the mechanical response. Thus, the size of the RVE does not affect the analysis and size effects are neglected.

The assumption of the uniformity of the macroscopic deformation field over the RVE is not valid, when zones of large deformation gradients are developed in the macroscopic structure. According to the descriptions presented in section 3.3.3 of chapter 3, localization phenomena arise in this case and the principle of separation of scales cannot be adopted. In addition, as mentioned in section 3.3.3, when strain softening of the macroscopic structure appears, it leads to loss of ellipticity of the governing equations which in turn results in the pathological dependence of the solution on the discretization size. For all these cases, the first-order computational homogenization cannot be applied. Instead, advanced multi-scale schemes have been proposed to capture localization of damage, as discussed in section 7.3.

The main steps of the multi-scale computational homogenization are summarized below [112], [206]:

**Microstructural formulation:** A microstructural RVE model is formulated, considering a known constitutive description for every constituent of the composite structure.

**Microstructural boundary conditions:** Chosen boundary conditions are applied to the boundaries of the RVE, formulating a corresponding *boundary value problem.*

**Effective macroscopic quantities:** The RVE is solved and the effective macroscopic quantities, thus, the stress and the stiffness, are determined.

**Building numerically the effective response:** The effective constitutive stress versus strain response is built numerically and the simulation of the composite, heterogeneous structure is evolved.

The mathematical formulation of the method is provided in the next section.

---

## 7.2 Formulation of first-order computational homogenization

### 7.2.1 Averaging relations

A core point of the mathematical formulation of the first-order, multi-scale computational homogenization is the introduction of averaging relations. The relevant descriptions are based initially on the assumption of small displacement gradients, for which the reference and deformed configurations coincide. At the end of the section, a more general formulation which is appropriate for large displacement analysis is briefly presented. For every quantity in the macroscopic and microscopic length scale, the letters $M$ and $m$ are used, respectively.

The main energy principle which is adopted within computational homogenization to provide a coupling between the macroscopic and microscopic work, is the *Hill-Mandel* principle or energy averaging theorem. This theorem states that the macroscopic volume average of the variation of work is equal to the local work variation, assigned to the RVE [90]:

$$\boldsymbol{\sigma}^M : \delta\boldsymbol{\varepsilon}^M = \frac{1}{V_m}\int_{V_m} \boldsymbol{\sigma}^m : \delta\boldsymbol{\varepsilon}^m dV_m \tag{7.1}$$

According to the concept of this multi-scale scheme, one microstructural RVE corresponds to every Gauss point of the macroscopic structure. Hence, there is an underlying macro-to-micro transition, elaborated from the microscopic boundary conditions which are derived using a macroscopic input field, such as the strain field. For the energetic validity of the method, these boundary conditions on the RVE should be such, that the Hill-Mandel principle is satisfied. It can be proved that among others, three widely used types of boundary conditions satisfy the Hill-Mandel principle, namely: a) prescribed (or linear) displacements, b) prescribed tractions and c) periodic boundary conditions [112].

According to prescribed displacements, a displacement loading $\mathbf{u}$ is applied to the boundaries $\partial V_m$ of the RVE, as provided below:

$$\mathbf{u}|_{\partial V_m} = \boldsymbol{\varepsilon}^M \cdot \mathbf{x} \tag{7.2}$$

where $\boldsymbol{\varepsilon}^M$ is the average macroscopic strain and $\mathbf{x}$ denotes the undeformed coordinates of the boundary nodes of the RVE. An example of applying prescribed boundary conditions on a two-dimensional plane stress RVE is given in section 8.2.

The prescribed tractions are formulated as follows:

$$\mathbf{t} = \mathbf{n} \cdot \boldsymbol{\sigma}^M \tag{7.3}$$

In general, traction boundary conditions do not fit within the displacement-driven finite element analysis which is widely used. However, a Lagrange multipliers scheme is proposed in [143] to compute the tractions on the boundary of the RVE, within a strain-controlled process.

According to periodic boundary conditions, periodic displacements as well as antiperiodic tractions, should appear in the opposite boundaries of the RVE. The relations between the displacements of the opposite boundaries are provided below:

$$\mathbf{u}_T - \mathbf{u}_B = \mathbf{u}_4 - \mathbf{u}_1 \tag{7.4a}$$

$$\mathbf{u}_R - \mathbf{u}_L = \mathbf{u}_2 - \mathbf{u}_1 \tag{7.4b}$$

where the displacements in the top (T), bottom (B), right (R) and left (L) boundary are determined using the prescribed displacements of three corner nodes of the RVE, namely 1, 2 and 4, given by relation (7.2). Practically, as indicated by relations (7.4), the difference in the displacement between two opposite boundary nodes should be equal, for every pair of nodes which belong to these boundaries. This imposes an inherent restriction on the discretization of the RVE, since pairs of opposite nodes should be identified in the mesh. In figure 7.2 is shown a two-dimensional RVE of a fibre reinforced composite. Squares and circles have been used in the figure, to depict how opposite boundary nodes are coupled within periodic boundary conditions. In figure 7.3 is shown the deformation contour of this RVE, under periodic boundary conditions. It is noticed that debonding between the matrix and the fibre arises, as explained analytically in [65].

By conducting a brief comparison between these boundary conditions, it is noticed that the prescribed displacements overestimate, while the prescribed tractions underestimate the effective response [227]. Thus, for problems where no localization phenomena arise, periodic boundary conditions seem to offer a reasonably well solution.

When localization appears, the prescribed displacements are too restrictive and cannot be used. This is due to the fact that all boundary displacements are prescribed, indicating that localization of damage in the form of cracks or damage zones crossing the boundaries of the RVE, cannot be depicted.

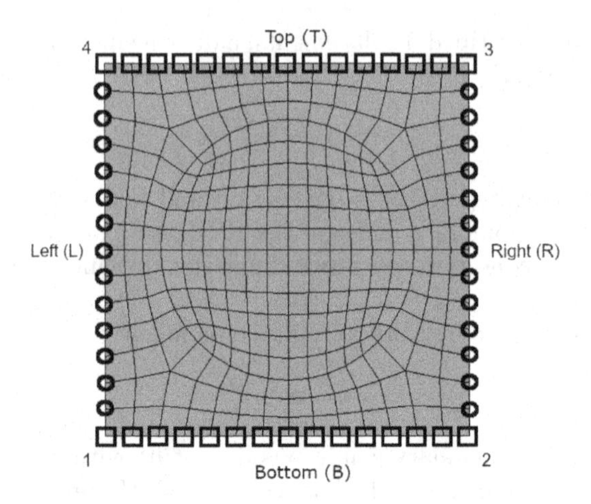


**FIGURE 7.2**: Two-dimensional fibre reinforced composite RVE.

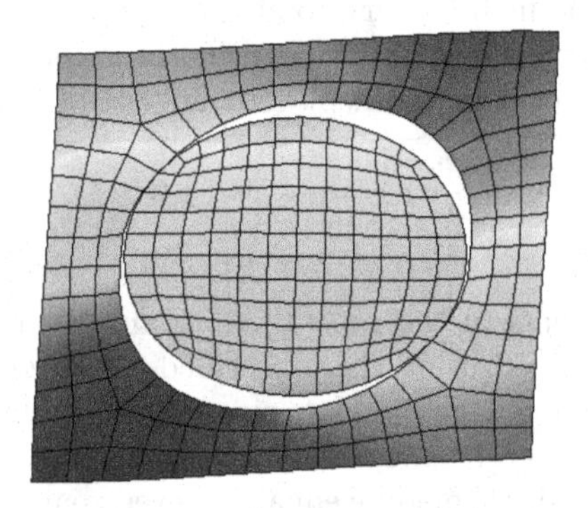

**FIGURE 7.3**: Deformation of RVE under periodic boundary conditions.

Periodic boundary conditions can also be proved restrictive in this case. For this reason, alternative boundary conditions have been proposed to capture localization, as it will be discussed in the following sections.

Next, the volume average quantities are determined for both the microscopic strain and stress. The general averaging relations, are provided below:

$$< \boldsymbol{\varepsilon} >_{V_m} = \frac{1}{V_m}\int_{V_m} \boldsymbol{\varepsilon}^m dV_m, \quad < \boldsymbol{\sigma} >_{V_m} = \frac{1}{V_m}\int_{V_m} \boldsymbol{\sigma}^m dV_m \tag{7.5}$$

Towards their numerical implementation, equations (7.5) can be further simplified. The volume average microscopic strain $< \boldsymbol{\varepsilon} >_{V_m}$ is equal to the macroscopic strain $\boldsymbol{\varepsilon}^M$ which has been applied as loading, to the boundaries of the RVE:

$$< \boldsymbol{\varepsilon} >_{V_m} = \boldsymbol{\varepsilon}^M \tag{7.6}$$

For the case of prescribed displacements, the following simplified formulation for the macroscopic stress can be used [112], [143]:

$$<\boldsymbol{\sigma}>_{V_m} = \frac{1}{V_m}\mathbf{f}\mathbf{x} = \boldsymbol{\sigma}^M \tag{7.7}$$

where $\mathbf{f}$ is the matrix of the resulting external, reaction forces in the undeformed coordinates $\mathbf{x}$ of the boundary nodes of the RVE, after the microscopic analysis has been completed. These reaction forces appear due to the displacement loading which is applied to the boundaries of the RVE.

A similar relation is applied for the macroscopic stress, in case periodic boundary conditions are used [112], [143]:

$$<\boldsymbol{\sigma}>_{V_m} = \frac{1}{V_m}\mathbf{f}_s\mathbf{x}_s = \boldsymbol{\sigma}^M \tag{7.8}$$

where $\mathbf{f}_s$ denotes the external, reaction forces in the three corner nodes with supported (prescribed) displacements and $\mathbf{x}_s$ represents the undeformed coordinates of these nodes.

The described multi-scale scheme can be developed within a more general framework, allowing for large displacement gradients. In this case, the deformation gradient tensor $\mathbf{F}$ defined in section 1.2.4.2 is used instead of the infinitesimal strain tensor $\boldsymbol{\varepsilon}$ and the first Piola-Kirchhoff stress tensor $\mathbf{P}$ defined in section 1.2.5.2 is used instead of the Cauchy stress $\boldsymbol{\sigma}$, as it is presented in [112]. The reason for using the deformation gradient tensor and the first Piola-Kirchhoff stress tensor is that these tensors are energetically conjugated. In addition, initial and current configurations are identified and the corresponding initial and current coordinates $\mathbf{X}$ and $\mathbf{x}$ are used. For completeness, the formulation is briefly provided below, in the following sequence: prescribed displacements (equation 7.9), prescribed tractions (equation 7.10), periodic boundary conditions (equations 7.11), averaging relations (equation 7.12), effective stress for prescribed displacements (equation 7.13) and effective stress for periodic boundary conditions (equation 7.14).

$$\mathbf{x}|_{\partial V_m} = \mathbf{F}^M \cdot \mathbf{X} \tag{7.9}$$

$$\mathbf{T} = \mathbf{P}^M \cdot \mathbf{N} \tag{7.10}$$

$$\mathbf{x}_T - \mathbf{x}_B = \mathbf{x}_4 - \mathbf{x}_1 \tag{7.11a}$$

$$\mathbf{x}_R - \mathbf{x}_L = \mathbf{x}_2 - \mathbf{x}_1 \tag{7.11b}$$

$$\mathbf{F}^M = \frac{1}{V_m}\int_{V_m} \mathbf{F}^m dV_m, \quad \mathbf{P}^M = \frac{1}{V_m}\int_{V_m} \mathbf{P}^m dV_m \tag{7.12}$$

$$\mathbf{P}^M = \frac{1}{V_m}\mathbf{f}\mathbf{X} \tag{7.13}$$

$$\mathbf{P}^M = \frac{1}{V_m}\mathbf{f}_s\mathbf{X}_s \tag{7.14}$$

In the above equations, **N** is the unit normal to the undeformed boundary of the RVE. In the next section is shown how the average stress-strain relation is built numerically, leading to the determination of the consistent tangent stiffness matrix which is used in the overall multi-scale scheme.

## 7.2.2 The overall multi-scale formulation

As presented in the previous sections of this chapter, two nested boundary value problems are solved, representing the effective macro and microstructural models. Within finite element analysis, classical formulations can be used to provide the solution to these models. Since the problems involve non-linearities, the incremental-iterative Newton-Raphson scheme is adopted for both the macro and the microscopic simulation. For the coarse scale no constitutive description is a priori chosen, indicating that the core concept of the overall formulation is to provide the effective macroscopic behaviour numerically, using the microstructural calculations.

For the RVE all the constituents are explicitly simulated and corresponding material laws are considered. After the strain (displacement) load derived from every macro Gauss point is applied to the RVE model, the latter is solved and once convergence of the non-linear solution is achieved, the effective, macroscopic stress and consistent tangent stiffness are numerically determined and sent back to the coarse scale simulation, as it is presented in this section. Briefly, the macroscopic stress is determined using equations given in section 7.2.1 while the consistent tangent stiffness is obtained by expressing the relation between the variations of the average microscopic (e.g. the macroscopic) stress and strain. Once these are calculated for every macroscopic Gauss point, the tangent stiffness matrix and the internal force vector, which are needed for the implementation of the non-linear analysis, are determined.

The formulation of the multi-scale scheme will be provided for both prescribed (linear) displacements and periodic boundary conditions, applied to the boundaries of the RVE. As it is given in section 2.4.3, when prescribed displacement loading is applied to a structural system, the solution is obtained by partitioning the system of equilibrium equations into free, "$f$" and supported, "$s$" degrees of freedom, noticing that the latter corresponds to prescribed nodal displacements. Since both prescribed displacements and periodic boundary conditions constitute a displacement loading for the RVE, partitioning will take place.

### 7.2.2.1 Prescribed displacements

In case prescribed displacements are used, simulation of the RVE is conducted and once convergence criteria of the Newton-Raphson scheme are satisfied, the microstructural analysis is complete. At this stage, the system of equilibrium equations is partitioned, as shown below:

$$\begin{bmatrix} \mathbf{K}_{ff} & \mathbf{K}_{fs} \\ \mathbf{K}_{sf} & \mathbf{K}_{ss} \end{bmatrix} \begin{bmatrix} \delta\mathbf{u}_f \\ \delta\mathbf{u}_s \end{bmatrix} = \begin{bmatrix} 0 \\ \delta\mathbf{f}_s \end{bmatrix} \tag{7.15}$$

With the letter $f$ are denoted the free and with $s$ the supported degrees of freedom. Equations (7.15) lead to:

$$\mathbf{K}^M \delta\mathbf{u}_s = \delta\mathbf{f}_s \tag{7.16a}$$

$$\mathbf{K}^M = \mathbf{K}_{ss} - \mathbf{K}_{sf}(\mathbf{K}_{ff})^{-1}\mathbf{K}_{fs} \tag{7.16b}$$

According to the next step of the procedure, the consistent tangent stiffness of the macroscopic Gauss point should be obtained. This is achieved by formulating the relation between the variation of the macroscopic stress and strain. First, the expression for the prescribed displacements on the whole RVE boundaries is given by:

$$\delta\mathbf{u}_s = \mathbf{B}\delta\boldsymbol{\varepsilon}^M \tag{7.17}$$

where $\mathbf{B}$ is the matrix with the undeformed positions on the boundary RVE nodes and $\delta\boldsymbol{\varepsilon}^M$ is the variation of the macroscopic strain.

The average stress of the RVE, which is equal to the macro stress, is given for prescribed displacements by the following relation:

$$\delta\boldsymbol{\sigma}^M = \frac{1}{V_m}\mathbf{B}^T\delta\mathbf{f}_s \tag{7.18}$$

where $\delta\mathbf{f}_s$ is the external, reaction force vector in the boundaries of the RVE. By substitution of (7.16a) into (7.18), it is derived that:

$$\delta\boldsymbol{\sigma}^M = \frac{1}{V_m}\mathbf{B}^T\mathbf{K}^M\delta\mathbf{u}_s \tag{7.19}$$

Relation (7.17) is then substituted into (7.19), resulting in the stress-strain relation for prescribed displacements and the corresponding consistent tangent stiffness of the macroscopic level, $\mathbf{D}^M$:

$$\delta\boldsymbol{\sigma}^M = \frac{1}{V_m}\mathbf{B}^T\mathbf{K}^M\mathbf{B}\delta\epsilon^M \tag{7.20}$$

$$\mathbf{D}^M = \frac{1}{V_m}\mathbf{B}^T\mathbf{K}^M\mathbf{B} \tag{7.21}$$

#### 7.2.2.2 Periodic boundary conditions

In case periodic boundary conditions are applied to the boundaries of the RVE, constraint equations (7.4) should be taken into account and included in the analysis, in addition to the system of equilibrium equations. There are two widely used methods, which can be adopted to consider equations of constraints. The first, is the method of Lagrange multipliers, which is used in [143] in the context of multi-scale computational homogenization. The second, is a method which relies on the transformation of the equilibrium equations, as shown in [50] and adopted within multi-scale analysis in [112]. This method is analytically presented here.

According to the main concept of the transformation, the initial system of equilibrium equations is properly transformed into a new system, where "retained" and "condensed" degrees of freedom are recognized. The constraint equations, which are used to implement the periodic boundary conditions, are also included in the formulation as it shown below. Eventually, the condensed degrees of freedom will be eliminated from the total system of equations. For the discussed problem, these degrees of freedom belong to two out of the four RVE boundaries, for instance to the bottom and left boundary. All the remaining degrees of freedom will be retained.

First, the constraint equations (7.4) are expressed in the discretized domain as follows:

$$\mathbf{C}\delta\mathbf{u} = \mathbf{0} \tag{7.22}$$

noticing that matrix $\mathbf{C}$ is properly assigned the values 0, 1 and -1, such that when multiplied by the vector of nodal displacements $\delta\mathbf{u}$, it provides the constraint equation (7.4) for every pair of opposite boundary nodes. Relation (7.22) is then partitioned to retained ($r$) and condensed ($c$) degrees of freedom:

$$\begin{bmatrix}\mathbf{C}_r & \mathbf{C}_c\end{bmatrix}\begin{bmatrix}\delta\mathbf{u}_r \\ \delta\mathbf{u}_c\end{bmatrix} = \mathbf{0} \tag{7.23}$$

which leads to:

$$\delta\mathbf{u}_c = \mathbf{C}_{rc}\delta\mathbf{u}_r \tag{7.24a}$$

$$\mathbf{C}_{rc} = -\mathbf{C}_c^{-1}\mathbf{C}_r \tag{7.24b}$$

Then, the following formulation is derived:

$$\begin{bmatrix}\delta\mathbf{u}_r \\ \delta\mathbf{u}_c\end{bmatrix} = \mathbf{T}\delta\mathbf{u}_r, \quad \mathbf{T} = \begin{bmatrix}\mathbf{I} \\ \mathbf{C}_{rc}\end{bmatrix} \tag{7.25}$$

where it is apparent that by using $\mathbf{T}$ as defined in (7.25), the identities $\delta\mathbf{u}_r = \delta\mathbf{u}_r$ and $\delta\mathbf{u}_c = \mathbf{C}_{rc}\delta\mathbf{u}_r$ arise, noticing that the second appears also in relation (7.24a).

Next, the initial system of equilibrium equations is partitioned into retained and condensed degrees of freedom:

$$\begin{bmatrix}\mathbf{K}_{rr} & \mathbf{K}_{rc} \\ \mathbf{K}_{cr} & \mathbf{K}_{cc}\end{bmatrix}\begin{bmatrix}\delta\mathbf{u}_r \\ \delta\mathbf{u}_c\end{bmatrix} = \begin{bmatrix}\delta\mathbf{f}_r \\ \delta\mathbf{f}_c\end{bmatrix} \tag{7.26}$$

By multiplying with $\mathbf{T}^T$ both parts of (7.26) and conducting the calculations, the final, reduced system of equations with the retained degrees of freedom is obtained:

$$\mathbf{K}^* \delta \mathbf{u}_r = \delta \mathbf{f}^* \tag{7.27}$$

where:

$$\mathbf{K}^* = \mathbf{K}_{rr} + \mathbf{C}_{rc}^T \mathbf{K}_{cr} + \mathbf{K}_{rc} \mathbf{C}_{rc} + \mathbf{C}_{rc}^T \mathbf{K}_{cc} \mathbf{C}_{rc} \tag{7.28}$$

and

$$\delta \mathbf{f}^* = \delta \mathbf{f}_r + \mathbf{C}_{rc}^T \delta \mathbf{f}_c \tag{7.29}$$

This system is solved and after convergence of the Newton-Raphson process is achieved, it is partitioned into prescribed and free degrees of freedom, noticing that now the three corner nodes of the RVE correspond to the prescribed degrees of freedom:

$$\begin{bmatrix} \mathbf{K}_{ff}^* & \mathbf{K}_{fs}^* \\ \mathbf{K}_{sf}^* & \mathbf{K}_{ss}^* \end{bmatrix} \begin{bmatrix} \delta \mathbf{u}_f \\ \delta \mathbf{u}_s \end{bmatrix} = \begin{bmatrix} 0 \\ \delta \mathbf{f}_s^* \end{bmatrix} \tag{7.30}$$

Equations (7.30) lead to:

$$(\mathbf{K}^{M*}) \delta \mathbf{u}_s = \delta \mathbf{f}_s^* \tag{7.31a}$$

$$(\mathbf{K}^{M*}) = \mathbf{K}_{ss}^* - \mathbf{K}_{sf}^* (\mathbf{K}_{ff}^*)^{-1} \mathbf{K}_{fs}^* \tag{7.31b}$$

For a two-dimensional problem with two degrees of freedom per node, $\delta \mathbf{u}_s$ is a $6 \times 1$ vector and $(\mathbf{K}^{M*})$ a $6 \times 6$ matrix. In addition, the prescribed displacements are provided by equation (7.17), where matrix $\mathbf{B}$ denotes in this case the undeformed positions of the three prescribed corner nodes of the RVE.

The average stress of the RVE, which is equal to the macro stress, is given for periodic boundary conditions by:

$$\delta \boldsymbol{\sigma}^M = \frac{1}{V_m} \mathbf{B}^T \delta \mathbf{f}_s^* \tag{7.32}$$

where $\delta \mathbf{f}_s^*$ is the external, reaction force vector in the three corner nodes of the RVE. Finally, the stress-strain relation and the corresponding consistent tangent stiffness are obtained respectively, by substituting relations (7.31a) and (7.17) into (7.32):

$$\delta \boldsymbol{\sigma}^M = \frac{1}{V_m} \mathbf{B}^T (\mathbf{K}^{M*}) \mathbf{B} \delta \boldsymbol{\varepsilon}^M \tag{7.33}$$

$$\mathbf{D}^M = \frac{1}{V_m} \mathbf{B}^T (\mathbf{K}^{M*}) \mathbf{B} \tag{7.34}$$

## 7.3 An introduction to multi-scale schemes adopted to simulate localization of damage

During the last years, several research efforts focus on the simulation of localization phenomena, in the framework of multi-scale analysis. As it is presented analytically in section 3.3.3, localization of damage arises when narrow regions (bands) are developed in structures, where all further deformations localize, as the external loading increases monotonically. These phenomena lead to material instability due to the appearance of softening constitutive laws, which are used to describe damage of composite materials such as concrete. Material instability may result in structural instability, accompanied by the loss of positive-definiteness for the tangent stiffness tensor, which becomes singular in this case. The most crucial consequence of this loss of positive-definiteness of the tangent stiffness is that it can result in loss of ellipticity of the governing equations. Within finite element analysis, this leads to pathological dependence of the solution on the discretization size.

From another point of view, once the size of the damage (for instance in the form of cracks) in fine scale is of the order of magnitude of the microstructure, the principle of separation of scales is not yet valid. For all these reasons, the first-order computational homogenization which is presented in section 7.2, cannot be used to provide an accurate representation of the mechanical response once localization of damage arises.

Different models have been proposed in the past, within the general framework of multi-scale analysis, for the investigation of localization phenomena in composite structures. In [112] a second-order computational homogenization scheme is proposed, resulting in a higher-order macroscopic continuum model. In [46], [47], [48] a new type of boundary conditions is developed, aiming to provide an alignment of the imposed boundary conditions with the evolving localization bands. For the extraction of both the traction-separation response of the macroscopic discontinuity and the stress-strain response of the surrounding bulk material, a single microscopic analysis is needed. For the representation of the macroscopic discontinuity, the *embedded discontinuities method* is adopted. Since a discontinuous scheme is used to represent damage in the macroscopic level, this approach belongs to the category of *continuous-discontinuous* models.

Continuous-discontinuous models are also proposed for the majority of multi-scale schemes, which are developed to capture localization. The reason for this, is that as mentioned in section 3.3.3, when loss of ellipticity arises for the governing equations, the corresponding solution depicts a discontinuity. Then, the multi-scale models introduce macroscopic discontinuities in a way that the response of the discontinuity, such as for instance the effective traction-separation law, is obtained using the microstructural simulations, within the multi-scale scheme.

Among the approaches which have been used to simulate discontinuities representing localization of damage in the context of multi-scale analysis, are the method of embedded discontinuities and the Extended Finite Element Method (XFEM). In [24] is presented an XFEM method which excludes unstable sub-domains of the microstructure exhibiting loss of material stability, using "perforated" unit cells. The method is further developed in [25] using concepts of nucleating cracks. The XFEM method is also used in [195] to simulate the cohesive behaviour of macroscopic cracks while cohesive zone models are introduced in the RVE. An extension of the work presented in [46], [47], [48], based on the usage of the XFEM method is found in [31].

In [134], [135], the anisotropy induced damage of masonry, is depicted by introducing a masonry meso scale structure, consisting of masonry blocks, mortar joints and a continuous damage model for the evaluation of failure. The method of *embedded localization bands* is chosen for the simulation of the macroscopic damage, based on the meso scale analysis. Both the emergence and the orientation of these bands rely on the meso scale computations. This work is extended to describe failure of thin planar masonry shells in [140], [141].

A different computational homogenization method is proposed in [223], to depict cohesive and adhesive failure. The multi-scale concept is only applied to the macroscopic integration points located on the macro cracks while for the microscopic scale, discrete failure is considered. Within a similar concept, a multi-scale approach is presented in [157], [158], [159] introducing a continuous damage model in the microstructure.

---

## 7.4 A multi-scale model proposed to simulate localization of damage in composite structures

In this section a model proposed in [63], which can be used to simulate localization of damage in composite materials, is presented. Within a multi-scale scheme, discontinuities (cracks) are introduced in the macroscopic level using the XFEM method, deriving their cohesive traction-separation response from mesoscopic simulations. The details for the numerical implementation of the XFEM method, emphasizing in simulation of cohesive cracks, can be found in section 5.3 of chapter 5.

At every integration point of the macroscopic discontinuities, a mesoscopic Representative Volume Element (RVE) is assigned, consisting of unilateral contact interfaces (joints) which are used to simulate the contact condition between discrete blocks. In addition, a simplified cohesive zone model is developed in the RVE model, to simulate tensile failure of the joints. The work focuses on applications in masonry structures, but it can be extended to different composite materials depicting localization of damage.

The overall multi-scale scheme is shown in figure 7.4. After loading is applied to the macroscopic structure, a crack pattern is developed along the joints. To depict this type of damage, a homogeneous macroscopic model is defined, where failure of the joints is simulated by introducing cohesive XFEM cracks. The displacement jump $||\mathbf{u}^M||$ calculated at each Gauss point of the cracks, is used as loading for the mesoscopic structure, as shown in figure 7.4. After the mesoscopic RVE is solved, the effective cohesive tangent tensor and the effective traction are calculated, as presented in the following lines of this section. These will then be introduced in the linearized cohesive XFEM analysis at the macro scale, as discussed in section 5.3.2.

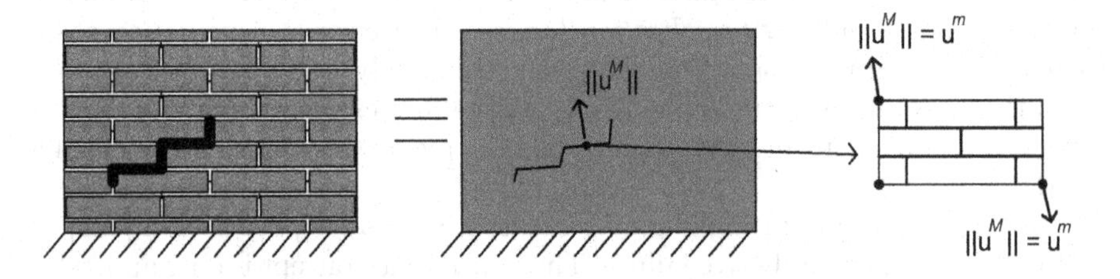


**FIGURE 7.4**: Schematic representation of the overall multi-scale model.

The mesoscopic structure consists of interfaces used to simulate the bond conditions between adjacent blocks. A unilateral contact law is applied to the interfaces, for the implementation of the non-penetration condition between the blocks. For the solution of the contact problem, an *Augmented Lagrangian formulation* is adopted, as presented in [65]. A simplified traction-separation law representing tensile failure in the normal direction of the interfaces is also used, as shown in figure 7.5.

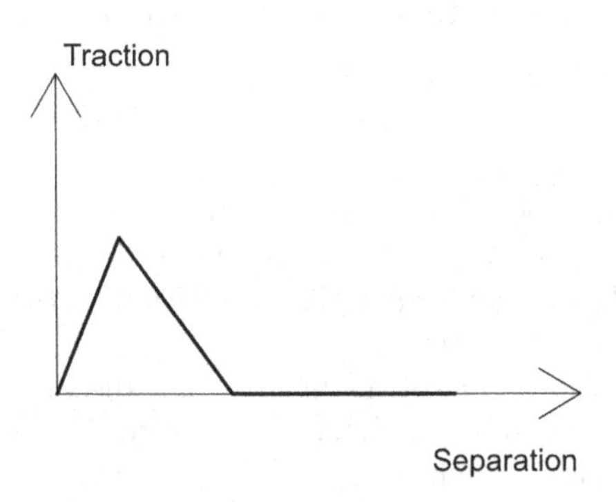


**FIGURE 7.5**: The tensile traction-separation law at the interfaces of the mesoscopic scale.

Loading for the mesoscopic structure is the displacement jump $||\mathbf{u}^M||$ which is calculated at the macroscopic crack. This displacement jump is applied to the corner nodes of the RVE as shown in the right part of figure 7.4, noticing that the direction of the macroscopic jump is derived at the macroscopic level and acts as loading direction for the mesoscopic model. Periodic boundary conditions are also applied to the boundaries of the RVE.

The proposed formulation allows the description of damaged induced anisotropy at the mesoscale, which arises due to failure at the interfaces. In addition, since the interfaces cross the boundaries of the RVE and a softening traction-separation law is applied to them, localization of damage arises in the fine scale. Thus, horizontal, vertical and diagonal damage patterns along the RVE boundaries, can be depicted by the proposed model.

Next, the formulation of this multi-scale scheme is presented. First, the Hill-Mandel energy principle for the mentioned problem is expressed as follows:

$$\delta||\mathbf{u}^M||^T\mathbf{t}^M = \frac{1}{V_m}\int_{V_m} \delta(\boldsymbol{\epsilon}^m)^T\boldsymbol{\sigma}^m dV_m \tag{7.35}$$

where the displacement jump $||\mathbf{u}^M||$ is determined at the integration points of the macroscopic cracks and the average traction $\boldsymbol{t}^M$ is obtained from RVE calculations. These quantities will be used to formulate the cohesive traction-separation law using the following relation:

$$\delta\boldsymbol{t}^M = \mathbf{D}^M_{coh}\delta||\mathbf{u}^M|| \tag{7.36}$$

in which $\mathbf{D}^M_{coh}$ is the cohesive tangent stiffness. For the determination of the average macroscopic traction the equation shown below is used:

$$\boldsymbol{t}^M = \boldsymbol{\sigma}^M\mathbf{n} \tag{7.37}$$

where $\mathbf{n}$ is the unit normal vector on the macroscopic crack and $\boldsymbol{\sigma}^M$ the average stress which is provided by equation (7.8). By substituting (7.8) into (7.37), and properly reformulating matrix $\mathbf{f}_s$ of (7.8) representing the external forces in the corner nodes with prescribed displacements and matrix $\mathbf{x}_s$ of (7.8) with the undeformed coordinates of these nodes, it is derived that:

$$\boldsymbol{t}^M = [\frac{1}{V_m}\mathbf{f}_s\mathbf{x}_s]\mathbf{n} = \frac{1}{V_m}\begin{bmatrix}\mathbf{x}_{br}^T\mathbf{n} & 0 & \mathbf{x}_{tl}^T\mathbf{n} & 0\\ 0 & \mathbf{x}_{br}^T\mathbf{n} & 0 & \mathbf{x}_{tl}^T\mathbf{n}\end{bmatrix}\mathbf{f}_s^* \tag{7.38}$$

noticing that $\mathbf{x}_{br}$ and $\mathbf{x}_{tl}$ are vectors representing the undeformed coordinates of the corner bottom-right and top-left nodes of the RVE where prescribed displacements are applied (dimensions 2x1 for each of these vectors), $\mathbf{n}$ is the unit normal vector to the macro crack (dimensions 2x1) and $\mathbf{f}_s^*$ is the vector of the external (reaction) forces at these nodes (dimensions 4x1).

Since periodic boundary conditions are applied to the boundaries of the RVE, constraint equations (7.4) should also be considered in the analysis. This is achieved by using the method of transformation of the initial system of equilibrium and constraint equations into a new system with "retained" and "condensed" degrees of freedom, as presented in section 7.2.2.2 for the case of first-order computational homogenization. The condensed degrees of freedom, which are in this case the bottom and left boundaries of the RVE, are eliminated while the remaining degrees of freedom are retained.

The system of retained degrees of freedom is then solved and after convergence of the Newton-Raphson process, it is partitioned into supported (prescribed) and free degrees of freedom, as shown in relations (7.30) and (7.31) at section 7.2.2.2. For completeness, relations (7.31) are also provided here:

$$(\mathbf{K}^{M*})\delta\mathbf{u}_s = \delta\mathbf{f}_s^* \tag{7.39a}$$

$$(\mathbf{K}^{M*}) = \mathbf{K}_{ss}^* - \mathbf{K}_{sf}^*(\mathbf{K}_{ff}^*)^{-1}\mathbf{K}_{fs}^* \tag{7.39b}$$

By substituting equation (7.39a) into the variation of equation (7.38), it is obtained that:

$$\delta t^M = \frac{1}{V_m}\begin{bmatrix} \mathbf{x}_{br}^T\mathbf{n} & 0 & \mathbf{x}_{tl}^T\mathbf{n} & 0 \\ 0 & \mathbf{x}_{br}^T\mathbf{n} & 0 & \mathbf{x}_{tl}^T\mathbf{n} \end{bmatrix}(\mathbf{K}^{M*})\delta\mathbf{u}_s \tag{7.40}$$

where the prescribed displacement loading $\mathbf{u}_s$ applied to the RVE is equal to the displacement jump obtained at the macroscopic level:

$$\mathbf{u}_s = \begin{bmatrix} u_{sx}^{br} \\ u_{sy}^{br} \\ u_{sx}^{tl} \\ u_{sy}^{tl} \end{bmatrix} = \frac{1}{2}\begin{bmatrix} ||u||_x \\ -||u||_y \\ -||u||_x \\ ||u||_y \end{bmatrix} = \begin{bmatrix} 1/2 & 0 \\ 0 & -1/2 \\ -1/2 & 0 \\ 0 & 1/2 \end{bmatrix} ||\mathbf{u}^M|| \tag{7.41}$$

By substituting the variation of equation (7.41) into (7.40), the final effective, traction-separation law is derived as follows:

$$\begin{aligned} \delta t^M &= \frac{1}{V_m}\begin{bmatrix} \mathbf{x}_{br}^T\mathbf{n} & 0 & \mathbf{x}_{tl}^T\mathbf{n} & 0 \\ 0 & \mathbf{x}_{br}^T\mathbf{n} & 0 & \mathbf{x}_{tl}^T\mathbf{n} \end{bmatrix}(\mathbf{K}^{M*})\begin{bmatrix} 1/2 & 0 \\ 0 & -1/2 \\ -1/2 & 0 \\ 0 & 1/2 \end{bmatrix}\delta||\mathbf{u}^M|| \\ &= \mathbf{D}_{coh}^M\delta||\mathbf{u}^M|| \end{aligned} \tag{7.42}$$

The term $\mathbf{D}_{coh}^M$ in equation (7.42) denotes the cohesive tangent stiffness which is eventually introduced in the macroscopic cohesive crack. It is noted that the bulk integration points surrounding the macroscopic discontinuities are assigned material properties which are obtained from linear numerical homogenization, chosen a priori for the undamaged material.

# Chapter 8

# *Data-driven analysis*

## 8.1 Introduction to data-driven numerical simulation

*Data-driven* structural analysis belongs to a broader category of a cutting edge concept called *digital twin*. Digital twin is the digital counterpart of a real-world system, which is designed to monitor and control its functionality. Data is exchanged between both the digital and the real-world system and feedback is provided, aiming in predicting and optimizing the response of both systems. This connection between the digital and the physical system is crucial for the successful implementation of the concept. It is expected that as the amount of feedback increases during time, the learning capability and capacity of the digital twin to provide improved predictions of the response of the physical infrastructure also increases.

Data-driven numerical modelling is defined within this general framework, as an alternative approach which can be used for the structural evaluation of composite heterogeneous structures and materials, in respect to the more classical methods presented in the previous chapters of this book. The core idea of this approach, is to replace the constitutive description needed for the implementation of a non-linear structural simulation, by databases containing elements of the same constitutive law. The databases may have been obtained experimentally, from real physical tests or computationally, by conducting numerical simulations. The concept has been developed during the last years, after the significant improvement of modern computers and cloud computing systems, which allow for manipulating large amounts of data. Data-driven techniques also profit from the existence of huge amount of experimental data and reply to the need to exploit this information directly within computational mechanics.

Several methodologies have been proposed for the implementation of data-driven techniques by adopting different concepts and tools. One of the key points refers to the length scale in which data-driven analysis is applied. Thus, these methodologies may either be used directly in a structural length scale or within a multi-scale framework. Another core point is related to the tool which is adopted in the structural code, to make use of the database. Since the database includes discrete values representing the constitutive description, for instance discrete stress-strain pairs, a process should be adopted to conduct

DOI: 10.1201/9781003017240-8

an interpolation in the database, of the values which satisfy compatibility and equilibrium of the structure under investigation. One technique which is extensively used, relies on the exploitation of *machine learning* and related tools, such as *artificial neural networks.* Another, more recent idea proposes the formulation of a minimum distance optimization problem, between data set points describing the constitutive behaviour and points satisfying the conservation laws of the underlying system. More concepts have also been proposed, adopting different interpolation schemes, such as for instance the *Delaunay triangulation* approach.

One of the main advantages of data-driven analysis is the fact that the incorporation of the database in the code results in more efficient simulations, in terms of the required computational time and cost, since, no "online" simulations are needed. For instance, contrary to the traditional multi-scale computational homogenization, for which a Representative Volume Element needs to be solved incrementally, at every increment-iteration of the Newton-Raphson process and at every Gauss point of the macroscopic structure, within the data-driven approach a database is used to provide the constitutive description, leading to reduced computational cost. Even if the database has been obtained numerically from simulations conducted on a RVE, these simulations are held "offline" and thus, independently of the overall multi-scale scheme. Another advantage of the method, is that the direct usage of databases offers the opportunity to "bypass" the constitutive description, eliminating uncertainties derived from the need to assume a material law and to calibrate its parameters. In the following sections of this chapter, some representative data-driven numerical schemes are presented.

---

## 8.2 Data-driven analysis using direct interpolation from databases

A concept of data-driven analysis relies on the direct usage of databases, in order to provide the material law which is needed for the numerical simulation. Since the databases will contain discrete values, an *interpolation scheme* is required to provide the constitutive description for any load level and direction. For the creation of the databases, numerical experiments are conducted on a Representative Volume Element (RVE) of the heterogeneous structure and values of the effective response, such as pairs of the effective strain-stress, strain-tangent stiffness or strain-energy, are derived.

This approach leads to the reduction of the computational cost of multi-scale analysis schemes, since there is no need for the simulation of a RVE at each increment-iteration of the overall macroscopic analysis, which otherwise would be the case. Thus, the "online" part of the simulation becomes computationally efficient. A high computational cost still arises, when the databases are derived numerically, with independent simulations on the RVE. However, these simulations take place "offline" and only once, for every microstructural pattern. In addition, some assumptions or extensions may need to be introduced in the numerical description, depending on the nature of the local problem. Another significant aspect, is the requirement for an interpolation scheme, used to provide a mapping between the effective strain (input) and output (stress, stiffness) quantities.

In [210] a continuous interpolation is implemented in order to characterize the constitutive behaviour of macroscopically orthotropic composites, depicting a non-linear elastic response. The test points used to build the database are considered as nodes and finite element analysis shape functions are used to provide the interpolation. In [245] the effective strain-energy potential of non-linear elastic heterogeneous materials is determined in a number of points (nodes) discretizing the macroscopic strain space. Two approaches are then adopted for the interpolation of the effective quantities. The first approach uses multidimensional cubic splines and the second considers an outer product decomposition of multidimensional data into rank-one tensors, avoiding high-rank data. In [246] the response of heterogeneous hyperelastic materials is determined, by using a database describing the effective strain energy density function. To provide a continuous representation of the energy, an interpolation approach is adopted, such that the full database is reduced by a tensor product approximation. In both [245], [246], the effective strain-stress relation, as well as the tangent stiffness, are calculated in a numerically explicit way, using the strain energy density which is determined from the interpolation scheme.

An alternative data-driven scheme, which is proposed in [62] for the study of heterogeneous structures in the framework of multi-scale, computational homogenization ($FE^2$), is presented below analytically. According to this concept, two databases containing the effective response of the composite microstructure are created, using independent simulations on a selected RVE. The first database provides the effective strain-stress response and the second, the effective strain-tangent stiffness response. These databases are then incorporated in the overall, macroscopic numerical analysis and an interpolation method is used to provide the solution of the simulation. The scheme is developed for two-dimensional, plane stress elasticity, although extension to three-dimensional problems is also possible. The proposed approach, which is schematically shown in figure 8.1, can be applied for any microstructural pattern, such as masonry or composite materials with cracks and is formulated within a small displacements assumption. It is summarized by the following steps:

**1:** A finite element analysis model is developed for a non-linear RVE. Any constitutive description can be assigned to depict damage of the constituents, taken for instance from elastoplasticity or contact mechanics.

**2:** Within plane stress elasticity, a strain load tensor is applied to the boundaries of the RVE, using either prescribed (linear) displacements or periodic boundary conditions. Three independent parameters representing the components of the loading strain tensor are then introduced, for the implementation of any possible strain load case.

**3:** For every load level and direction, the RVE is solved, the effective strains and stresses are calculated and the corresponding database is built.

**4:** At every strain load level, three test incremental loading strain tensors are introduced and steps 2, 3 are repeated. Then, by developing the incremental strain-stress relation, the instantaneous, effective tangent tensor is obtained for each strain load level and the second database is built.

**5:** An overall, macroscopic multi-scale homogenization model is developed, which uses the two databases for the derivation of the constitutive description. Within this model, also called *metamodel*, the required constitutive quantities (stress and stiffness) are interpolated from the databases.

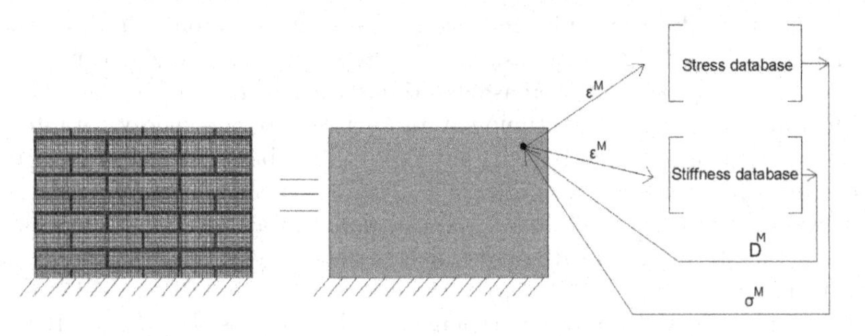


**FIGURE 8.1**: Schematic representation of the data-driven computational homogenization approach.

The prescribed displacements and the periodic boundary conditions, which are adopted in the framework of computational homogenization, have already been described in chapter 7. Since they are also adopted in the present formulation for the explicit loading of the RVE, are provided below for clarity. The prescribed displacement boundary conditions are expressed as given in equation (8.1):

$$\mathbf{u}|_{\partial V_m} = \boldsymbol{\varepsilon}^M \cdot \mathbf{x} \tag{8.1}$$

where $\boldsymbol{\varepsilon}^M$ is the strain load tensor, $\partial V_m$ indicates the boundaries of the RVE and $\mathbf{x}$ is a vector representing the undeformed coordinates of the boundary nodes. For a two-dimensional, plane stress problem, equation (8.1) is further elaborated as given in equations (8.2):

$$u_x = \varepsilon_{xx} x + 0.5\gamma_{xy} y \tag{8.2a}$$

$$u_y = \varepsilon_{yy} y + 0.5\gamma_{xy} x \tag{8.2b}$$

noticing that $u_x$, $u_y$ are the displacement loads along the $X$, $Y$ axes at the boundaries of the RVE and $\varepsilon_{xx}$, $\varepsilon_{yy}$, $\gamma_{xy}$ are the components of the loading strain tensor. For prescribed displacement boundary conditions, the displacement loading is applied at each boundary node of the RVE, with $x$, $y$ being the undeformed coordinates of these nodes.

Periodic boundary conditions require periodic displacements, as well as antiperiodic tractions, in the opposite boundaries of the RVE. The displacements on the opposite top ($T$) - bottom ($B$) and right ($R$) - left ($L$) boundaries of the RVE are determined using the prescribed displacements of three corner nodes on the boundaries, namely 1, 2 and 4, as follows:

$$\mathbf{u}^T - \mathbf{u}^B = \mathbf{u}^4 - \mathbf{u}^1 \tag{8.3a}$$

$$\mathbf{u}^R - \mathbf{u}^L = \mathbf{u}^2 - \mathbf{u}^1 \tag{8.3b}$$

By substituting relation (8.1) in the right-hand part of equations (8.3), it is derived that the periodic boundary conditions are provided by:

$$\mathbf{u}^T - \mathbf{u}^B = \boldsymbol{\varepsilon}^M \cdot (\mathbf{x}^4 - \mathbf{x}^1) \tag{8.4a}$$

$$\mathbf{u}^R - \mathbf{u}^L = \boldsymbol{\varepsilon}^M \cdot (\mathbf{x}^2 - \mathbf{x}^1) \tag{8.4b}$$

where $\mathbf{x}^1$, $\mathbf{x}^2$ and $\mathbf{x}^4$ are vectors denoting the undeformed coordinates of the corner nodes 1, 2 and 4, respectively.

For the simplest case of a single rectangular (four-node) element, with dimensions $L_x$, $L_y$ and origin of the coordinate system at the bottom-left node, the prescribed displacements at the four corner nodes 1 - 4 defined anti-clockwise, are obtained using relations (8.1) and (8.2) as follows:

$$u_x^1 = 0, \quad u_y^1 = 0 \tag{8.5a}$$

$$u_x^2 = \epsilon_{xx} L_x, \quad u_y^2 = 0.5\gamma_{xy} L_x \tag{8.5b}$$

$$u_x^3 = \epsilon_{xx} L_x + 0.5\gamma_{xy} L_y, \quad u_y^3 = \epsilon_{yy} L_y + 0.5\gamma_{xy} L_x \tag{8.5c}$$

$$u_x^4 = 0.5\gamma_{xy} L_y, \quad u_y^4 = \epsilon_{yy} L_y \tag{8.5d}$$

After the strain load is applied to the boundaries of the RVE, a numerical simulation for every strain load is conducted and the effective strain and stress tensors are determined, using the concepts presented in chapter 7. Briefly, the volume average strain of the microstructure is equal to the known, load strain tensor:

$$< \boldsymbol{\varepsilon} >_{V_m} = \boldsymbol{\varepsilon}^M \tag{8.6}$$

When prescribed displacements are considered, the following relation is used for the average stress [112], [143]:

$$< \boldsymbol{\sigma} >_{V_m} = \frac{1}{V_m} \mathbf{f}\mathbf{x} = \boldsymbol{\sigma}^M \tag{8.7}$$

where $\mathbf{f}$, $\mathbf{x}$ are matrices representing the reaction forces in the boundary nodes and the undeformed coordinates of these nodes, respectively.

In case periodic boundary conditions are considered, the following relation for the average stress is used [112], [143]:

$$< \boldsymbol{\sigma} >_{V_m} = \frac{1}{V_m} \mathbf{f}_p \mathbf{x}_p = \boldsymbol{\sigma}^M \tag{8.8}$$

where $\mathbf{f}_p$, $\mathbf{x}_p$ are matrices representing the reaction forces in the three corner nodes of the RVE with prescribed displacements and the undeformed coordinates of these nodes, respectively.

To obtain the stiffness of the overall macroscopic model, for every strain load case considered in each microscopic analysis, three test, incremental strain tensors are applied to the boundaries of the RVE as given in relation (8.9a). The solution of the RVE models loaded with the incremental strain tensors, leads to three incremental effective stress tensors, according to relation (8.9b). Then, the instantaneous, effective tangent tensor is estimated by using the system of equations given in (8.9c). This effective tangent tensor is used to build the tangent stiffness matrix in the overall macroscopic analysis.

$$[\delta \boldsymbol{\epsilon}^M] = [\delta \boldsymbol{\epsilon}_1^M \quad \delta \boldsymbol{\epsilon}_2^M \quad \delta \boldsymbol{\epsilon}_3^M] \tag{8.9a}$$

$$[\delta \boldsymbol{\sigma}^M] = [\delta \boldsymbol{\sigma}_1^M \quad \delta \boldsymbol{\sigma}_2^M \quad \delta \boldsymbol{\sigma}_3^M] \tag{8.9b}$$

$$[\delta \boldsymbol{\sigma}^M] = \mathbf{D}^M [\delta \boldsymbol{\epsilon}^M] \Rightarrow \mathbf{D}^M = [\delta \boldsymbol{\sigma}^M][\delta \boldsymbol{\epsilon}^M]^{-1} \tag{8.9c}$$

It is noticed that a vector notation is used in equations (8.9) for the incremental strain and stress tensors $\delta \boldsymbol{\epsilon}^M$, $\delta \boldsymbol{\sigma}^M$ and a matrix notation for the tangent tensor $\mathbf{D}^M$. In plane stress problems, $\mathbf{D}^M$ is a 3x3 matrix, while each of the $\delta \boldsymbol{\epsilon}_1^M$, $\delta \boldsymbol{\epsilon}_2^M$, $\delta \boldsymbol{\epsilon}_3^M$, $\delta \boldsymbol{\sigma}_1^M$, $\delta \boldsymbol{\sigma}_2^M$, $\delta \boldsymbol{\sigma}_3^M$ is a 3x1 vector. Thus, each of $\delta \boldsymbol{\epsilon}^M$ and $\delta \boldsymbol{\sigma}^M$ are 3x3 matrices, respectively.

The final step of this approach, is the implementation of the data-driven scheme in the overall, macroscopic simulation. Within the incremental-iterative Newton-Raphson process, equilibrium of the system according to the formulation presented in section 2.4.5 for material non-linearity with small displacements assumption, is expressed as follows:

$$\mathbf{K}_T \Delta \mathbf{U} = \mathbf{F}_{ext} - \mathbf{F}_{int} \tag{8.10}$$

where, by substituting the tangent stiffness matrix $\mathbf{K}_T$ as well as the external and internal force vectors $\mathbf{F}_{ext}$ and $\mathbf{F}_{int}$, it becomes:

$$\begin{gathered} \left( \sum_{i=1}^{N_e} \int_{V_e} \mathbf{B}^T \mathbf{D}^M \mathbf{B} dV \right) \Delta \mathbf{U} = \\ \sum_{i=1}^{N_e} \int_{V_e} \rho \mathbf{N}^T \mathbf{f} dV + \sum_{i=1}^{N_e} \int_{S_e} \mathbf{N}^T \mathbf{t} dS - \sum_{i=1}^{N_e} \int_{V_e} \mathbf{B}^T \boldsymbol{\sigma}^M dV \end{gathered} \tag{8.11}$$

To build the tangent stiffness matrix $\mathbf{K}_T$, the tangent stiffness tensor $\mathbf{D}^M$ is calculated for every Gauss point of the macroscopic structure, using the stiffness database. The internal force vector is determined for every Gauss point, after the effective macroscopic stress tensor $\boldsymbol{\sigma}^M$ is calculated using the stress database.

To obtain these quantities from the databases, an interpolation technique should be used. According to the approach which is proposed in [62], a scattered data set defined by locations $X$ and corresponding values $Y$ is interpolated using a Delaunay triangulation of $X$. Then, a surface which has the form $Y = f(X)$ is built. The interpolation function $f$ is used for the evaluation of the surface at any query location $qX$, using $qY = f(qX)$, where $qX$ lies within the convex hull of $X$. For the implementation of the mentioned interpolation a Matlab toolbox is used [136].

Contrary to the concept presented in [245], [246], where the constitutive description is built using the strain energy which is included in the database and determined from an interpolation scheme, the approach which is developed above determines the constitutive response using a direct interpolation on stress and stiffness databases.

It is also mentioned that according to the formulation which is developed in this section, only the averaging relations in the microscopic scale are used for the estimation of the necessary information for the final metamodel (stress and stiffness). Thus, no manipulation of the final tangent stiffness matrix of the RVE is needed for the representation of the effective stress-strain law, contrary to classical computational homogenization [112]. This aspect, makes the implementation of the proposed scheme possible even by using a commercial software package, provided that volume average quantities can be calculated and a simple finite element metamodel incorporating the relation (8.9c) is developed.

---

## 8.3 Data-driven analysis incorporating machine learning

An approach which is widely used recently for the implementation of data-driven analysis in computational mechanics, introduces *machine learning* elements in the numerical code, to provide the constitutive description. According to a general definition, machine learning is a numerical scheme, which uses an optimization algorithm for the determination of the parameters of a function, that minimizes the approximation error over a set of values [36].

A simplified example of a unidirectional structural element can be used to provide a further insight on machine learning methodology. For a rod of a truss structure, it is assumed that experimental investigation provides the strain versus stress response, in the form of strain-stress pairs, included in a database $A$. Thus, $A = (\varepsilon_i, \sigma_i)$ contains the known, strain-stress pairs for the rod. For displacement controlled experiments, every strain value $\varepsilon_i$ is the input and every stress $\sigma_i$ is the corresponding output on the database $A$. A machine learning algorithm introduces some functions $f$, for instance polynomial functions of a given order, which are used to predict the output stress values $\sigma$. According to the process which is followed, an optimization algorithm is adopted, to determine the parameters of the functions $f$, such that the distance between the output values $\sigma$ from the database $A$ and the predicted output values using the functions $f$, thus the error, is minimized. For the case of polynomial functions $f$, the parameters determined from the optimization process are the coefficients of the polynomials.

The process of the error minimization and the determination of the parameters of the functions $f$ is called *training*. Once the training is complete, the machine learning algorithm incorporating the known parameters of functions $f$, can be used for the calculation of the output stress values, for any, arbitrarily chosen, input values of strains.

Several machine learning algorithms have been developed, relying on different numerical formulations and using different error functions and optimization schemes. One of the simplest machine learning schemes, is summarized by the concept of *linear regression*. Linear regression is the process of determining the coefficients of a linear, first-order polynomial function, such that the error distance between the discrete output values from a database and the predicted values is minimized. This problem is solved by adopting the *least squares* fitting scheme presented in section 8.3.1.1, according to which the coefficients of the linear polynomial are calculated by minimizing the sum of the squares of the offsets, also called *residuals*, of the database points from the polynomial prediction. For the solution of the least squares minimization problem, a closed form relation is provided. The method can also be used with non-linear functions, such as higher-order polynomials, by adopting an iterative scheme applied to a linearized form of the functions, using linear least squares fitting, until convergence is achieved.

More machine learning methods have been developed, such as *artificial neural networks*, *ridge regression*, *lasso regression*, *support vector machines*, *k-nearest neighbors* and others, as well as combinations of them. Emphasis in this section is given to artificial neural networks, since this approach is widely used for the determination of the structural response of composite materials, within data-driven, computational mechanics applications.

### 8.3.1 Soft computing, artificial intelligence and neural networks

The success of neural networks in machine learning tasks, from classical approximation to more complicated like computer vision, signal and speech processing, attracted the interest of computational mechanics community. Recent developments like the availability of specialized hardware designed to efficiently execute neural networks, like the Tensor Processing Units of Google and the Neural Engine of Apple, and the fusion of hardware and software leading to the notion of digital twins, accelerate this process. The emergence of scientific machine learning and embedded systems requires effective combination of results from various research fields. A short introduction on machine learning, emphasizing on neural network applications relevant to the topic of the present book, is given here.

#### 8.3.1.1 Feed-forward neural networks

In order to give an example of a neural network, it is considered the most common and simplest neural network which is called *feed-forward neural network* or *multi-layer perceptron* or *multi-layer feed-forward neural network.* Every element of a neural network which holds an input is called "neuron" or "node", noticing that the basic element of a single node is known as *single perceptron.* The information is forwarded from the previous nodes, multiplied by suitable synaptic weights $w_{ij}$ to the node at hand. After addition of a bias $b$, the information is locally processed by a non-linear *activation function* $\sigma$. Finally, each perceptron realizes an affine linear transformation $W^l\mathrm{z} + b^l$, which is followed by a non-linear activation function $\sigma^l$. The same pattern is followed, as it is indicated in the multi-layer feed-forward network of figure 8.2 (section 8.3.2.1), such that at the end for each input $z$ the output is written as:

$$y(z) = \mathbf{W}^L\sigma^L[\mathbf{W}^{L-1}\sigma^{L-1}(\ldots\sigma^1(\mathbf{W}^0\mathbf{z} + \mathbf{b}^0)) + \mathbf{b}^{L-1}] + \mathbf{b}^L \tag{8.12}$$

where $\mathbf{W}^l$ and $\mathbf{b}^l$ are weight matrices and bias vectors of layer $l$ and $z$ is the input vector to the neural network. The most common activation function is the *sigmoidal* one, as provided by:

$$\sigma(x) = \frac{1}{1 + e^{-x}} \tag{8.13}$$

Other commonly used activation functions are the hyperbolic tangent *tanh* and the rectified linear unit $ReLU = max\{x, 0\}$.

Usage of a suitable topology, i.e. number of layers and nodes within each layer, as well as suitable values for the weights and the biases, allows a feed-forward neural network with the highly non-linear structure of the approximation function given in (8.12), to approximate any continuous mapping from the input to the output space. These theoretical results led to practical applications after the emergence of the *backpropagation learning algorithm* which made possible the training of neural networks in order to approximate a given mapping by using examples. Available couples of input-output values, the examples, are used in order to adjust the values of the parameters until convergence to the required accuracy is achieved. This is done by adjusting the variables layerwise, through backpropagation of the error between predicted and known responses.

Backpropagation is an iterative, distributed algorithm adopted for the solution of the training process, used to solve a large scale optimization problem.

It can be solved using *gradient-based optimization* such as *stochastic gradient descent*, or the *Adam optimizer*, a variant of stochastic gradient descent based on an adaptive estimation of first-order and second-order derivatives, adopted to improve the speed of convergence.

Next, it is assumed a random set of known data points, which are realizations of the unknown mapping to be approximated, i.e. $M$ couples of points $z_1, z_2, \ldots, z_M$ and the corresponding outputs $f(z_1), f(z_2), \ldots, f(z_M)$. Provided all $M$ examples are used for training, the parameters (weight and bias) of the neural network $\mathbf{W}$, $\mathbf{b}$ can be identified by minimizing the least squares error function:

$$E = \left( \frac{1}{M} \sum_{j=1}^{M} \left( y(z^j) - f(z^j) \right)^2 \right)^{\frac{1}{2}} \tag{8.14}$$

with $y$ being the prediction for the outputs, which has been defined in equation (8.12).

The flexibility of neural network approximations can be reduced due to the effort required to define an effective neural network architecture, to choose appropriate training sets and accomplish the training. Unfortunately, the size and topology of the network cannot be determined by using a specified theory, indicating that some experimentation is required. Similarly, the preparation of the training data as well as the effectiveness of the training process, depend on user's experience. Some attempts have been made in order to tailor the neural network for every specific problem by using *global optimization* or *genetic programming*. These techniques may be useful for given tasks.

#### 8.3.1.2 Deep learning

Recently, the extension to more complicated networks, the so-called *deep neural networks* appeared, as a universal technique for function approximation and image correlation. This development comes at the price of a large number of parameters to be determined in the supervised learning phase, accompanied by the demand for a large volume of training data. In fact, in a fully connected feed-forward neural network, the determination of all involved parameters requires the availability of proportional number of examples for training. In addition, very large fully connected feed-forward neural networks may lead to numerical difficulties during training.

Deep learning imitates biological networks, where some processing is realized between the layers while many connections are not used. Omitting certain links between nodes leads to compression of the transmitted information and gives to the network deep learning properties. These tools are currently under development, noticing that many software packages have already been released to provide computer implementation.

*Convolutional neural networks* have also been proposed and used, for multidimensional data processing. The most common applications are picture and video processing, scene recognition and feature extraction. In view of possible applications in the field of material science and computational mechanics, correlation between two- and three-dimensional images with microstructural information or composite structures layout and the corresponding mechanical behaviour, is possible. This task can, in principle, be accomplished by means of general-purpose image processing tools, as it has been shown in applications of convolutional, deep neural networks. In particular, one more pre-processing step may be added, using a graph neural network that introduces graph information from the microstructure. For instance, the loading paths created due to contact interactions in a granular medium can be seen as graphs within this context. Another possible application could be the graphs created from topology optimization by following a ground structure approach.

#### 8.3.1.3 Available tools

The procedure of developing integrated systems with artificial intelligence components, mainly trained deep neural networks, becomes an emerging field in the industry. Creating and teaching a computer vision tool can be completed off-line. The trained neural network needs to be deployed at an onboard computer in order to work, with restricted resources, energy consumption and real-time performance (edge computing). For the time being these steps are being followed for the development of electronic components to assist autonomous driving. Even the preparation of pre-trained neural networks, which can be a starting point for further, maybe online fine tuning, constitutes a meaningful step within this development.

From the ability to integrate artificial intelligence constitutive models within finite element computations, as is has been demonstrated in the present book, one may predict that a similar development will be followed in this area as well. Suitably trained neural networks will exploit the *big data* which is available from experiments and will be embedded to large-scale computations. Software and hardware will work together, as it is already the case with experimental devices interconnected and controlled by computational algorithms.

### 8.3.2 Machine learning applications in engineering mechanics

Neural networks and soft computing tools have been applied for the approximation of *direct* and *inverse* problems in mechanics at both material and structural level. According to some early attempts, the usage of *Hoopfield* neural networks with internal energy which is minimized during unsupervised training, is proposed for the solution of structural analysis problems including cracks, contact and elastoplasticity [111], [215], [216].

Neural networks have also been used for the modelling of material constitutive laws in composites [171], the description of failure surfaces in masonry [174] and the vulnerability study of masonry structures under seismic actions [13]. Machine learning concepts have been applied for the analysis of reinforced concrete structural elements in [41] and for the preliminary investigation of multi-storey frame structures in [21]. They have also been used for post-processing of experimental measurements, like the determination of elastoplasticity laws from indentation tests in [93], [94]. These concepts can be extended to cover time-dependent relations, using a moving window technique [102].

Moreover, neural networks are able to approximate, directly, the relation between assumed damage parameters and mechanical responses of a structure. Therefore, they have been used for the solution of inverse and identification problems in mechanics. Parameter identification problems have been studied by using backpropagation neural networks for elastoplasticity problems in [1], [214] and for structural elements like semi-rigid connections of steel structures in [203]. Results of crack identification problems have been published in [198], [200] and of damage detection in curtain-wall facades in [68].

Further reviews and applications of neural networks in structural analysis and computational mechanics can be found in [29], [82], [199], [219], [226] and in the monograph [238].

Within non-linear finite element analysis, the response of the material which is traditionally specified by a mathematical, mechanical constitutive model, can alternatively be described by a suitably trained neural network. Early implementations of this concept as well as technical details are found in [88], [233]. These works can be considered to be the first studies of data-driven finite element analysis. In particular, a finite element scheme can be created, wherein the constitutive equations are replaced by an artificial neural network. This neural network is trained with data about stresses, strains, strain rates and hardening or other internal variables depending on the adopted model, in a range of values which are expected to appear in the finite element simulations [97], [168], [205]. The same technique can be used at every level of engineering mechanics models (rods, beams, shells) as well as in three-dimensional elasticity.

Machine learning concepts can also be adopted in multi-scale analysis and especially within the computational homogenization or FE$^2$ method presented in chapter 7 of this book, in the sense that neural networks are integrated as an estimator of the result expected by a local, microscopic, RVE analysis [127], [220]. In particular, *data-driven computational homogenization* has demonstrated the potential to drastically reduce the computational time comparing to traditional multi-scale FE$^2$ computations. Simplified formulations of this concept, using directly an one-dimensional experimentally obtained stress-strain curve for the extraction of the required material data, have also been proposed and tested on elastoplasticity problems in [207]. Alternatively, evolutionary polynomial regression for *on the fly* tunning of material parameters within the finite element method has been proposed in [70]. Finally, multiple metamodels can be introduced and used, selectively, depending on the requirements of each point at every loading stage [77].

It should be noted that although it is usually believed that machine learning is restricted to interpolation and does not perform well in extrapolation tasks, there have been studies that demonstrate unexpected performance for predicting the behaviour of extraordinary materials, i.e. for predicting the response of materials with properties that extend beyond previously seen values [106]. Relevant selected applications and proposals for future work are briefly discussed in the next sections.

#### 8.3.2.1 Constitutive material modelling

Within engineering mechanics, the relation between stresses and strains, possibly including information from previous loading path or time history, is required. The classical approach is based on modelling of this constitutive relation, by adopting principles of elasticity, plasticity, or viscoelasticity, among others. The raw material, which is available in the form of experimentally measured data, is used in order to approximate the parameters of each constitutive law.

Using an analytical constitutive relation has certainly many advantages. For example, symmetry conditions or thermodynamic requirements leading to convexity requirements for the potential energy are, in most cases, easily enforced. On the other hand, this approach poses the question of the choice of a suitable model, which is not easily answered in some cases. An alternative approach is the usage of raw data for the training of a neural network constitutive *metamodel*, which in turn can be combined with the other elements of engineering mechanics like equilibrium and compatibility equations, to form the complete mathematical description of the problem. The question of selecting appropriate inputs and outputs still remains, on top of the choice of proper parameters for the neural network. Figures 8.2, 8.3 and 8.4 provide schematic representations of suitable approximations of stress-strain laws, using neural networks as metamodels.

In a more general setting, machine learning and hybrid techniques constitute powerful approaches among data-driven methods. Neural network based metamodels can be combined with computational homogenization within finite element analysis, for the realization of efficient multi-scale schemes, as it has been discussed in this book. Constitutive modelling has also been studied by means of neural networks in [88], [95], [225]. Further information can be found in edited volumes and review articles, for instance in [29], [61], [178].

Computational homogenization can be considered as an extension of a constitutive approximation at the level of a Representative Volume Element. Machine learning based on experimentally or numerically generated data can be used in order to predict the effective material properties in linear and non-linear cases, as presented in [28], [38], [122], [185], [241]. Early attempts rely on correlation of unstructured data by using backpropagation neural networks. Convolutional neural networks (CNNs), which take images of microstructures as input, have been adopted in recent studies to construct microstructure-property linkages and to predict the macroscopic properties. Artificial neural networks (ANNs) and deep neural networks (DNNs), which are trained to construct complex non-linear relations between predefined features (e.g. strain components, volume fractions) and mechanical response data (e.g. average stress or elastic modulus), have been coupled with finite element simulations to accelerate multi-scale schemes for bone remodelling [85], non-linear elastic composites [117] and other microstructures [120].

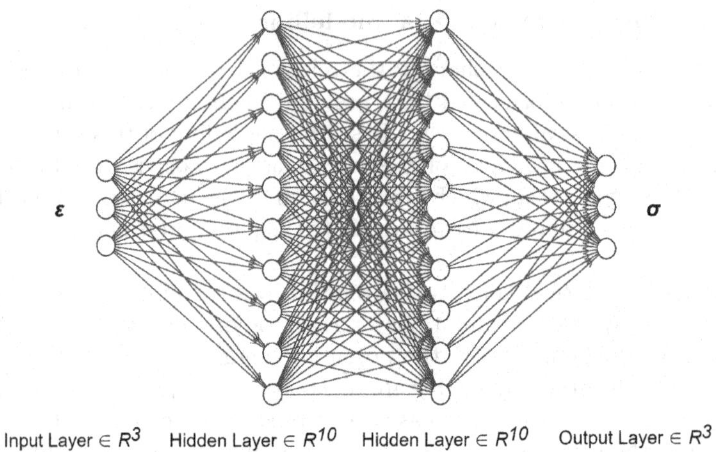


**FIGURE 8.2**: Learning a strain-stress relation in two-dimensional elasticity by using a feed-forward neural network.

#### 8.3.2.2 Consideration of thermodynamic restrictions

The blind approximation of a constitutive relation from a neural network has a main drawback, which is attributed to the fact that the predictions do not necessarily obey the laws of physics. Hence, the predictions may become physically inconsistent, with obvious negative consequences on the convergence of every computational mechanics scheme based on them and the quality of the results.

A thermodynamics-based neural network for the approximation of strain rate independent processes has been proposed in [133]. In this work, the two basic principles of thermodynamics are encoded in the network's architecture by taking advantage of automatic differentiation to compute the numerical derivatives of the network with respect to its inputs. In this way, derivatives of the free-energy, the dissipation rate and their relation with the stress and internal state variables are hardwired in the architecture of the neural network. In the same direction, a new neural-network architecture, called the *Cholesky-factored symmetric positive definite neural network*, has been proposed in [235] for modelling constitutive relations in computational mechanics. In this case,

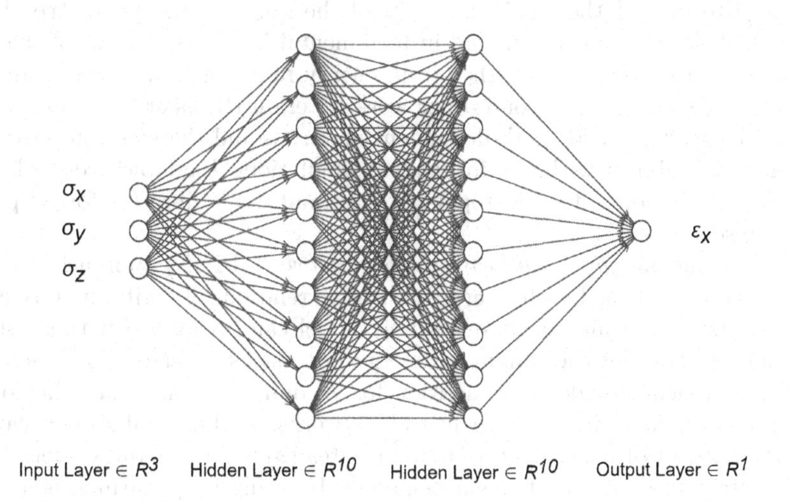


**FIGURE 8.3**: Learning a strain-stress relation in two-dimensional elasticity by using a feed-forward neural network that separately approximates each element of the output tensor.

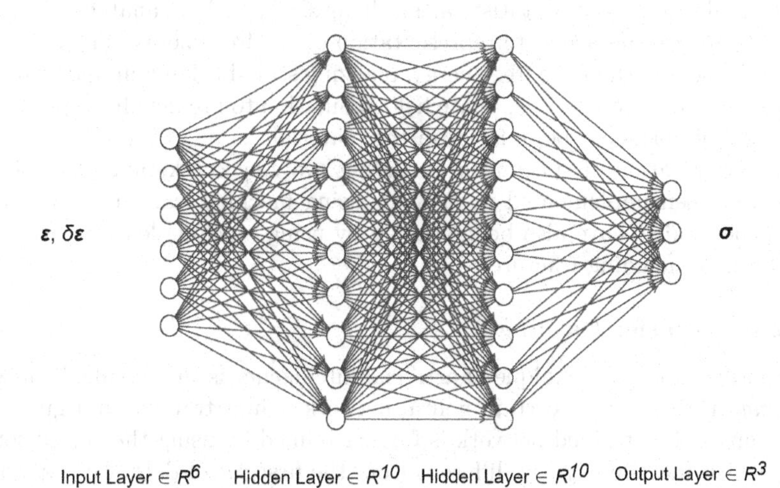


**FIGURE 8.4**: Learning a strain rate-stress relation in two-dimensional elasticity by using a feed-forward neural network.

instead of directly predicting the stress of the material, the neural network is trained to predict the Cholesky factor of the tangent stiffness matrix, based on which the stress is calculated in incremental form. As a result of this special structure, convexity on the strain energy function is imposed, leading to the satisfaction of the second-order work criterion (Hill's criterion) as well as to time consistency for path-dependent materials and therefore, improves the numerical stability, within finite element simulations. Examples from solid mechanics on hyperelastic, elastoplastic, and multi-scale fibre reinforced plates, are presented.

By using the *physics-informed neural network* (PINN) approach, which is introduced later in this chapter, constitutive relations of digital materials can be constructed. This machine learning model can be trained without supervision by adopting the minimum energy criteria as its error (loss) function. This approach provides a foundation for encoding the physical behaviour of digital materials directly into neural networks, enabling label-free learning for the design of next-generation composites [249]. In general, correlation of microstructure with mechanical behaviour by using deep learning is a newly proposed tool in material design and topology optimization [242].

#### 8.3.2.3 Multi-scale modelling and scale-bridging

When dealing with composites, the notion of free energy is used in many homogenized material models. Although free energy data is not generally found directly, its derivatives can be observed or calculated. In [209], an *Integrable Deep Neural Network*, (IDNN) has been trained by data obtained from atomic scale models and statistical mechanics. Then, it is analytically integrated to recover an accurate representation of the free energy. The resulting DNN representation of the free energy, realizing scale-bridging in material systems, is used in a mesoscopic, phase field simulation to predict the appropriate formation of antiphase boundaries in the material.

Homogenization of specimens, including evolving cracks and microstructures, has been implemented in [248] by using machine learning techniques. Mesoscale models have also been studied by using machine learning in [167], for continua including fracture.

#### 8.3.2.4 Transfer learning

Another modern technique of artificial intelligence is the so-called *transfer learning*. In this case, a certain neural network is first trained on a given set of examples. The trained network is further refined by using the actual set of data, which may be slightly different from the first set of data. For example, the first round of training can be based on numerically generated data, while fine tuning can rely on additional experimental measurements.

A novel multi-fidelity physics-informed deep neural network is presented in [39]. The framework proposed is particularly suitable when the physics of the problem is known in an approximate sense (low-fidelity physics) and only a few high-fidelity data are available. The machine learning system blends physics-informed and data-driven deep learning techniques by using the concept of transfer learning. The approximate governing equation is first used to train a low-fidelity physics informed deep neural network. This is followed by transfer learning where the low-fidelity model is updated by using the available high-fidelity data. This approach is able to encode useful information on the physics of the problem from the approximate governing differential equation and hence, provides accurate prediction even in zones with no data.

An interesting application of transfer learning, by using bidirectional recurrent neural networks on a finite element analysis problem, has been reported in [101]. A recently proposed hybrid methodology, combines classical constitutive laws (model-based), with a data-driven correction component, within computational multi-scale analysis. A model-based material representation is locally improved with data from lower scales, obtained by means of a non-linear numerical homogenization procedure, leading to a model-data-driven approach. Therefore, macro scale simulations explicitly incorporate the true micro scale response, maintaining the same level of accuracy that would be obtained with online micro-macro simulations but with a computational cost comparable to classical model-driven approaches [78], [81], [95]. This technique can be compared with transfer learning, where the model-based prediction constitutes the first approximation of the response, which is further enhanced with the usage of data-based tools. A discussion providing parallel developments in the field of digital twin can be found in [42].

#### 8.3.2.5 Postprocessing and zooming

Within multi-scale modelling of composite materials, an important step is the estimation of local fields from the solution of the homogeneous continuum. A neural network solution to this *zooming analysis problem* has been proposed in [239] for the simulation of carbon fibre reinforced composite materials (CFRP). Zooming is used to simplify a finite element model by dividing it into global coarse meshes and local fine meshes. When traditional shape functions are adopted for the approximation of the displacement field, displacements of boundary nodes of the local model cannot be obtained accurately, if the nodes are outside the global model. Hence, the zooming method is performed by a neural network that learns the relationship between nodal coordinates and nodal displacements of a global model. Then, the boundary displacements of a local model are obtained using the trained neural network.

#### 8.3.2.6 Physics-informed neural networks

The previous descriptions mainly focus on *supervised* neural networks, which require existence of input-output data for training. Although models of unsupervised neural networks have also been proposed, their applicability on the solution of mechanical problems is still restricted. Another proposal, which allows for the automatic generation of training data using the governing equations, relies on the usage of the so-called *physics-informed neural networks* (PINNs). In this case, the governing partial differential equations are used, together with the approximation of necessary derivatives from the same neural network, in order to define the error required for training. A schematic representation of the method is provided in figure 8.5.

Physics-informed neural networks constitute a scientific machine learning technique for solving partial differential equation (PDE) problems, since they use only the information provided by the PDEs, rather than a large number of pairs of the input and output variables. Therefore, they generate and approximate solutions to PDEs by training a neural network, in order to minimize a loss function, consisting of terms representing the misfit of the predicted and wished governing equations at each point of the domain as well as the misfit of the boundary conditions along the boundary.

Physics-informed neural networks have been proposed in the pioneer work of Lagaris and co-workers [113], [114]. The development of computer tools and hardware made them popular in the last years, as a promising alternative of solving complicated direct and inverse problems related to ordinary and partial differential equations, as well as initial and boundary value problems [86], [175]. They remind the *mesh-free* collocation methods and the least square finite elements of computational mechanics. The development of various applications of PINNs, is supported by software which is used to implement automatic differentiation of neural networks.

Within the context of multi-scale modelling, certain steps related to the evaluation at the RVE level could be realized with PINNs. For example, the response of the lower-level, RVE boundary value problem could be derived from a physics-informed neural network, used to determine the homogenized coefficients of the continuum [249].

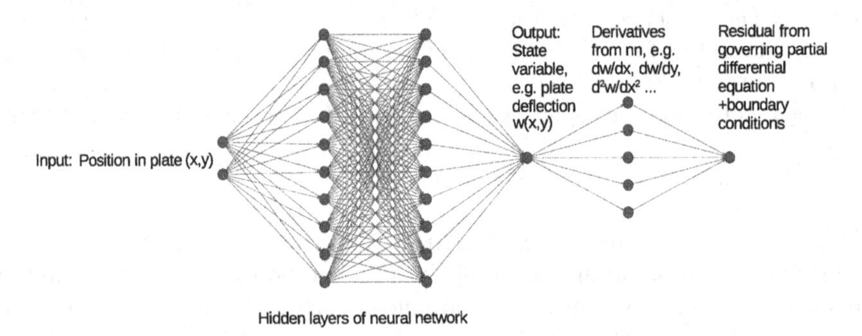


**FIGURE 8.5**: Physics-informed neural network for the approximation of a plate bending problem [151].

#### 8.3.2.7 Integration of machine learning tools in multi-scale $FE^2$-like procedures

The original multi-scale computational homogenization or $FE^2$ approach which is presented in chapter 7, requires the evaluation of a local, RVE model with all possible non-linearities and microstructural information at each Gauss integration point of the upper-level, homogenized continuum, in order to calculate the mechanical response at every given loading step and iteration. This procedure does not require an initial guess for the constitutive response of the macroscopic homogenized continuum, since this is being built gradually during the iterations of the algorithm. On the other hand, it is clear that the original implementation of the $FE^2$ is very expensive computationally, even if classical tools such as parallelization are employed.

Correlation of micromechanics with homogenized material properties can be based on classical approaches, which are described in various papers, for instance in [211] or in monographs [34], [251]. Usage of analytical or second-order asymptotic homogenization may increase the effectiveness of the procedure [217]. Refinement of the technical implementation has also brought to this approach a certain maturity [26], [115]. Further information can be found in the recent review paper [176].

The replacement of the lower-level, local finite element solution with some approximation leads to considerable reduction of the computational cost. This approximation of the mechanical behaviour on the fine scale, can be based on numerical or experimental results of microstructures which cover the whole space of interest, with respect to parameter and load variation, followed by subsequent interpolation, using classical or machine learning schemes.

Data-driven techniques relying on numerical polynomial interpolation of databases and machine learning tools, are implemented within Matlab in [62], [64]. Usage of these tools is proved to be very helpful in the case of RVEs with contact interactions, where the differentiability of the solution cannot be guaranteed [62]. Neural network and machine learning interpolation, based on the data of repeated RVE solutions, have also been presented in [2], [117], [190].

It should be noted that hybrid model-supported, data-driven techniques can be adopted. Further information is found in [137], [183], [244], while a comparison of various approaches is presented in [179].

#### 8.3.2.8 Emerging fields

As it has previously been mentioned, usage of deep learning such as convolutional neural networks, can be adopted in order to correlate the picture of the microstructure, including possible cracks and other defects, with the mechanical properties.

A challenging task relies on the extension of constitutive material and multi-scale approximation methods, towards dynamic actions. First, in frequency domain, material properties are frequency-dependent, which makes the procedure iterative. Furthermore, wave propagation effects interact with the microstructure and lead to *band-gaps*. Recent attempts to study these effects can be found in [58], [123], [196], [222], [250].

From the practical point of view, modern Python-driven scripts enable the implementation of multi-scale and machine learning tools, with commercial finite element packages (e.g. [243]).

---

## 8.4 Data-driven analysis based on the minimum distance among points of the data set

According to a concept recently presented in [108], a data-driven analysis scheme can be formulated, by considering the distance minimization problem between every value of a data set, representing the constitutive description of a structure obtained from physical or numerical experiments and a corresponding value which satisfies equilibrium and compatibility. The data set may describe a non-linear material law in a structural length scale. For instance, the non-linear stress versus strain response of a truss element is adopted in [108], with pairs of stress-strain values being included in the data set. The scheme can also be used to describe the non-linear response of a composite material, in the framework of multi-scale computational homogenization [236]. The data set in this case includes pairs of the average macroscopic stress versus average strain tensor, which characterize the response of the composite material on the macroscopic length scale. This data is derived from experiments or numerical simulations on the microstructural scale of the material. The formulation which is presented below applies to both cases.

The first step of this data-driven approach, is the consideration of a penalty function $F$, used to define the distance minimization problem between all pairs of the strain and stress tensors $(\boldsymbol{\varepsilon}', \boldsymbol{\sigma}')$ in the data set and the pair which satisfies compatibility and equilibrium, $(\boldsymbol{\varepsilon}, \boldsymbol{\sigma})$. Using tensor notation, this penalty function can be expressed as:

$$F(\boldsymbol{\varepsilon}, \boldsymbol{\sigma}) = \min_{(\boldsymbol{\varepsilon}', \boldsymbol{\sigma}') \in D} \frac{1}{2} \int_{\Omega} \left( (\boldsymbol{\varepsilon} - \boldsymbol{\varepsilon}') : \mathbf{C} : (\boldsymbol{\varepsilon} - \boldsymbol{\varepsilon}') + (\boldsymbol{\sigma} - \boldsymbol{\sigma}') : \mathbf{C}^{-1} : (\boldsymbol{\sigma} - \boldsymbol{\sigma}') \right) d\Omega \tag{8.15}$$

where $D$ represents the data set and $\mathbf{C}$ is a known numerical tensor, chosen only to build the formulation and thus, without any physical meaning. Equation (8.15) can be rewritten using matrix and vector notation as follows:

$$F(\boldsymbol{\varepsilon}, \boldsymbol{\sigma}) = \min_{(\boldsymbol{\varepsilon}', \boldsymbol{\sigma}') \in D} \frac{1}{2} \int_{\Omega} \left( (\boldsymbol{\varepsilon} - \boldsymbol{\varepsilon}')^T \mathbf{C} (\boldsymbol{\varepsilon} - \boldsymbol{\varepsilon}') + (\boldsymbol{\sigma} - \boldsymbol{\sigma}')^T \mathbf{C}^{-1} (\boldsymbol{\sigma} - \boldsymbol{\sigma}') \right) d\Omega \tag{8.16}$$

Each of the terms on the right-hand side of both equations resembles the strain energy of the system [169]. Still, these terms are introduced only to define the minimum distance problem and hence, they are not related to any material property. To formulate the overall optimization problem, equilibrium and compatibility conditions will be expressed as constraints. The principle of virtual work given in relation (2.21) is rewritten here using tensor notation, by neglecting body forces and considering small displacement analysis:

$$\int_{\Omega} \boldsymbol{\sigma} : \delta\boldsymbol{\varepsilon} d\Omega = \int_{S} \mathbf{t} \cdot \delta\mathbf{u} dS \tag{8.17}$$

The strain-nodal displacement relation, which expresses the compatibility condition in the framework of finite element analysis, is provided below:

$$\boldsymbol{\varepsilon} = \mathbf{B} \cdot \mathbf{u} \Rightarrow \delta\boldsymbol{\varepsilon} = \mathbf{B} \cdot \delta\mathbf{u} \tag{8.18}$$

By substituting equation (8.18) in (8.17), it is obtained:

$$\int_{\Omega} \boldsymbol{\sigma} : \mathbf{B} d\Omega = \int_{S} \mathbf{t} dS \Rightarrow \int_{\Omega} \boldsymbol{\sigma} : \mathbf{B} d\Omega - \int_{S} \mathbf{t} dS = 0 \tag{8.19}$$

The overall optimization problem can now be formulated, by adopting the minimization of the objective function (8.15), the equilibrium constraint (8.19) and the compatibility constraint (8.18), as follows:

$$\begin{aligned} &\text{minimize}\, F \,\text{subject to} \\ &\int_{\Omega} \boldsymbol{\sigma} : \mathbf{B} d\Omega - \int_{S} \mathbf{t} dS = 0 \\ &\boldsymbol{\varepsilon} = \mathbf{B} \cdot \mathbf{u} \end{aligned} \tag{8.20}$$

For the consideration of the equilibrium constraint, the method of Lagrange multipliers is used. In particular, the Lagrange multipliers $\boldsymbol{\lambda}$ are introduced and the functional $\Pi$ which is shown below is built:

$$\Pi = F(\boldsymbol{\varepsilon}, \boldsymbol{\sigma}) - \boldsymbol{\lambda} \left( \int_{\Omega} \boldsymbol{\sigma} : \mathbf{B} d\Omega - \int_{S} \mathbf{t} dS \right) \tag{8.21}$$

The optimization problem is then solved by considering the following stationary problem:

$$\delta\Pi = 0 \Rightarrow \delta \left( F(\boldsymbol{\varepsilon}, \boldsymbol{\sigma}) - \boldsymbol{\lambda} \left( \int_{\Omega} \boldsymbol{\sigma} : \mathbf{B} d\Omega - \int_{S} \mathbf{t} dS \right) \right) = 0 \tag{8.22}$$

The compatibility condition expressed by relation (8.18) is substituted in equation (8.22) and the variations in terms of $\mathbf{u}$, $\boldsymbol{\sigma}$ and $\boldsymbol{\lambda}$ are considered, leading to the following equations:

$$\delta\mathbf{u} \rightarrow \int_{\Omega} \mathbf{C} : (\mathbf{B} \cdot \mathbf{u} - \boldsymbol{\varepsilon}^*) : \mathbf{B} d\Omega = 0 \tag{8.23a}$$

$$\delta\boldsymbol{\sigma} \rightarrow \boldsymbol{\sigma} - \boldsymbol{\sigma}^* = \mathbf{C} : \mathbf{B} \cdot \boldsymbol{\lambda} \tag{8.23b}$$

$$\delta\boldsymbol{\lambda} \rightarrow \int_{\Omega} \boldsymbol{\sigma} : \mathbf{B} d\Omega - \int_{S} \mathbf{t} dS = 0 \tag{8.23c}$$

noticing that equation (8.23b) can be rewritten in matrix-vector notation as follows:

$$\boldsymbol{\sigma} = \boldsymbol{\sigma}^* + \mathbf{CB}\boldsymbol{\lambda} \tag{8.24}$$

In equations (8.23), $(\boldsymbol{\varepsilon}^*, \boldsymbol{\sigma}^*)$ are the optimal data points which minimize the function $F$, among all data points $(\boldsymbol{\varepsilon}', \boldsymbol{\sigma}')$ of the data set $D$. Thus, for these optimal data points, the following condition is satisfied:

$$\begin{aligned} &\int_\Omega \left((\boldsymbol{\varepsilon} - \boldsymbol{\varepsilon}^*) : \mathbf{C} : (\boldsymbol{\varepsilon} - \boldsymbol{\varepsilon}^*) + (\boldsymbol{\sigma} - \boldsymbol{\sigma}^*) : \mathbf{C}^{-1} : (\boldsymbol{\sigma} - \boldsymbol{\sigma}^*)\right) d\Omega \leq \\ &\int_\Omega \left((\boldsymbol{\varepsilon} - \boldsymbol{\varepsilon}') : \mathbf{C} : (\boldsymbol{\varepsilon} - \boldsymbol{\varepsilon}') + (\boldsymbol{\sigma} - \boldsymbol{\sigma}') : \mathbf{C}^{-1} : (\boldsymbol{\sigma} - \boldsymbol{\sigma}')\right) d\Omega \end{aligned} \tag{8.25}$$

When all optimal data points are determined for the structure under consideration, equations (8.23) lead to the system (8.26) of linear equations which is shown below. In particular, equation (8.23a) leads to (8.26a) and substitution of $\boldsymbol{\sigma}$ from (8.23b) to (8.23c) leads to equation (8.26b).

$$\int_\Omega \mathbf{B} : \mathbf{C} : \mathbf{B} \cdot \mathbf{u} d\Omega = \int_\Omega \mathbf{C} : \boldsymbol{\varepsilon}^* : \mathbf{B} d\Omega \tag{8.26a}$$

$$\int_\Omega \mathbf{B} : \mathbf{C} : \mathbf{B} \cdot \boldsymbol{\lambda} d\Omega = \int_S \mathbf{t} dS - \int_\Omega \boldsymbol{\sigma}^* : \mathbf{B} d\Omega \tag{8.26b}$$

Equations (8.26) are used for the solution of the described data-driven simulation. Thus, the first equation is used to provide the solution of the nodal displacement vector $\mathbf{u}$ and the second equation to provide the solution of the vector $\boldsymbol{\lambda}$ of Lagrange multipliers. It should be emphasized that the solution of the equations is only possible, when the optimal data points $(\boldsymbol{\varepsilon}^*, \boldsymbol{\sigma}^*)$ have previously been defined. In the following lines of this section, it will be shown that this can be done within an iterative process.

It is worth noticing that the form of equations (8.26) resembles the formulation which is used in the framework of finite element analysis. In particular, the term in the left-hand part of both equations represents the stiffness matrix, which is provided in relation (2.52) of section 2.3.2. The second component in the right-hand part of equation (8.26b) is the internal force vector, which is used in the solution of non-linear finite element analysis, for both geometric and material non-linearity, as presented in sections 2.4.4.3 and 2.4.5, respectively. Based on these descriptions, equations (8.26) can be rewritten using a matrix-vector notation:

$$\left(\int_\Omega \mathbf{B}^T \mathbf{C} \mathbf{B} d\Omega\right) \mathbf{u} = \int_\Omega \mathbf{B}^T \mathbf{C} \boldsymbol{\varepsilon}^* d\Omega \tag{8.27a}$$

$$\left(\int_\Omega \mathbf{B}^T \mathbf{C} \mathbf{B} d\Omega\right) \boldsymbol{\lambda} = \int_S \mathbf{t} dS - \int_\Omega \mathbf{B}^T \boldsymbol{\sigma}^* d\Omega \tag{8.27b}$$

Using a simplified notation, equations (8.27) can be expressed as follows:

$$\mathbf{K}\mathbf{u} = \mathbf{F} \tag{8.28a}$$

$$\mathbf{K}\boldsymbol{\lambda} = \mathbf{F}_{ext} - \mathbf{F}_{int} \tag{8.28b}$$

Equation (8.28a) is similar to the final formulation of linear finite element analysis, if someone replaces the force vector $\mathbf{F}$ by the component in the right-hand part of equation (8.27a). Relation (8.28b) resembles the equations of non-linear finite element analysis. However, in the framework of data-driven formulation, no tangent stiffness is introduced. Next, it will be discussed how this particular point influences the solution of data-driven analysis, comparing to classical formulations of non-linear mechanics which make use of the tangent stiffness matrix.

The next step which completes the presented data-driven formulation, focuses on the determination of the optimal pair $(\boldsymbol{\varepsilon}^*, \boldsymbol{\sigma}^*)$ from the data set $D$, which results in the closest possible satisfaction of equilibrium and compatibility. Thus, the goal of the process will be, to identify for a given strain-stress pair $(\boldsymbol{\varepsilon}^{(0)}, \boldsymbol{\sigma}^{(0)})$ which satisfies equilibrium and compatibility, which pair from the data set is the closest. This description leads to the formulation of the minimum distance problem, which is provided by the norm:

$$\|\text{distance}\| = \left((\boldsymbol{\varepsilon}^{(0)} - \boldsymbol{\varepsilon}^*)^T \mathbf{C}(\boldsymbol{\varepsilon}^{(0)} - \boldsymbol{\varepsilon}^*) + (\boldsymbol{\sigma}^{(0)} - \boldsymbol{\sigma}^*)^T \mathbf{C}^{-1}(\boldsymbol{\sigma}^{(0)} - \boldsymbol{\sigma}^*)\right)^{\frac{1}{2}} \tag{8.29}$$

This problem can be solved by developing the *Voronoi diagram* or *Dirichlet tessellation* [108], for the data set points. When the points of a data set are considered, a Voronoi diagram is built by partitioning a plane into convex polygons, such that each polygon contains exactly one generating point from the data set. Then, every other point located in a polygon, will be closer to the corresponding generating point of this polygon, than to any other point from the data set [230]. A Voronoi diagram created in Matlab for points of a data set is shown in figure 8.6 [136].

The solution of the overall data-driven scheme can now be presented. As mentioned above an iterative process is adopted, according to the following steps:

**1:** For every material point of the structural system, a randomly chosen $(\boldsymbol{\varepsilon}^{*(0)}, \boldsymbol{\sigma}^{*(0)})$ pair is chosen for the initiation of the process. It is noted that within the solution of isoparametric finite element analysis, the material points of a structure are represented by the Gauss (or integration) points.

**2:** Equations (8.27) are solved to determine the nodal displacement vector $\mathbf{u}^{(0)}$ and the Lagrange multipliers vector $\boldsymbol{\lambda}^{(0)}$.

**3:** The pair $(\boldsymbol{\varepsilon}^{(0)}, \boldsymbol{\sigma}^{(0)})$ which satisfies compatibility and equilibrium is calculated for every Gauss point, using relations (8.18) and (8.24).

**4:** Using the concept of Voronoi diagrams, for every $(\boldsymbol{\varepsilon}^{(0)}, \boldsymbol{\sigma}^{(0)})$ pair, is determined the corresponding pair $(\boldsymbol{\varepsilon}^{*(1)}, \boldsymbol{\sigma}^{*(1)})$ in the data set $D$, closest to $(\boldsymbol{\varepsilon}^{(0)}, \boldsymbol{\sigma}^{(0)})$.

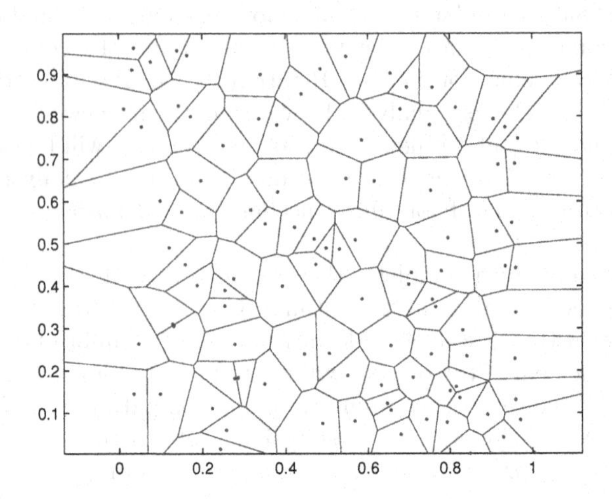


**FIGURE 8.6**: Voronoi diagram for a number of points from a data set.

**5:** Convergence for the iterative algorithm is checked, by comparing the pairs $(\boldsymbol{\varepsilon}^{*(0)}, \boldsymbol{\sigma}^{*(0)})$ and $(\boldsymbol{\varepsilon}^{*(1)}, \boldsymbol{\sigma}^{*(1)})$.

If $(\boldsymbol{\varepsilon}^{*(1)}, \boldsymbol{\sigma}^{*(1)}) = (\boldsymbol{\varepsilon}^{*(0)}, \boldsymbol{\sigma}^{*(0)})$ for every Gauss point, then the displacement vector and strain-stress pairs are the ones calculated above: $\mathbf{u}^{(0)}, (\boldsymbol{\varepsilon}^{(0)}, \boldsymbol{\sigma}^{(0)})$ and iterations are completed. It is noticed that equality $(\boldsymbol{\varepsilon}^{*(1)}, \boldsymbol{\sigma}^{*(1)}) = (\boldsymbol{\varepsilon}^{*(0)}, \boldsymbol{\sigma}^{*(0)})$ indicates that the pair from the data set used to calculate the strain-stress pair satisfying compatibility and equilibrium, is the same with the pair from the data set which is the closest to the calculated strain-stress pair satisfying compatibility and equilibrium.

If not, iterations are repeated from step **2**, using the pairs $(\boldsymbol{\varepsilon}^{*(1)}, \boldsymbol{\sigma}^{*(1)})$.

In section A.2 of the appendix, Matlab codes which implement the data-driven scheme of this section, are described.

The present concept has recently been extended in applications related to dynamic analysis of structures [109] and fracture mechanics [37].

# Appendix A

## Matlab codes on numerical modelling of composite heterogeneous structures

Several Matlab codes have been developed by the authors, for the implementation of a number of topics related to computational mechanics applications for composite heterogeneous structures. In this chapter, analytical descriptions are presented, explaining the topics under investigation and key aspects for using the codes. Thus, after reading this material, the user will be able to visit the suggested sites, download the codes and run them. Any further development of the codes will then be possible.

The problems which are discussed within this framework, focus on mechanical descriptions for fibre reinforced composites as well as for masonry structures. However, the relevant codes can be extended further towards the simulation of different composite materials, such as concrete or for three-dimensional problems and other multiphysics applications. In this sense, these descriptions can be considered as cases studies, since they represent applications for concepts which have been discussed analytically in the main body of the book.

The Matlab codes which are briefly described in the next sections of the Appendix, are used for applications in the following topics:

**1:** Multi-scale computational homogenization or $FE^2$ for fibre reinforced composites, using the formulations which are provided in chapter 7 of the book. This work focuses on debonding failure at the microscopic level. Except the overall multi-scale scheme, the reader will be able to implement a simple but efficient contact law, which is used in the microscopic level to provide the matrix-fibre debonding.

**2:** Data-driven computational homogenization ($FE^2$), using the approach presented in section 8.4 of chapter 8 of the book. This approach, which relies on the concept of minimization of a distance function, is recently developed and used in literature for several applications, including among others fracture mechanics and dynamic analysis of structures.

**3:** Data-driven multi-scale computational homogenization ($FE^2$), using artificial neural networks for the regression of stress and stiffness information from databases. This work is relevant to the discussions presented in

DOI: 10.1201/9781003017240-A

section 8.3 of chapter 8 of the book. In the Matlab codes are included the implementation of the training of a feed-forward, backpropagation neural network, and the incorporation of the trained neural network in the overall $FE^2$ scheme.

**4:** Multi-scale analysis for depicting localization in composite structures. The Extended Finite Element Method is used in the macroscopic scale, to simulate damage in the form of cracks, depicting a cohesive response. The corresponding softening traction-separation laws, at every Gauss point of the macro cracks, are derived from microscopic simulations on a Representative Volume Element (RVE), in the context of $FE^2$ analysis. Unilateral contact interfaces crossing the boundaries of the RVE are adopted to describe failure of (mesoscopic) masonry structures.

---

## A.1 Multi-scale computational homogenization ($FE^2$)

Matlab codes have been developed, to provide the overall multi-scale computational homogenization scheme for a fibre reinforced composite. The macroscopic level represents a rectangular homogeneous structure, at every integration point of which, a Representative Volume Element (RVE) of a fibre reinforced composite is assigned. A single fibre and the surrounding matrix have been considered in this case, as shown in figure 7.2 of chapter 7. The macro and micro structures are shown in figure A.1. More details for the models and related results can be found in [65].

The interaction between the matrix and the fibre is simulated using a unilateral contact interface. The simplest, node to node discretization is adopted in this case. For the implementation of the contact problem, an *Augmented Lagrangian* formulation is used in the microstructure, within the Newton-Raphson incremental-iterative process.

Periodic boundary conditions described in section 7.2.2.2 have also been assigned as loading on the RVE. For the implementation of this type of boundary conditions, the strain at each macro integration point is used as input.

In a first load step, total compression is applied to the macrostructure. A vertical shear action is applied at the right side, in a second load step, as shown in figure A.1. This macroscopic loading, allows for the activation of the contact condition between the matrix and the fibre in the microstructure. A fixed support in the left side is also assigned (figure A.1).

For both the macro and the micro analysis plane stress conditions are used, but plane strain conditions can equivalently be applied. According to the principles of multi-scale computational homogenization, no material properties are introduced in the macroscopic model. These are derived at every load increment, after the solution of the RVE is obtained.

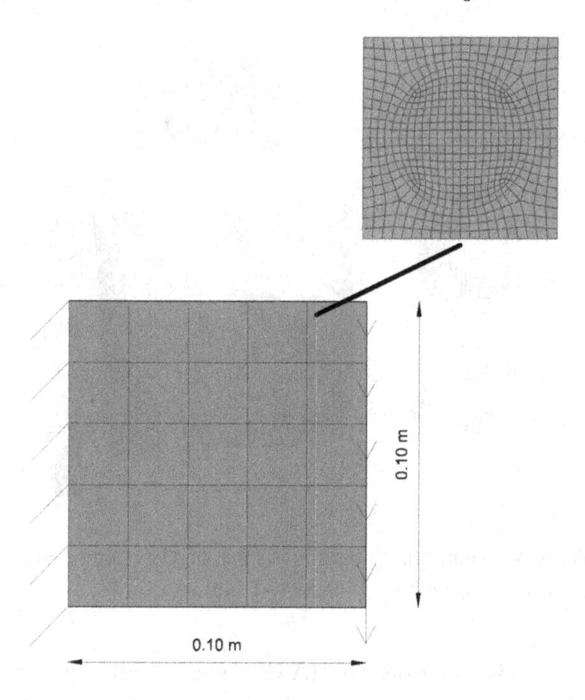


**FIGURE A.1**: The macro and micro structure simulated within the $FE^2$ approach in the provided Matlab codes.

In particular, when convergence of the non-linear RVE model arises, the effective stress and the consistent tangent stiffness are calculated. These are sent to the macro model and used in the main equilibrium equation, in order to provide the internal force vector and the tangent stiffness matrix, within the Newton-Raphson process. Parallel analysis is used in the macro model using the *parfor* Matlab function, to provide a faster solution. Figure A.2 shows the vertical displacements of the macro structure and the corresponding micro structure, at selected Gauss points of the last increment of the overall, multi-scale simulation.

The main Matlab files and functions are provided below, with some brief descriptions:

- *central_multiscale.m* is the central Matlab file providing the overall multi-scale simulation for the macroscopic structure. A number of Matlab functions providing the microstructure (RVE) as well as the necessary descriptions within non-linear finite element analysis, are also found in this file.

- *stiffness_internalForce.m* is the Matlab function which is used in the *central_multiscale.m* file, to provide the tangent stiffness and the internal force vector per element. These will be used to build the global tangent stiffness matrix and internal force vector, within the Newton-Raphson process.

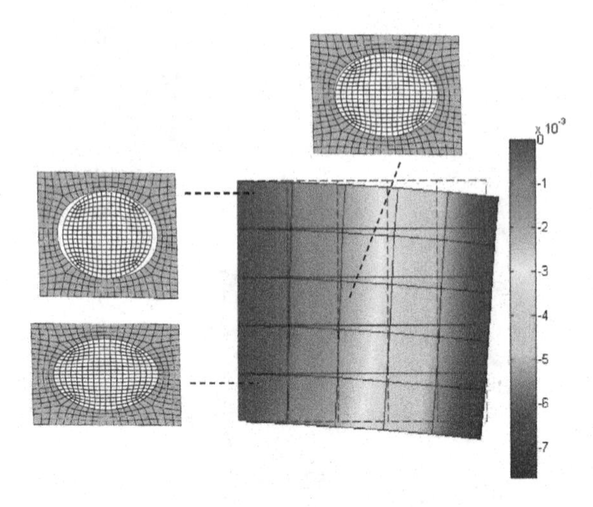


**FIGURE A.2**: Vertical displacements (m) of the macro structure and the corresponding micro structure at selected Gauss points.

Furthermore, in this function, the RVE model is called and solved at every integration point of the macrostructure.

- *RVE.m* is the Matlab function found within the *stiffness_ internalForce.m*, which calls and solves the model of the RVE. Input of this function is the strain vector (of dimensions $3 \times 1$ under plane stress conditions), and output the stress vector *Sigma_ M* ($3 \times 1$) and the consistent tangent stiffness $C_T$ ($3 \times 3$). These are used to build the element tangent stiffness and internal force vector in *stiffness_ internalForce.m* as well as the global stiffness matrix and internal force vector in *central_ multiscale.m.* In turn, these are used to solve the main Newton-Raphson equation in the central Matlab file.

The mesh of the RVE is created in a commercial finite element package. Relevant *.txt* files describing the node coordinates and element connectivity are exported from the commercial software and imported in the *RVE.m* function. It is noted that all the constituent material properties are provided in the *RVE.m.* In addition, periodic boundary conditions are applied to the boundaries of the RVE, using the approach of the transformation of the equilibrium equations which is discussed in section 7.2.2.2. When convergence of the non-linear RVE analysis is achieved, the effective stress vector and consistent tangent stiffness are calculated using relations (7.8) and (7.34), respectively, as presented in chapter 7.

## A.2 Data-driven computational homogenization (FE$^2$) using the concept of minimization of a distance function

Matlab codes which can be used to implement the concept of data-driven structural analysis, relying on the minimization of a distance function, are discussed in this section. Relevant descriptions are provided in section 8.4 of chapter 8 in this book. Further details can also be found in [108].

For the formulation of this data-driven scheme, the distance minimization problem between every value of a database, representing the constitutive law of a structure and a corresponding value which satisfies equilibrium and compatibility, is considered. The overall description is provided by the optimization formulation (8.20) and the equations (8.27) or (8.28), which are the main relations of the data-driven scheme. These equations are used to determine the unknown quantities of the problem, which are the nodal displacement vector and the Lagrange multipliers vector. The method of Lagrange multipliers has been used to consider the equilibrium constraint of the optimization problem (8.20).

The Matlab codes which are presented here, can be used to apply the mentioned concept in the framework of multi-scale computational homogenization (FE$^2$). In particular, a database has been developed to provide the effective strain-stress response for a heterogeneous masonry microstructure (RVE). This database has been built using numerical simulations on a non-linear RVE model. The overall data-driven code is then developed to represent the non-linear structural response of a masonry wall. Below are provided basic descriptions for the relevant Matlab files:

- *central_datadriven.m* is the central Matlab file which provides the overall data-driven, multi-scale simulation for the macroscopic structure. The database is loaded and used to implement this data-driven scheme. Equations (8.27) are solved in this file and the function *datadriven_iter.m* is called and used to apply the minimum distance optimization.

- *datadriven_iter.m* is the Matlab function which is developed to implement the minimum distance problem. The Matlab function *knnsearch* is used here, to provide the strain-stress pair $(\boldsymbol{\varepsilon}^*, \boldsymbol{\sigma}^*)$ from the database, which is closest to the pair $(\boldsymbol{\varepsilon}^{(0)}, \boldsymbol{\sigma}^{(0)})$, satisfying compatibility and equilibrium. In this function, the right-hand part terms of relations (8.27) are also calculated.

- *stiff_ini.m* is the Matlab function which is used to provide the stiffness, as well as initial values for the right-hand part terms of relations (8.27).

## A.3 Data-driven computational homogenization (FE$^2$) using machine learning principles

In this section are provided descriptions which are related to Matlab codes developed for the implementation of data-driven structural analysis, using machine learning principles. Two different concepts can be recognized within this framework. The first, is relevant to machine learning algorithms, adopted to contribute the data-driven aspect on the structural analysis. In particular, artificial neural networks are developed, trained and used to handle databases containing the constitutive response. The second concept refers to multi-scale computational homogenization, which is used to determine the non-linear response of a composite structure. The trained neural networks are incorporated in the multi-scale scheme to provide the microstructural response. The overall concept is schematically shown in figure A.3.

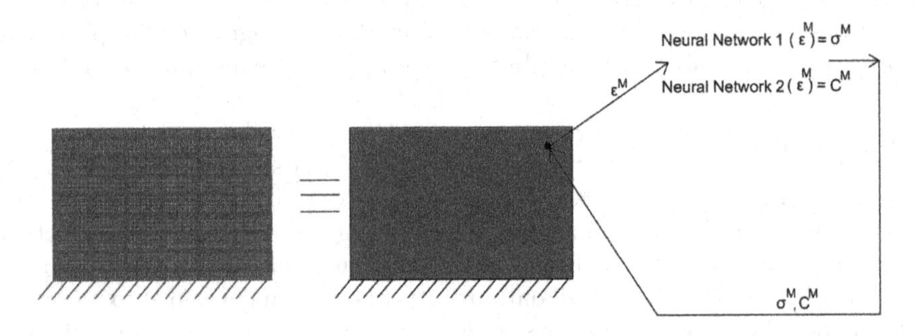


**FIGURE A.3**: Data-driven computational homogenization (FE$^2$) using artificial neural networks.

Two databases have been built using independent numerical simulations on a masonry RVE model. The first, corresponds strain to effective stress and the second corresponds strain to effective elasticity. These databases are used to train two neural networks, one using as input the strain and output the effective stress and the other using as input the strain and output the effective elasticity. The trained neural networks can then predict the effective stress and elasticity, for arbitrary strain values. In figures A.4 and A.5 are provided examples of the outcome of the training process, comparing the desired values in the database and the corresponding predicted values by the trained neural network, for a stress and an elasticity component, respectively.

After training is complete, the trained neural networks are introduced in the multi-scale (FE$^2$) scheme, to provide the effective microstructural response. Thus, they are used to substitute the RVE simulation, for every load increment, at each Gauss point of the macroscopic structure. It is noted that the computing time to predict the stress and elasticity per Gauss point using

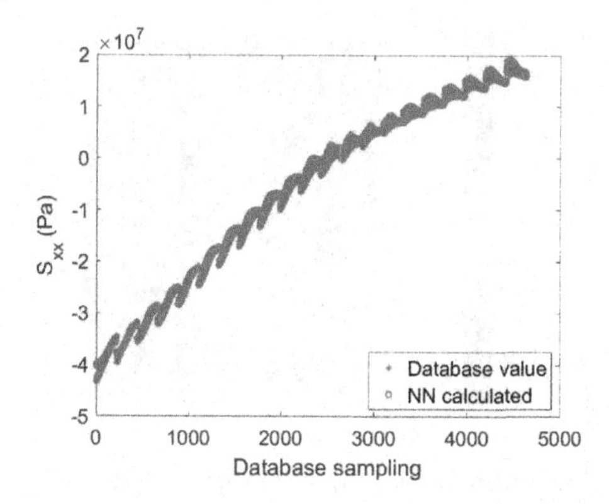


**FIGURE A.4**: Desired and predicted values of a stress component using the trained artificial neural network.

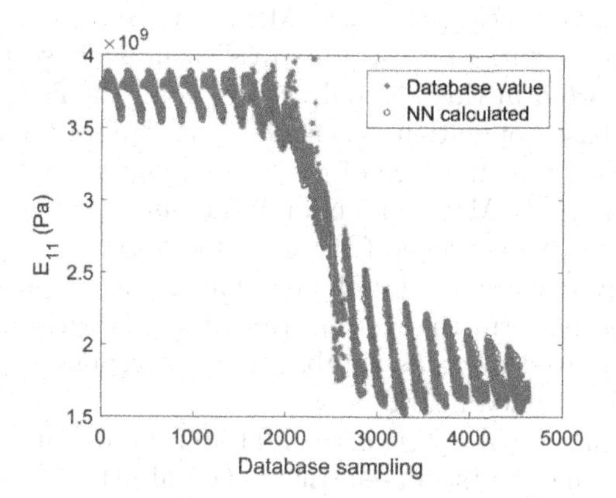


**FIGURE A.5**: Desired and predicted values of an elasticity component using the trained artificial neural network.

the trained neural networks is extremely lower than the time to conduct one RVE simulation, as presented in [64]. Thus, this data-driven approach contributes to the computational efficiency of the multi-scale formulation. Figure A.6 shows the maximum principal stress distribution, as obtained from the overall, data-driven approach, for a masonry wall with openings (windows).

In the following lines are provided the details of the developed Matlab codes. The folder "Neural Networks" includes all the necessary Matlab files and functions for the training process. After this is complete, the overall multi-scale simulation can be conducted, using the Matlab files of the folder "Multi_scale_FE2".

Relevant descriptions are initially provided for the training of the neural networks. The Matlab codes which are mentioned below, should be run in the order they appear in the text:

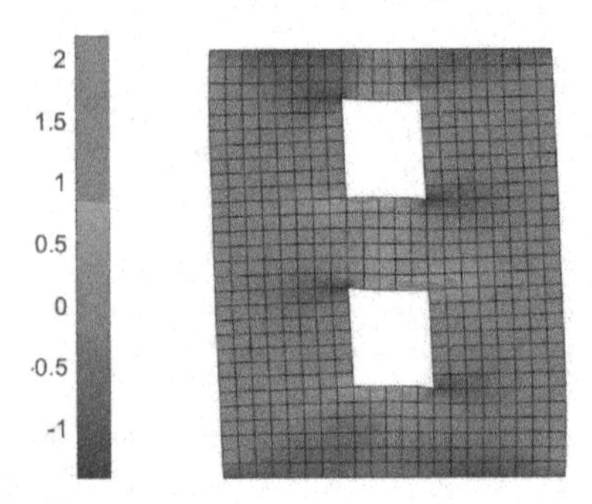


**FIGURE A.6**: Maximum principal stress distribution (MPa) for a masonry wall with openings, obtained from the data-driven multi-scale approach.

1) *nn_training_testing.m* is the Matlab code which can be used for the training and testing of a "strain-stress" neural network. A strain-stress database is imported in this file and used to train the neural network, using half of the database population. Then, the other half of the database population is used to test the efficiency of the trained neural network.

- *bbackprop.m* is the Matlab function inside *nn_training_testing.m*, which is used to perform the training of the neural network.

- *usebbackprop.m* is the Matlab function inside *nn_training_testing.m*, which is used to perform the testing, thus, the comparison between desired output values, derived from the database and corresponding predicted values, derived using the trained neural network.

2) *nn_training_testing_b.m* is the Matlab code which can be used for the training and testing of a "strain-elasticity" neural network. A strain-elasticity database is imported in this file and used to train the neural network, using half of the database population. Then, the other half of the database population is used to test the efficiency of the trained neural network.

The function *bbackprop.m* which is mentioned above, is also used in this file to perform the training of the neural network.

- *usebbackprop_b.m* is the Matlab function inside *nn_training_testing_b.m*, which is used to perform the testing, thus, the comparison between desired output values, derived from the database and corresponding predicted values, derived using the trained neural network.

3) *nn_use.m* is the Matlab file which calls the trained strain-stress neural network and predicts the stress for any randomly assigned strain.

- *usebbackprop_nn.m* is the Matlab function which is used in this code to perform the regression through the trained neural network.

4) *nn_use_b.m* is the Matlab file which calls the trained strain-elasticity neural network and predicts the elasticity tensor for any randomly assigned strain.

- *usebbackprop_nn_b.m* is the Matlab function which is used in this code to perform the regression through the trained neural network.

After all these steps have been completed, the user can proceed towards the data-driven multi-scale simulation. In particular, the following Matlab files should be transferred from the "Neural Networks" to the "Multi_scale_FE2" folder:

- *minmax.mat*
- *minmax_b.mat*
- *TrainedNetwork.mat*
- *TrainedNetwork_elasticity.mat*

These files will provide all the necessary information for the trained neural networks. Furthermore, within the "Multi_scale_FE2" folder, the following Matlab files are found:

- *multiscale_macro_DisplLoad.m* is the central Matlab file which can be used to perform the overall data-driven, multi-scale simulation of a masonry wall with displacement loading, by adopting the trained neural networks.
- *multiscale_macro_ForceLoad.m* is the central Matlab file which can be used to perform the overall data-driven, multi-scale simulation of a masonry wall with force loading, by adopting the trained neural networks.
- *multiscale_macro_DisplLoad_openings.m* is the central Matlab file which can be used to perform the overall data-driven, multi-scale simulation of a masonry wall with openings (windows) and displacement loading, by adopting the trained neural networks.

Inside those three files, are included the following Matlab functions:

- *RVE_stress_NN.m* is the Matlab function which calls the trained strain-stress neural network, uses as input the strain per Gauss integration point and returns the corresponding effective stress.
- *RVE_stiffness_NN.m* is the Matlab function which calls the trained strain-elasticity neural network, uses as input the strain per Gauss integration point and returns the corresponding effective consistent elasticity.

Both the stress and the consistent elasticity are used to produce the global tangent stiffness matrix and the internal force vector within the Newton-Raphson process, in the central multi-scale codes.

---

## A.4 A multi-scale scheme proposed to simulate localization of damage in composite structures using the XFEM method and principles of contact mechanics

In this section, Matlab codes and some representative results are provided, from the multi-scale, $FE^2$ scheme presented in section 7.4 of chapter 7. This multi-scale scheme introduces cohesive XFEM cracks in the macroscopic scale, with the corresponding traction-separation response of those cracks being derived from mesoscopic simulations. The model is able to predict localization of damage in composite materials such as masonry, as discussed analytically in [63].

In particular, one or more cracks are introduced in the macro model using the XFEM method, which is presented in chapter 5 of the book. The cohesive, traction-separation response at every Gauss point of the macro cracks, is derived numerically, by solving the corresponding RVE model. This mesoscopic RVE represents a masonry structure, consisting of stone blocks and mortar interfaces, simulated as unilateral contact interfaces crossing the RVE boundaries. Softening traction-separation laws are also assigned to the interfaces, to capture tensile failure.

The response of the RVE is first studied here, by conducting independent meso scale simulations. As mentioned in section 7.4, the anisotropy damage pattern induced by the interfaces crossing the boundaries of the RVE, can be represented by this model. Therefore, failure of the RVE may involve the development of vertical, diagonal and horizontal displacement jumps, as shown in figures A.7, A.8 and A.9. It is also noticed that the effective traction-separation diagrams depict a softening response for each of these cases.

The proposed model can then be applied to a masonry wall with an opening, the geometry, load steps and experimental failure pattern of which are depicted in figure A.10. A compressive load is assigned at the top of the structure in a first load step, followed by a uniformly distributed shear displacement load applied in a second load step, as shown in figure A.10.

The propagation of damage in the macroscopic level and the corresponding meso scale failure pattern obtained at the integration points of the macro cracks, are shown in figures A.11 and A.12. Comparison between the experimental and the numerical response indicates that a similar crack path is gradually depicted in both cases.

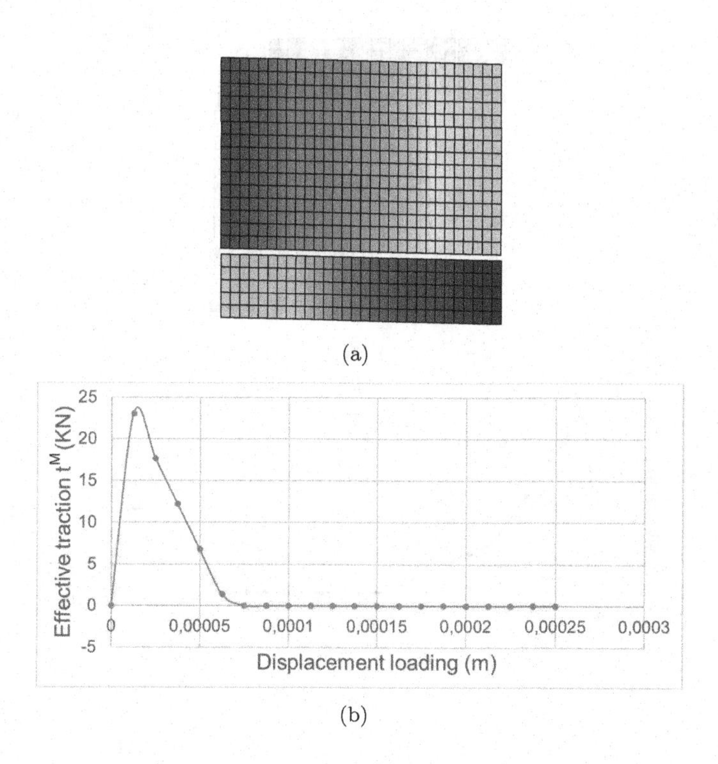


**FIGURE A.7**: a) Vertical displacement jump on the RVE and b) the effective, tensile traction-separation diagram *(Reprinted by permission from Springer Nature Customer Service Centre GmbH: Nature Springer, Archive of Applied Mechanics, A computational homogenization approach for the study of localization of masonry structures using the XFEM, Georgios A. Drosopoulos, Georgios E. Stavroulakis, Volume 88, pages 2135–2152 (2018)).*

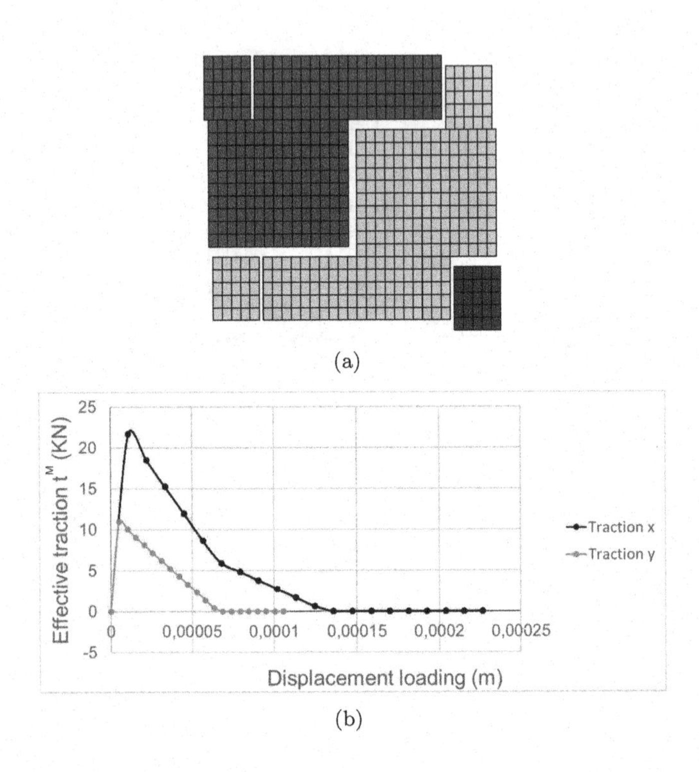


**FIGURE A.8**: a) Displacement jump on the RVE developed along a diagonal direction and b) the effective, tensile traction-separation diagram *(Reprinted by permission from Springer Nature Customer Service Centre GmbH: Nature Springer, Archive of Applied Mechanics, A computational homogenization approach for the study of localization of masonry structures using the XFEM, Georgios A. Drosopoulos, Georgios E. Stavroulakis, Volume 88, pages 2135–2152 (2018)).*

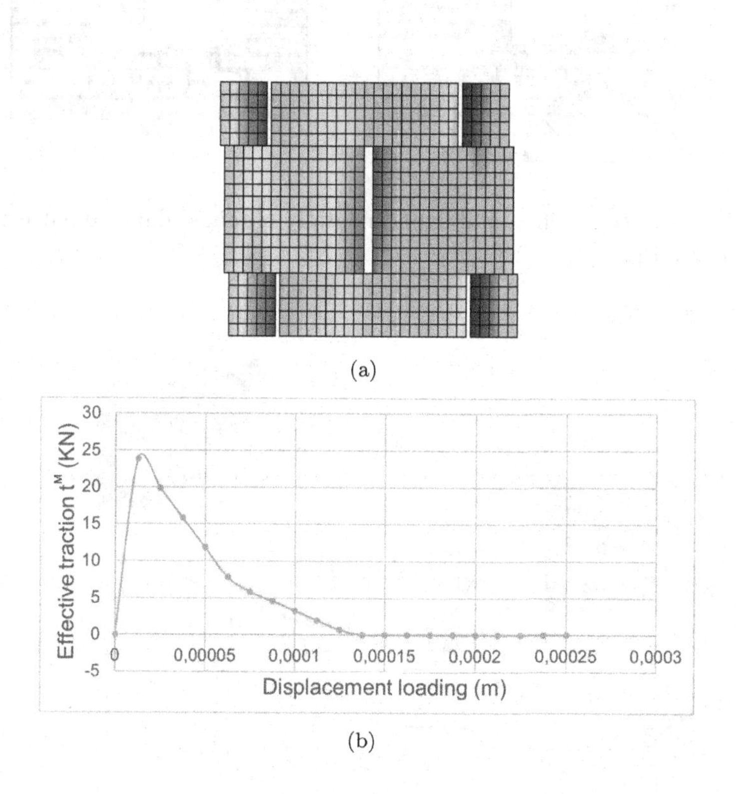


**FIGURE A.9**: a) Horizontal displacement jump on the RVE and b) the effective, tensile traction-separation diagram *(Reprinted by permission from Springer Nature Customer Service Centre GmbH: Nature Springer, Archive of Applied Mechanics, A computational homogenization approach for the study of localization of masonry structures using the XFEM, Georgios A. Drosopoulos, Georgios E. Stavroulakis, Volume 88, pages 2135–2152 (2018)).*

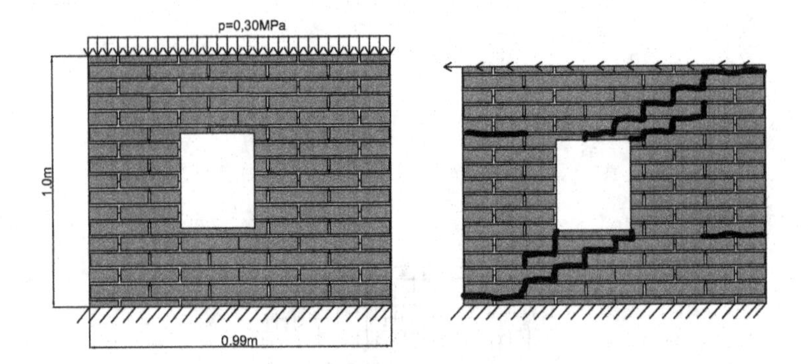


**FIGURE A.10**: Loading and experimentally obtained damage pattern for a masonry wall [224].

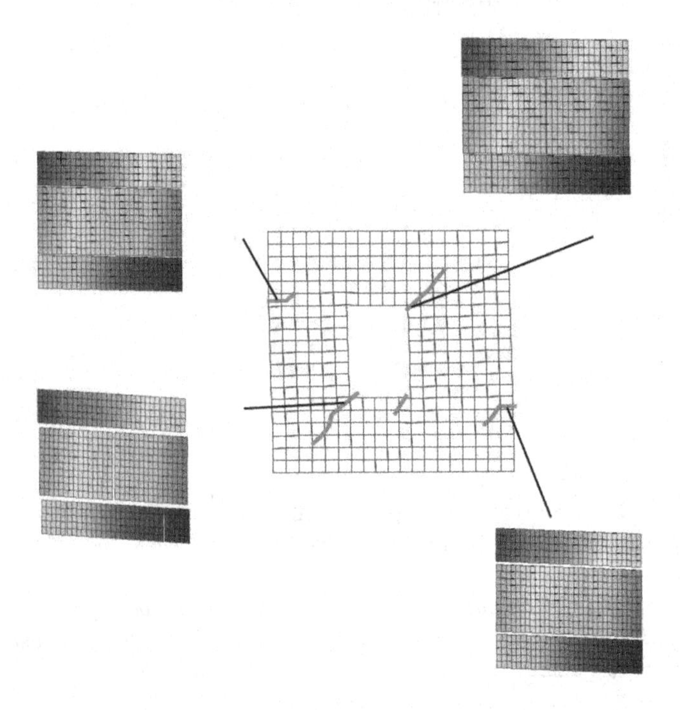

**FIGURE A.11**: Macroscopic and mesoscopic failure obtained by numerical simulation at an intermediate load step *(Reprinted by permission from Springer Nature Customer Service Centre GmbH: Nature Springer, Archive of Applied Mechanics, A computational homogenization approach for the study of localization of masonry structures using the XFEM, Georgios A. Drosopoulos, Georgios E. Stavroulakis, Volume 88, pages 2135–2152 (2018)).*

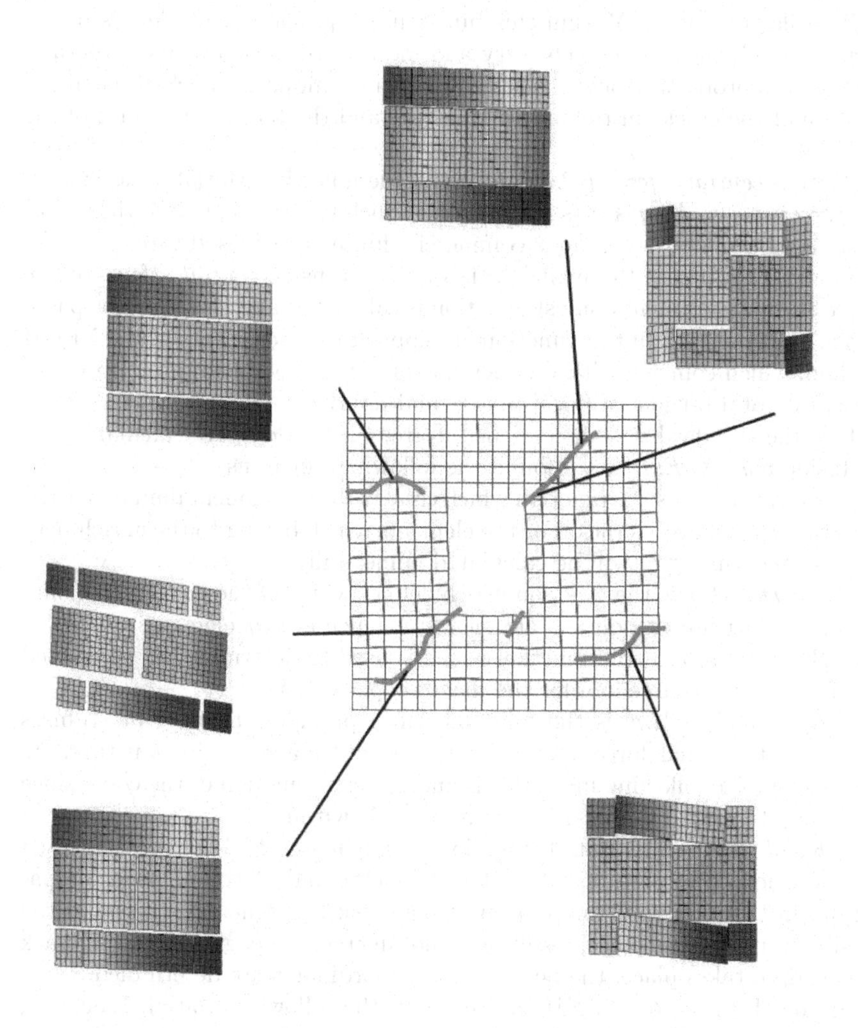

**FIGURE A.12**: Macroscopic and mesoscopic failure obtained by numerical simulation at the final load step *(Reprinted by permission from Springer Nature Customer Service Centre GmbH: Nature Springer, Archive of Applied Mechanics, A computational homogenization approach for the study of localization of masonry structures using the XFEM, Georgios A. Drosopoulos, Georgios E. Stavroulakis, Volume 88, pages 2135–2152 (2018)).*

The main Matlab files and functions which are used to implement the above scheme are provided below:

- *central_xfem_fe2.m* is the central Matlab file which runs the overall multi-scale simulation. Within this function are provided core details of the overall model, such as the geometry and mesh of the macroscopic structure, the material properties for the *bulk* elements (not containing cracks), the initial position of the cracks in the macro structure and the tensile resistance of the material.

Within *central_xfem_fe2.m*, are found the following Matlab functions:

- *createmesh_Abaqus.m* is the function which is used to provide the macro mesh, as obtained from Abaqus commercial finite element software.
- *mainXFEM.m* is the main Matlab function inside *central_xfem_fe2.m*, where the finite element analysis solution is calculated and the crack propagation is implemented. In this function are applied the increments of the XFEM method, which coincide with the increments of the Newton-Raphson process. Thus, the total tangent stiffness matrix and internal force vector, which both include the standard and the enriched degrees of freedom, are calculated.

Inside *mainXFEM.m*, are found the following functions:

- *crackDetect.m* is the function which creates the enrichment domain for the two crack tips of every crack. For the elements which belong to the enrichment domain, the J-integral will be calculated numerically.
- *nodeDetect.m* is the function which defines which of the elements belonging to the enrichment domain, are *split* and which are *tip* elements.
- *BCs_macro.m* is the function which is used to determine the supported and free degrees of freedom for the macroscopic structure.
- *Kg_Rg_XFEM.m* is the function which provides the tangent stiffness matrix and internal force vector for the overall macroscopic structure. As discussed in the following lines, this is another core function of the code, since the connection with the microstructure is built within it.
- *KcalJint.m* is the Matlab function within *mainXFEM.m* which is used for the calculation of the stress intensity factor and of the angle of propagation. In this function, comparison of the crack tip principal stress and the tensile strength takes place, resulting or not in crack propagation. When crack propagation takes place, the new crack tip coordinates are determined.

Inside the *Kg_Rg_XFEM.m*, are found the following Matlab functions, which are used to determine the total tangent stiffness matrix and internal force vector of the structure. It should be emphasized, that within *Kg_Rg_XFEM.m* the cohesive response at every Gauss point of each macroscopic crack is evaluated, by conducting the RVE simulations, in the context of $FE^2$.

- *xfemBmatcoh.m* is the function which is used to calculate the strain-displacement matrix for standard and enriched (split and tip) degrees of freedom.
- *Kcoh_Rcoh.m* is the function which is used to calculate the tangent stiffness and the internal force vector obtained from the cohesive response of

the macroscopic cracks. In this function the RVE is called and the multi-scale, $FE^2$ concept is implemented.

As mentioned above, the RVE represents a mesoscopic masonry structure. The corresponding RVE model is found inside *Kcoh_Rcoh*:

- *RVE.m* is the Matlab function which implements the microstructural simulation, conducted at every Gauss point of the macroscopic cracks. It represents a masonry RVE, which consists of several blocks. In the interfaces between these blocks, unilateral contact interfaces are considered, using a node-to-node approximation. Tensile traction-separation laws depicting softening, are also assigned in a direction normal to the interfaces. In the tangential direction of the interfaces, some shear springs with small (linear) stiffness are used, to provide stability to this direction.

It is noted that this numerical scheme may present some instability. The reasons for this instability, is the presence of the softening law and the fact that the interfaces cross the RVE boundaries, as explained in section 3.3.3 of the book. Hence, numerical singularities may arise, depending on several parameters, such as the level of the macro displacement load, the slope of the softening branch of the traction-separation diagram assigned in the RVE interfaces, the stiffness of the linear, shear springs at the interfaces or even the number of the macro cracks. The interested reader may need to conduct some parametric investigation, in order to calibrate the model.

Finally, interesting extensions of the model would involve its reformulation within a data-driven framework, according to concepts which are presented in chapter 8 of the book.

# Bibliography


[1] KM Abdalla and GE Stavroulakis. A backpropagation neural network model for semi-rigid steel connections. *Computer-Aided Civil and Infrastructure Engineering*, 10(2):77–87, 1995.

[2] M Abendroth, G Hütter, C Settgast, A Malik, B Kiefer, and M Kuna. A hybrid approach to describe the elastic-plastic deformation behavior of porous media including damage effects. *Technische Mechanik*, 40(1):5–14, 2020.

[3] V Acary and B Brogliato. *Numerical methods for nonsmooth dynamical systems. Applications in mechanics and electronics.* Springer, 2008.

[4] S Adly. *A variational approach to nonsmooth dynamics. Applications in unilateral mechanics and electronics.* Springer, 2017.

[5] EC Aifantis. On the microstructural origin of certain inelastic models. *Journal of Engineering Materials and Technology (ASME)*, 106:326–330, 1984.

[6] AM Al-Fahed and GE Stavroulakis. A complementarity problem formulation of the frictional grasping problem. *Computer Methods in Applied Mechanics and Engineering*, 190:941–952, 2000.

[7] AM Al-Fahed, GE Stavroulakis, and PD Panagiotopoulos. Hard and soft fingered robot grippers. *ZAMM, Journal of Applied Mathematics and Mechanics*, 71:257–266, 1991.

[8] AM Al-Fahed, GE Stavroulakis, and PD Panagiotopoulos. A linear complementarity approach to the frictionless gripper problem. *International Journal of Robotics Research*, 11:112–122, 1992.

[9] H Antes and PD Panagiotopoulos. *The boundary integral approach to static and dynamic contact problems. Equality and inequality methods.* Birkhäuser, Basel - Boston - Berlin, 1992.

[10] A Anthoine. Derivation of the in-plane elastic characteristics of masonry through homogenization theory. *International Journal of Solids and Structures*, 32(2):137–163, 1995.

[11] PMA Areias and T Belytschko. Non-linear analysis of shells with arbitrary evolving cracks using XFEM. *International Journal for Numerical Methods in Engineering*, 62:384–415, 2005.

[12] A Asadpoure and S Mohammadi. Developing new enrichment functions for crack simulation in orthotropic media by the extended finite element method. *International Journal for Numerical Methods in Engineering*, 69:2150–2172, 2007.

[13] PG Asteris, A Moropoulou, AD Skentou, M Apostolopoulou, A Mohebkhah, L Cavaleri, H Rodrigues, and H Varum. Stochastic vulnerability assessment of masonry structures: Concepts, modeling and restoration aspects. *Applied Sciences*, 9(2), 2019.

[14] C Baggio. Collapse behaviour of three-dimensional brick-block systems using non-linear programming. *Structural Engineering and Mechanics*, 10:181–195, 2000.

[15] A Bagirov, N Karmitsa, and M Mäkelä. *Introduction to nonsmooth optimization. Theory, practice and software.* Springer, 2014.

[16] GI Barenblatt. The mathematical theory of equilibrium cracks in brittle fracture. *Advances in Applied Mechanics*, 7:55–129, 1962.

[17] KJ Bathe. *Finite element procedures.* Prentice Hall, Upper Saddle River, New Jersey 07458, 1996.

[18] KJ Bathe and PA Bouzinov. On the constraint function method for contact problems. *Computers and Structures*, 64:1069–1086, 1997.

[19] ZP Bažant, T Belytschko, and TP Chang. Continuum theory for strain-softening. *Journal of Engineering Mechanics (ASCE)*, 110:1666–1692, 1984.

[20] ZP Bažant and G Pijaudier-Cabot. Measurement of the characteristic length of nonlocal continuum. *Journal of Engineering Mechanics (ASCE)*, 115:755–767, 1989.

[21] GK Bekas and GE Stavroulakis. Machine learning and optimality in multi storey reinforced concrete frames. *Infrastructures*, 2(2), 2017.

[22] T Belytschko and T Black. Elastic crack growth in finite elements with minimal remeshing. *International Journal for Numerical Methods in Engineering*, 45(5):601–620, 1999.

[23] T Belytschko, J Fish, and BE Engelman. A finite element with embedded localization zones. *Computer Methods in Applied Mechanics and Engineering*, 70:59–89, 1988.

[24] T Belytschko, S Loehnert, and JH Song. Multiscale aggregating discontinuities: A method for circumventing loss of material stability. *International Journal for Numerical Methods in Engineering*, 73:869–894, 2008.

[25] T Belytschko and JH Song. Coarse-graining of multiscale crack propagation. *International Journal for Numerical Methods in Engineering*, 81:537–563, 2010.

[26] V Beng Chye Tan, K Raju, and H Pueh Lee. Direct $FE^2$ for concurrent multilevel modelling of heterogeneous structures. *Computer Methods in Applied Mechanics and Engineering*, 360:112694, 2020.

[27] A Bensoussan, JL Lions, and G Papanicolaou. *Asymptotic analysis for periodic structures.* North-Holland, Amesterdam, Netherlands, 1978.

[28] M Bessa, R Bostanabad, Z Liu, A Hu, DW Apley, C Brinson, W Chen, and WK Liu. A framework for data-driven analysis of materials under uncertainty: Countering the curse of dimensionality. *Computer Methods in Applied Mechanics and Engineering*, 320:633–667, 2017.

[29] FE Bock, RC Aydin, CJ Cyron, N Huber, SR Kalidindi, and B Klusemann. A review of the application of machine learning and data mining approaches in continuum materials mechanics. *Frontiers in Materials*, 6, 2019.

[30] G Bolzon. Complementarity problems in structural engineering: an overview. *Archives of Computational Methods in Engineering*, 24:23–36, 2017.

[31] E Bosco, VG Kouznetsova, and MGD Geers. Multi-scale computational homogenization-localization for propagating discontinuities using X-FEM. *International Journal for Numerical Methods in Engineering*, 102:496–527, 2015.

[32] F Brezzi and M Fortin. *Mixed and hybrid finite element methods.* Springer, New York, 1991.

[33] B Brogliato. *Nonsmooth mechanics models, dynamics and control.* Springer, 2016.

[34] V Buryachenko. *Micromechanics of heterogeneous materials.* Springer Publishing Company, Incorporated, 1st edition, 2007.

[35] GT Camacho and M Ortiz. Computational modeling of impact damage in brittle materials. *International Journal of Solids and Structures*, 33:2899–2938, 1996.

[36] G Capuano. *Smart finite elements: An application of machine learning to reduced-order modeling of multi-scale problems.* PhD thesis, Georgia Institute of Technology, 2019.

[37] P Carrara, L De Lorenzis, L Stainier, and M Ortiz. Data-driven fracture mechanics. *Computer Methods in Applied Mechanics and Engineering*, 372:113390, 2020.

[38] A Cecen, H Dai, YC Yabansu, SR Kalidindi, and S Le. Material structure-property linkages using three-dimensional convolutional neural networks. *Acta Materialia*, 146:76–84, 2018.

[39] S Chakraborty. Transfer learning based multi-fidelity physics informed deep neural network. *Journal of Computational Physics*, 426:109942, 2021.

[40] TR Chandrupatla and AD Belegundu. *Introduction to finite elements in engineering.* Prentice Hall, Upper Saddle River, New Jersey 07458, 2002.

[41] AE Charalampakis and VK Papanikolaou. Machine learning design of R/C columns. *Engineering Structures*, 226:111412, 2021.

[42] F Chinesta, E Cueto, E Abisset-Chavanne, JL Duval, and FEI Khaldi. Virtual, digital and hybrid twins: A new paradigm in data-based engineering and engineered data. *Archives of Computational Methods in Engineering*, 27:105–134, 2020.

[43] PW Christensen and A Klarbring. Newton's method for frictional contact problems. In P Wunderlich, editor, *ECCM'99 European Conference on Computational Mechanics*, 1999.

[44] PW Christensen, A Klarbring, JS Pang, and N Strömberg. Formulation and comparison of algorithms for frictional contact problems. *International Journal for Numerical Methods in Engineering*, 42:145–173, 1998.

[45] BG Christoff, H Brito-Santana, R Talreja, and V Tita. Development of an ABAQUS$^{TM}$ plug-in to evaluate the fourth-order elasticity tensor of a periodic material via homogenization by the asymptotic expansion method. *Finite Elements in Analysis and Design*, 181:103482, 2020.

[46] EWC Coenen, VG Kouznetsova, E Bosco, and MGD Geers. A multi-scale approach to bridge microscale damage and macroscale failure: A nested computational homogenization-localization framework. *International Journal of Fracture*, 178:157–178, 2012.

[47] EWC Coenen, VG Kouznetsova, and MGD Geers. Multi-scale continuous-discontinuous framework for computational homogenization-localization. *Journal of the Mechanics and Physics of Solids*, 60:1486–1507, 2012.

[48] EWC Coenen, VG Kouznetsova, and MGD Geers. Novel boundary conditions for strain localization analyses in microstructural volume elements. *International Journal for Numerical Methods in Engineering*, 90:1–21, 2012.

[49] DA Colera and HG Kim. Asymptotic expansion homogenization analysis using two-phase representative volume element for non-periodic composite materials. *Multiscale Science and Engineering*, 1:130–140, 2019.

[50] RD Cook, DS Malkus, ME Plesha, and RJ Witt. *Concepts and applications of finite element analysis*. John Wiley & Sons, Inc. New York, NY, USA, 2002.

[51] G Dal Maso. *An introduction to C-convergence*. Birkhäuser, Boston, United States of America, 1993.

[52] C Dascalu, G Bilbie, and EK Agiasofitou. Damage and size effects in elastic solids: A homogenization approach. *International Journal of Solids and Structures*, 45:409–430, 2008.

[53] R de Borst. The zero-normal-stress condition in plane-stress and shell elasto-plasticity. *Communications in Applied Numerical Methods*, 7:29–33, 1991.

[54] R De Borst, MA Crisfield, JJC Remmers, and CV Verhoosel. *Non-linear finite element analysis of solids and structures*. John Wiley & Sons Ltd, West Sussex, PO19 8SQ, United Kingdom, 2012.

[55] R de Borst, J Pamin, RHJ Peerlings, and LJ Sluys. On gradient-enhanced damage and plasticity models for failure in quasi-brittle and frictional materials. *Computational Mechanics*, 17:130–141, 1995.

[56] R de Borst, JJC Remmers, and A Needleman. Mesh-independent discrete numerical representations of cohesive-zone models. *Engineering Fracture Mechanics*, 73:160–177, 2006.

[57] R de Borst, LJ Sluys, H-B Muhlhaus, and J Pamin. Fundamental issues in finite element analyses of localization of deformation. *Engineering Computations*, 10(2):99–121, 1993.

[58] EA de Souza Neto, PJ Blanco, PJ Sánchez, and RA Feijóo. An RVE-based multiscale theory of solids with micro-scale inertia and body force effects. *Mechanics of Materials*, 80:136–144, 2015.

[59] HPJ de Vree, WAM Brekelmans, and MAJ van Gils. Comparison of nonlocal approaches in continuum damage mechanics. *Computers and Structures*, 55:581–588, 1995.

[60] VF Demyanov, GE Stavroulakis, LN Polyakova, and PD Panagiotopoulos. *Quasidifferentiability and nonsmooth modelling in mechanics, engineering and economics.* Kluwer-Springer, 1996.

[61] DM Dimiduk, EA Holm, and SR Niezgoda. Perspectives on the impact of machine learning, deep learning, and artificial intelligence on materials, processes, and structures engineering. *Integrating Materials and Manufacturing Innovation*, 7:157–172, 2018.

[62] GA Drosopoulos, K Giannis, ME Stavroulaki, and GE Stavroulakis. Metamodelling-assisted numerical homogenization for masonry and cracked structures. *Journal of Engineering Mechanics (ASCE)*, 144:04018072, 2018.

[63] GA Drosopoulos and GE Stavroulakis. A computational homogenization approach for the study of localization of masonry structures using the XFEM. *Archive of Applied Mechanics*, 88:2135–2152, 2018.

[64] GA Drosopoulos and GE Stavroulakis. Data-driven computational homogenization using neural networks: $FE^2$-NN application on damaged masonry. *ACM Journal on Computing and Cultural Heritage*, 14(1), 2021.

[65] GA Drosopoulos, P Wriggers, and GE Stavroulakis. A multi-scale computational method including contact for the analysis of damage in composite materials. *Computational Materials Science*, 95:522–535, 2014.

[66] DS Dugdale. Yielding of steel sheets containing slits. *Journal of the Mechanics and Physics of Solids*, 8:100–108, 1960.

[67] EN Dvorkin, AM Cuitiño, and G Gioia. Finite elements with displacement interpolated embedded localization lines insensitive to mesh size and distortions. *International Journal for Numerical Methods in Engineering*, 30(3):541–564, 1990.

[68] Ch Efstathiades, CC Baniotopoulos, P Nazarko, L Ziemianski, and GE Stavroulakis. Application of neural networks for the structural health monitoring in curtain-wall systems. *Engineering Structures*, 29(12):3475–3484, 2007.

[69] F Facchinei, A Fischer, and C Kanzow. Inexact Newton methods for semismooth equations with applications to variational inequality problems. In G DiPillo and F Giannessi, editors, *Nonlinear optimization and applications*, pages 125–149, New York, 1996. Plenum Press.

[70] A Faramarzi, AA Javadi, and M Alani. An EPR-based self-learning approach to material modelling. *Computers and structures*, 137:63–71, 2014.

[71] MC Ferris and F Tin-Loi. Limit analysis of frictional block assemblies as a mathematical program with complementarity constraints. *International Journal of Mechanical Sciences*, 43:209–224, 2001.

[72] F Feyel. Multiscale $FE^2$ elastoviscoplastic analysis of composite structures. *Computational Materials Science*, 16:344–354, 1999.

[73] G Fichera. Boundary value problems in elasticity with unilateral constraints. In S Flügge, editor, *Enzyclopedia of Physics*, volume VI a/2. Springer Verlag, Berlin, 1972.

[74] A Fischer. A special Newton-type optimization method. *Optimization*, 24:269–284, 1992.

[75] J Fish, Z Yang, and Z Yuan. A second-order reduced asymptotic homogenization approach for nonlinear periodic heterogeneous materials. *International Journal for Numerical Methods in Engineering*, 119:469–489, 2019.

[76] M Fremond. *Nonsmooth thermomechanics*. Springer Verlag, Berlin, 2002.

[77] F Fritzen, M Fernández, and F Larsson. On-the-fly adaptivity for nonlinear twoscale simulations using artificial neural networks and reduced order modeling. *Frontiers in Materials*, 6, 2019.

[78] JN Fuhg, C Bóhm, N Bouklas, A Fau, P Wriggers, and Marino M. Model-data-driven constitutive responses: Application to a multiscale computational framework. *International Journal of Engineering Science*, 167:103522, 2021.

[79] TC Gasser and GA Holzapfel. Modeling 3D crack propagation in unreinforced concrete using PUFEM. *Computer Methods in Applied Mechanics and Engineering*, 194:2859–2896, 2005.

[80] MGD Geers. *Experimental analysis and computational modelling of damage and fracture*. PhD thesis, Eindhoven University of Technology, Eindhoven, The Netherlands, 1997.

[81] MGD Geers, VG Kouznetsova, K Matouš, and J Yvonnet. *Homogenization Methods and Multiscale Modeling: Nonlinear Problems*, pages 1–34. American Cancer Society, 2017.

[82] J Ghaboussi. *Advances in neural networks in computational mechanics and engineering*, pages 191–236. Springer, Vienna, Vienna, 2010.

[83] D Gross and T Seelig. *Fracture Mechanics: With an Introduction to Micromechanics*. Springer-Verlag, Berlin, Heidelberg, 2011.

[84] ME Gurtin, E Fried, and L Anand. *The mechanics and thermodynamics of continua.* Cambridge University Press, Cambridge, 2010.

[85] R Hambli, H Katerchi, and CL Benhamou. Multiscale methodology for bone remodelling simulation using coupled finite element and neural network computation. *Biomechanics and Modeling in Mechanobiology*, 10:133–145, 2011.

[86] J Han, A Jentzen, and E Weinan. Solving high-dimensional partial differential equations using deep learning. *Proceedings of the National Academy of Sciences of the United States of America*, 115(34):8505–8510, 2018.

[87] W Han and BD Reddy. *Plasticity: mathematical theory and numerical analysis.* Springer, New York, 1999.

[88] YMA Hashash, S Jung, and J Ghaboussi. Numerical implementation of a neural network based material model in finite element analysis. *International Journal for Numerical Methods in Engineering*, 59:989–1005, 2004.

[89] R Hill. A general theory of uniqueness and stability in elastic-plastic solids. *Journal of the Mechanics and Physics of Solids*, 6:236–249, 1958.

[90] R Hill. Elastic properties of reinforced solids: Some theoretical principles. *Journal of the Mechanics and Physics of Solids*, 11:357–372, 1963.

[91] A Hillerborg, M Modeer, and PE Petersson. Analysis of crack formation and crack growth in concrete by means of fracture mechanics and finite elements. *Cement and Concrete Research*, 6:773–782, 1976.

[92] M Hjiaj, J Fortin, and G de Sacxcé. A complete stress update algorithm for the non-associated Drucker-Prager model including the treatment of the apex. *International Journal of Engineering Science*, 41:1109–1143, 2003.

[93] N Huber and Ch Tsakmakis. Determination of constitutive properties from spherical indentation data using neural networks. Part I: The case of pure kinematic hardening in plasticity laws. *Journal of the Mechanics and Physics of Solids*, 47(7):1569–1588, 1999.

[94] N Huber and Ch Tsakmakis. Determination of constitutive properties from spherical indentation data using neural networks. Part II: Plasticity with nonlinear isotropic and kinematic hardening. *Journal of the Mechanics and Physics of Solids*, 47(7):1589–1607, 1999.

[95] R Ibáñez, E Abisset-Chavanne, D González, JL Duval, E Cueto, and F Chinesta. Hybrid constitutive modeling: Data-driven learning of corrections to plasticity models. *International Journal of Material Forming*, 12:717–725, 2019.

[96] JR Jain and S Ghosh. Damage evolution in composites with a homogenization-based continuum damage mechanics model. *International Journal of Damage Mechanics*, 18(6):533–568, 2009.

[97] A-A Javadi and M Rezania. Applications of artificial intelligence and data mining techniques in soil modeling. *Geomechanics and Engineering*, 1(1):53–74, 2009.

[98] Y Jeawon, GA Drosopoulos, G Foutsitzi, GE Stavroulakis, and S Adali. Optimization and analysis of frequencies of multi-scale graphene/fibre reinforced nanocomposite laminates with non-uniform distributions of reinforcements. *Engineering Structures*, 228:111525, 2021.

[99] M Jirásek. Damage and smeared crack models. In G Hofstetter and G Meschke, editors, *Numerical modeling of concrete cracking*, chapter 1, pages 1–49. Springer, 2011.

[100] M Jirásek and M Bauer. Numerical aspects of the crack band approach. *Computers and Structures*, 110-111:60–78, 2012.

[101] M Jokar and F Semperlotti. Finite element network analysis: A machine learning based computational framework for the simulation of physical systems. *Computers and Structures*, 247:106484, 2021.

[102] S Jung and J Ghaboussi. Neural network constitutive model for rate-dependent materials. *Computers and Structures*, 84(15-16):955–963, 2006.

[103] Z Kan, F Li, N Song, and H Peng. Novel nonlinear complementarity function approach for mechanical analysis of tensegrity structures. *AIAA Journal*, 59(4):1483–1495, 2021.

[104] C Kanzow. Some equation-based methods for the nonlinear complementarity problem. *Optimization Methods and Software*, 3:327–340, 1994.

[105] C Kanzow. Nonlinear complementarity as unconstrained optimization. *Journal of Optimization Theory and Applications*, 88:139–155, 1996.

[106] SK Kauwe, J Graser, R Murdock, and TD Sparks. Can machine learning find extraordinary materials? *Computational Materials Science*, 174:109498, 2020.

[107] AR Khoei. *Extended finite element method: Theory and applications.* John Wiley and Sons, 2015.

[108] T Kirchdoerfer and M Ortiz. Data-driven computational mechanics. *Computer Methods in Applied Mechanics and Engineering*, 304:81–101, 2016.

[109] T Kirchdoerfer and M Ortiz. Data-driven computing in dynamics. *International Journal for Numerical Methods in Engineering*, 113(11):1697–1710, 2018.

[110] A Klarbring and G Björkman. A mathematical programming approach to contact problem with friction and varying contact surface. *Computers and Structures*, 30:1185–1198, 1988.

[111] S Kortesis and PD Panagiotopoulos. Neural networks for computing in structural analysis: Methods and prospects of applications. *International Journal for Numerical Methods in Engineering*, 36(13):2305–2318, 1993.

[112] V Kouznetsova. *Computational homogenization for the multiscale analysis of multi-phase materials.* PhD thesis, Technical University of Eindhoven, The Netherlands, 2002.

[113] IE Lagaris, A Likas, and DI Fotiadis. Artificial neural networks for solving ordinary and partial differential equations. *IEEE Transactions on Neural Networks*, 9(5):987–1000, 1998.

[114] IE Lagaris, A Likas, and DG Papageorgiou. Neural-network methods for boundary value problems with irregular boundaries. *EEE Transactions on Neural Networks*, 11(5):1041–1049, 2000.

[115] N Lange, G Hütter, and B Kiefer. An efficient monolithic solution scheme for $FE^2$ problems. *Computer Methods in Applied Mechanics and Engineering*, 382:113886, 2021.

[116] TA Laursen. *Computational contact and impact mechanics.* Springer Verlag, Berlin, 2002.

[117] BA Le, J Yvonnet, and Q-C He. Computational homogenization of nonlinear elastic materials using neural networks. *International Journal for Numerical Methods in Engineering*, 104(12):1061–1084, 2015.

[118] J Lemaitre and JL Chaboche. *Mechanics of solid materials.* Cambridge University Press, Cambridge, 1994.

[119] AYT Leung, Ch Guoqing, and Ch Wanji. Smoothing Newton method for solving two- and three-dimensional frictional contact problems. *International Journal for Numerical Methods in Engineering*, 41:1001–1027, 1998.

[120] B Li and X Zhuang. Multiscale computation on feedforward neural network and recurrent neural network. *Frontiers of Structural and Civil Engineering*, 14:1285–1298, 2020.

[121] FZ Li, CF Shih, and A Needleman. A comparison of methods for calculating energy release rates. *Engineering Fracture Mechanics*, 21:405–421, 1985.

[122] X Li, Z Liu, S Cui, C Luo, C Li, and Z Zhuang. Predicting the effective mechanical property of heterogeneous materials by image based modeling and deep learning. *Computer Methods in Applied Mechanics and Engineering*, 347:735–753, 2019.

[123] C Liu and C Reina. Dynamic homogenization of resonant elastic metamaterials with space/time modulation. *Computational Mechanics*, 64:147–161, 2019.

[124] Y Liu, FP Van der Meer, LJ Sluys, and JT Fan. A numerical homogenization scheme used for derivation of a homogenized viscoelastic-viscoplastic model for the transverse response of fiber-reinforced polymer composites. *Composite Structures*, 252:112690, 2020.

[125] HR Lofti and PB Shing. Embedded representation of fracture in concrete with mixed finite-elements. *International Journal for Numerical Methods in Engineering*, 38:1307–1325, 1995.

[126] PB Lourenço. A matrix formulation for the elastoplastic homogenisation of layered materials. *Mechanics of Cohesive-Frictional Materials*, 1:273–294, 1996.

[127] X Lu, DG Giovanis, J Yvonnet, V Papadopoulos, F Detrez, and J Bai. A data-driven computational homogenization method based on neural networks for the nonlinear anisotropic electrical response of graphene/polymer nanocomposites. *Computational Mechanics*, 64:307–321, 2018.

[128] J Lubliner. On the thermodynamic foundations of non-linear solid mechanics. *International Journal of Non-Linear Mechanics*, 7:237–254, 1972.

[129] J Lubliner. *Plasticity theory.* Springer, New York, 1990.

[130] G Maier and T Hueckel. Nonassociated and coupled flow rules of elastoplasticity for rock-like materials. *International Journal of Rock Mechanics and Mining Sciences and Geomechanics Abstracts*, 16:77–92, 1979.

[131] G Maier and A Nappi. On the unified framework provided by mathematical programming to plasticity. In GF Dvorak and RT Shield, editors, *Mechanics of Material Behavior*, volume 6 of *Studies in Applied Mechanics*, pages 253–273. Elsevier, 1984.

[132] OL Manzoli and PB Shing. A general technique to embed non-uniform discontinuities into standard solid finite elements. *Computers and Structures*, 84:742–757, 2006.

[133] F Masi, I Stefanou, P Vannucci, and V Maffi-Berthier. Thermodynamics-based artificial neural networks for constitutive modeling. *Journal of the Mechanics and Physics of Solids*, 147:104277, 2021.

[134] TJ Massart, RHJ Peerlings, and MGD Geers. An enhanced multi-scale approach for masonry wall computations with localization of damage. *International Journal for Numerical Methods in Engineering*, 69:1022–1059, 2007.

[135] TJ Massart, RHJ Peerlings, and MGD Geers. Structural damage analysis of masonry walls using computational homogenization. *International Journal of Damage Mechanics*, 16:199–226, 2007.

[136] MATLAB. *Version 8.3.0.532 (R2014a)*. The MathWorks Inc., Natick, Massachusetts, 2014.

[137] K Matouš, MGD Geers, VG Kouznetsova, and A Gillman. A review of predictive nonlinear theories for multiscale modeling of heterogeneous materials. *Journal of Computational Physics*, 330:192–220, 2017.

[138] J Mazars and G Pijaudier-Cabot. Continuum damage theory-application to concrete. *ASCE Journal of Engineering*, 115:345–365, 1989.

[139] JM Melenk and I Babuška. The partition of unity finite element method: basic theory and applications. *Computer Methods in Applied Mechanics and Engineering*, 139:289–314, 1996.

[140] BCN Mercatoris, Ph Bouillard, and TJ Massart. Multi-scale detection of failure in planar masonry thin shells using computational homogenisation. *Engineering Fracture Mechanics*, 76:479–499, 2009.

[141] BCN Mercatoris and TJ Massart. A coupled two-scale computational scheme for the failure of periodic quasi-brittle thin planar shells and its application to masonry. *International Journal for Numerical Methods in Engineering*, 85:1177–1206, 2011.

[142] CD Meyer. *Matrix analysis and applied linear algebra*. SIAM: Society for Industrial and Applied Mathematics, 2004.

[143] C Miehe and A Koch. Computational micro-to-macro transitions of discretized microstructures undergoing small strains. *Archive of Applied Mechanics*, 72(4-5):300–317, 2002.

[144] C Miehe, J Schröder, and J Schotte. Computational homogenization analysis in finite plasticity simulation of texture development in polycrystalline materials. *Computer Methods in Applied Mechanics and Engineering*, 171:387–418, 1999.

[145] ES Mistakidis and GE Stavroulakis. *Nonconvex optimization in mechanics. Algorithms, heuristics and engineering applications by the F.E.M.* Kluwer Academic Springer, 1998.

[146] N Moës and T Belytschko. Extended finite element method for cohesive crack growth. *Engineering Fracture Mechanics*, 69:813–833, 2002.

[147] N Moës, J Dolbow, and T Belytschko. A finite element method for crack growth without remeshing. *International Journal for Numerical Methods in Engineering*, 46:131–150, 1999.

[148] J. Moreau. Application of convex analysis to the treatment of elastoplastic systems. In P Germain and B Nayroles, editors, *Applications of methods of functional analysis to problems in mechanics. Lecture Notes in Mathematics*, volume 503, chapter 4, pages 56–89. Springer, Berlin, Heidelberg, 1976.

[149] JJ Moreau, PD Panagiotopoulos, and G Strang. *Topics in nonsmooth mechanics*. Birkhäuser, Basel Switzerland, 1988.

[150] H-B Mühlhaus and I Vardoulakis. The thickness of shear bands in granular materials. *Géotechnique*, 37:271–283, 1987.

[151] AD Muradova and GE Stavroulakis. Physics-informed neural networks for elastic plate problems with bending and Winkler-type contact effects. *Journal of the Serbian Society for Computational Mechanics*. 15(2):45–54, 2021.

[152] F Murat and L Tartar. H-convergence. In A Cherkaev and R Kohn, editors, *Topics in the mathematical modelling of composite materials, series progress in nonlinear differential equations and their applications*, volume 31, pages 21–43. Birkhäuser, Boston, United States of America, 1997.

[153] KG Murty. *Linear complementarity, linear and nonlinear programming*. Heldermann, Berlin, 1988.

[154] A Nappi and F Tin-Loi. Numerical model for masonry implemented in the framework of a discrete formulation. *Structural Engineering and Mechanics*, 11:171–184, 2001.

[155] A Needleman. A continuum model for void nucleation by inclusion debonding. *Journal of Applied Mechanics*, 54:525–531, 1987.

[156] A Needleman. Material rate dependence and mesh sensitivity in localization problems. *Computer Methods in Applied Mechanics and Engineering*, 28:859–878, 1987.

[157] VP Nguyen, O Lloberas-Valls, M Stroeven, and LJ Sluys. Homogenization-based multiscale crack modelling: From micro-diffusive damage to macro-cracks. *Computer Methods in Applied Mechanics and Engineering*, 200:1220–1236, 2011.

[158] VP Nguyen, O Lloberas-Valls, M Stroeven, and LJ Sluys. Computational homogenization for multiscale crack modeling. implementational and computational aspects. *International Journal for Numerical Methods in Engineering*, 89:192–226, 2012.

[159] VP Nguyen, M Stroeven, and LJ Sluys. An enhanced continuous-discontinuous multiscale method for modelling mode-I failure in random heterogeneous quasibrittle materials. *Engineering Fracture Mechanics*, 79:78–102, 2012.

[160] A Nosier and F Fallah. Non-linear analysis of functionally graded circular plates under asymmetric transverse loading. *International Journal of Non-Linear Mechanics*, 44(8):928–942, 2009.

[161] JA Oliveira, J Pinho-da-Cruz, and F Teixeira-Dias. Asymptotic homogenisation in linear elasticity. Part II: Finite element procedures and multiscale applications. *Computational Materials Science*, 45:1081–1096, 2009.

[162] J Oliver. Modelling strong discontinuities in solid mechanics. *International Journal for Numerical Methods in Engineering*, 39:3575–3623, 1996.

[163] M Ortiz, Y Leroy, and A Needleman. A finite element method for localized failure analysis. *Computer Methods in Applied Mechanics and Engineering*, 61:189–214, 1987.

[164] M Paggi and P Wriggers. A nonlocal cohesive zone model for finite thickness interfaces - Part II: FE implementation and application to polycrystalline materials. *Computational Materials Science*, 50(5):1634–1643, 2011.

[165] PD Panagiotopoulos. *Inequality problems in mechanics and applications. Convex and nonconvex energy functions.* Birkhäuser, Basel - Boston - Stuttgart, 1985.

[166] PD Panagiotopoulos. *Hemivariational inequalities. Applications in mechanics and engineering.* Springer, Berlin - Heidelberg - New York, 1993.

[167] N Panda, D Osthus, G Srinivasan, D O'Malley, V Chau, D Oyen, and H Godinez. Mesoscale informed parameter estimation through machine learning: A case-study in fracture modeling. *Journal of Computational Physics*, 420:109719, 2020.

[168] V Papadopoulos, G Soimiris, D-G Giovanis, and M Papadrakakis. A neural network based surrogate model for carbon nanotubes with geometric nonlinearities. *Computer Methods in Applied Mechanics and Engineering*, 328:411–430, 2018.

[169] M Papadrakakis. *Analysis of structures using the finite element method (in Greek)*. Papasotiriou, 2001.

[170] RHJ Peerlings. *Enhanced damage modelling for fracture and fatigue*. PhD thesis, Eindhoven University of Technology, Eindhoven, The Netherlands, 1999.

[171] RMV Pidaparti and MJ Palakal. Material model for composites using neural networks. *AIAA Journal*, 31(8):1533–1535, 1993.

[172] S Pietruszczak and Z Mróz. Finite element analysis of deformation of strain-softening materials. *International Journal for Numerical Methods in Engineering*, 17:327–334, 1981.

[173] J Pinho-da-Cruz, JA Oliveira, and F Teixeira-Dias. Asymptotic homogenisation in linear elasticity. Part I: Mathematical formulation and finite element modelling. *Computational Materials Science*, 45:1073–1080, 2009.

[174] V Plevris and PG Asteris. Modeling of masonry failure surface under biaxial compressive stress using neural networks. *Construction and Building Materials*, 55:447–461, 2014.

[175] M Raissi, P Perdikaris, and GE Karniadakis. Physics-informed neural networks: A deep learning framework for solving forward and inverse problems involving nonlinear partial differential equations. *Journal of Computational Physics*, 378:686–707, 2019.

[176] K Raju, TE Tay, and VBC Tan. A review of the FE$^2$ method for composites. *Multiscale and Multidisciplinary Modeling, Experiments and Design*, 4:1–24, 2021.

[177] A Ramìrez-Torres, S Di Stefano, A Grillo, R RodrÃŋguez-Ramos, J Merodio, and R Penta. An asymptotic homogenization approach to the microstructural evolution of heterogeneous media. *International Journal of Non-Linear Mechanics*, 106:245–257, 2018.

[178] R Ramprasad, R Batra, G Pilania, A Mannodi-Kanakkithodi, and C Kim. Machine learning in materials informatics: Recent applications and prospects. *Computational Materials*, 3:54, 2017.

[179] M Raschi, O Lloberas-Valls, A Huespe, and J Oliver. High performance reduction technique for multiscale finite element modeling (HPR-FE$^2$): Towards industrial multiscale FE software. *Computer Methods in Applied Mechanics and Engineering*, 375:113580, 2021.

[180] JN Reddy. *Mechanics of laminated composite plates and shells: Theory and analysis*. CRC Press, 2003.

[181] JN Reddy. *An introduction to continuum mechanics*. Cambridge University Press, New York, USA, 2008.

[182] MTA Robinson and S Adali. Variational solution for buckling of nonlocal carbon nanotubes under uniformly and triangularly distributed axial loads. *Composite Structures*, 156:101–107, 2016.

[183] IBCM Rocha, P Kerfriden, and FP van der Meer. Micromechanics-based surrogate models for the response of composites: A critical comparison between a classical mesoscale constitutive model, hyper-reduction and neural networks. *European Journal of Mechanics - A/Solid*, 82:103995, 2020.

[184] G Romano, L Rosati, and F Marotti de Sciarra. Variational principles for a class of finite step elastoplastic problems with non-linear mixed hardening. *Computer Methods in Applied Mechanics and Engineering*, 109:293–314, 1993.

[185] Q Rong, H Wei, and H Bao. Deep learning methods based on cross-section images for predicting effective thermal conductivity of composites. *arXiv: Computational Physics*, 2019.

[186] JG Rots, P Nauta, GMA Kusters, and Blaauwendraad J. Smeared crack approach and fracture localization in concrete. *Heron*, 30:1–49, 1985.

[187] E Sacco. A nonlinear homogenization procedure for periodic masonry. *European Journal of Mechanics A/Solids*, 28:209–222, 2009.

[188] E Sanchez-Palencia. *Non-homogeneous media and vibration theory*. Lectures Notes in Physics, vol. 127, Springer-Verlag, Berlin, Germany, 1980.

[189] G Scalet and F Auricchio. Computational methods for elastoplasticity: An overview of conventional and less-conventional approaches. *Archives of Computational Methods in Engineering*, 25:545–589, 2018.

[190] C Settgast, G Hütter, M Kuna, and M Abendroth. A hybrid approach to simulate the homogenized irreversible elastic-plastic deformations and damage of foams by neural networks. *International Journal of Plasticity*, 126:102624, 2020.

[191] LC Silva, PB Lourenço, and G Milani. Numerical homogenization-based seismic assessment of an English-bond masonry prototype: Structural level application. *Earthquake Engineering and Structural Dynamics*, 49:841–862, 2020.

[192] JC Simo and TJR Hughes. *Computational inelasticity*. Springer, New York, 1998.

[193] H Singh and P Mahajan. Strain localization in reduced-order asymptotic homogenization. *Mathematics and Mechanics of Solids*, 25(4):913–936, 2020.

[194] RJM Smit, WAM Brekelmans, and HEH Meijer. Prediction of the mechanical behaviour of non-linear heterogeneous systems by multi-level finite element modeling. *Computer Methods in Applied Mechanics and Engineering*, 155:181–192, 1998.

[195] FV Souza and DH Allen. Modeling the transition of microcracks into macrocracks in heterogeneous viscoelastic media using a two-way coupled multiscale model. *International Journal of Solids and Structures*, 48:3160–3175, 2011.

[196] A Sridhar, VG Kouznetsova, and MGD Geers. Homogenization of locally resonant acoustic metamaterials towards an emergent enriched continuum. *Computational Mechanics*, 57:423–435, 2016.

[197] S Srividhya, P Raghu, A Rajagopal, and JN Reddy. Nonlocal nonlinear analysis of functionally graded plates using third-order shear deformation theory. *International Journal of Engineering Science*, 125:1–22, 2018.

[198] G Stavroulakis and H Antes. Nondestructive elastostatic identification of unilateral cracks through BEM and neural networks. *Computational Mechanics*, 20:439–451, 1997.

[199] G Stavroulakis, G Bolzon, Z Waszczyszyn, and L Ziemianski. Inverse analysis. In B Karihaloo, RO Ritchie, and I Milne, editors, *Comprehensive structural integrity, Vol. 3: Numerical and computational methods*, pages 685–718. Elsevier Science Ldt, 2003.

[200] GE Stavroulakis. *Inverse and crack identification problems in engineering mechanics*. Springer, 2001.

[201] GE Stavroulakis and H Antes. Nonlinear boundary equation approach for inequality 2- D elastodynamics. *Engineering Analysis with Boundary Elements*, 23(5-6):487–501, 1999.

[202] GE Stavroulakis and H Antes. Nonlinear equation approach for inequality elastostatics. A 2-D BEM implementation. *Computers and Structures*, 75(6):631–646, 2000.

[203] GE Stavroulakis, AV Avdelas, KM Abdalla, and PD Panagiotopoulos. A neural network approach to the modelling, calculation and identification of semi-rigid connections in steel structures. *Journal of Constructional Steel Research*, 44(1):91–105, 1997. Structural Steel Research in Greece.

[204] GE Stavroulakis, PD Panagiotopoulos, and AM Al-Fahed. On the rigid body displacements and rotations in unilateral contact problems and applications. *Computers and Structures*, 40:599–614, 1991.

[205] M Stoffel, F Bamer, and B Markert. Artificial neural networks and intelligent finite elements in non-linear structural mechanics. *Thin-Walled Structures*, 131:102–106, 2018.

[206] PM Suquet. Local and global aspects in the mathematical theory of plasticity. In A Sawczuk and G Bianchi, editors, *Plasticity today: modelling, methods and applications*, pages 279–310. Elsevier Applied Science Publishers, 1985.

[207] S Tang, Y Li, H Qiu, H Yang, S Saha, S Mojumder, WK Liu, and X Guo. Map123-ep: A mechanistic-based data-driven approach for numerical elastoplastic analysis. *Computer Methods in Applied Mechanics and Engineering*, 364, 2020.

[208] G Tefera, G Bright, and S Adali. Flexural and shear properties of CFRP laminates reinforced with functionalized multiwalled CNTs. *Nanocomposites*, 7(1):141–153, 2021.

[209] GH Teichert, AR Natarajan, A Van der Ven, and K Garikipati. Machine learning materials physics: Integrable deep neural networks enable scale bridging by learning free energy functions. *Computer Methods in Applied Mechanics and Engineering*, 353:201–216, 2019.

[210] I Temizer and P Wriggers. An adaptive method for homogenization in orthotropic nonlinear elasticity. *Computer Methods in Applied Mechanics and Engineering*, 35-36:3409–3423, 2007.

[211] I Temizer and TI Zohdi. A numerical method for homogenization in non-linear elasticity. *Computational Mechanics*, 40:281–298, 2007.

[212] K Terada and N Kikuchi. Nonlinear homogenization method for practical applications. In *Proceedings of the 1995 ASME International Mechanical Engineering Congress and Exposition*, pages 1–16, San Francisco, CA, USA, 1995. American Society of Mechanical Engineers, Applied Mechanics Division, AMD.

[213] K Terada and N Kikuchi. A class of general algorithms for multi-scale analyses of heterogeneous media. *Computer Methods in Applied Mechanics and Engineering*, 190:5427–5464, 2001.

[214] PS Theocaris, C Bisbos, and PD Panagiotopoulos. On the parameter identification problem for failure criteria in anisotropic bodies. *Acta Mechanica*, 123:37–56, 1997.

[215] PS Theocaris and PD Panagiotopoulos. Neural networks for computing in fracture mechanics. methods and prospects of applications. *Computer Methods in Applied Mechanics and Engineering*, 106(1):213–228, 1993.

[216] PS Theocaris and PD Panagiotopoulos. Generalised hardening plasticity approximated via anisotropic elasticity: A neural network approach. *Computer Methods in Applied Mechanics and Engineering*, 125(1):123–139, 1995.

[217] E Tikarrouchine, G Chatzigeorgiou, F Praud, B Piotrowski, Y Chemisky, and F Meraghni. Three-dimensional $FE^2$ method for the simulation of non-linear, rate-dependent response of composite structures. *Composite Structures*, 193:165–179, 2018.

[218] D Tsalis, G Chatzigeorgiou, and N Charalambakis. Homogenization of structures with generalized periodicity. *Composites Part B: Engineering*, 43(6):2495–2512, 2012.

[219] Y Tsompanakis, ND Lagaros, and GE Stavroulakis. Soft computing techniques in parameter identification and probabilistic seismic analysis of structures. *Advances in Engineering Software*, 39(39):612–624, 2008.

[220] JF Unger and C. Könke. Coupling of scales in a multiscale simulation using neural networks. *Computers and Structures*, 86(21-22):1994–2003, 2008.

[221] O Van der Sluis, PJG Schreurs, WAM Brekelmans, and HEH Meijer. Overall behaviour of heterogeneous elastoviscoplastic materials: effect of microstructural modelling. *Mechanics of Materials*, 32:449–462, 2000.

[222] TFW van Nuland, PB Silva, A Sridhar, MGD Geers, and VG Kouznetsova. Transient analysis of nonlinear locally resonant metamaterials via computational homogenization. *Mathematics and Mechanics of Solids*, 24(10):3136–3155, 2019.

[223] CV Verhoosel, JJC Remmers, MA Gutiérrez, and R de Borst. Computational homogenization for adhesive and cohesive failure in quasi-brittle solids. *International Journal for Numerical Methods in Engineering*, 83:1155–1179, 2010.

[224] AT Vermeltfoort, TMJ Raijmakers, and HJM Janssen. Shear tests on masonry walls. In *Proceedings of 6th North America Masonry Conference*, pages 1183–1193, Philadelphia, USA, 1993.

[225] K Wang and WC Sun. Meta-modeling game for deriving theory-consistent, microstructure-based traction-separation laws via deep reinforcement learning. *Computer Methods in Applied Mechanics and Engineering*, 346:216–241, 2019.

[226] Z Waszczyszyn and L Ziemianski. Neural networks in mechanics of structures and materials-new results and prospects of applications. *Computers and Structures*, 79:2261–2276, 2001.

[227] E Weinan, B Engquist, X Li, W Ren, and E. Vanden-Eijnden. Heterogeneous Multiscale Methods: A Review. *Communications in Computational Physics*, 2(3):367–450, 2007.

[228] GN Wells, R de Borst, and LJ Sluys. A consistent geometrically nonlinear approach for delamination. *International Journal for Numerical Methods in Engineering*, 54:1333–1355, 2002.

[229] GN Wells and LJ Sluys. A new method for modeling cohesive cracks using finite elements. *International Journal for Numerical Methods in Engineering*, 50:2667–2682, 2001.

[230] Inc. Wolfram Research. Voronoi diagram. `https://mathworld.wolfram.com/VoronoiDiagram.html`, April 2021.

[231] P Wriggers. *Computational contact mechanics*. J. Wiley and Sons, 2002.

[232] P Wriggers. *Nonlinear finite element methods*. Springer-Verlag, Berlin, Heidelberg, 2008.

[233] X Wu and J Ghaboussi. *Neural network-based material modeling. Civil Engineering Studies*. Structural Research Series No. 599, University of Illinois at Urbana Champaing, Civil Engineering Department, 1995.

[234] J Wuite and S Adali. Deflection and stress behaviour of nanocomposite reinforced beams using a multiscale analysis. *Composite Structures*, 71(3):388–396, 2005.

[235] K Xu, DZ Huang, and E Darve. Learning constitutive relations using symmetric positive definite neural networks. *Journal of Computational Physics*, 428:110072, 2021.

[236] R Xu, J Yang, W Yan, Q Huang, G Giunta, S Belouettar, H Zahrouni, TB Zineb, and H Hu. Data-driven multiscale finite element method: From concurrence to separation. *Computer Methods in Applied Mechanics and Engineering*, 363:112893, 2020.

[237] XP Xu and A Needleman. Numerical simulations of fast crack growth in brittle solids. *Journal of the Mechanics and Physics of Solids*, 42:1397–1434, 1994.

[238] G Yagawa and A Oishi. *Computational mechanics with neural networks*. Springer, 2021.

[239] T Yamaguchi and H Okuda. Zooming method for FEA using a neural network. *Computers and Structures*, 247:106480, 2021.

[240] Y Yang, FY Ma, CH Lei, YY Liu, and JY Li. Nonlinear asymptotic homogenization and the effective behavior of layered thermoelectric composites. *Journal of the Mechanics and Physics of Solids*, 61:1768–1783, 2013.

[241] Z Yang, YC Yabansu, R Al-Bahrani, W keng Liao, AN Choudhary, SR Kalidindi, and A Agrawal. Deep learning approaches for mining structure-property linkages in high contrast composites from simulation datasets. *Computational Material Science*, 151:278–287, 2018.

[242] Z Yang, C-H Yu, and MJ Buehler. Deep learning model to predict complex stress and strain fields in hierarchical composites. *Science Advances*, 7(15):eabd7416, 2021.

[243] Z Yuan and J Fish. Toward realization of computational homogenization in practice. *Journal for Numerical Methods in Engineering*, 73:361–380, 2008.

[244] J Yvonnet. *Computational homogenization of heterogeneous materials with finite elements.* Springer, 2019.

[245] J Yvonnet and D Gonzalez. Numerically explicit potentials for the homogenization of nonlinear elastic heterogeneous materials. *Computer Methods in Applied Mechanics and Engineering*, 198:2723–2737, 2009.

[246] J Yvonnet, E Monteiro, and Q-C He. Computational homogenization method and reduced database model for hyperelastic heterogeneous structures. *International Journal for Multiscale Computational Engineering*, 11(3):201–225, 2013.

[247] G Zavarise, P Wriggers, and BA Schrefler. A method for solving contact problems. *International Journal for Numerical Methods in Engineering*, 43:473–498, 1998.

[248] X Zhang and K Garikipati. Machine learning materials physics: Multiresolution neural networks learn the free energy and nonlinear elastic response of evolving microstructures. *Computer Methods in Applied Mechanics and Engineering*, 372:113362, 2020.

[249] Z Zhang and GX Gu. Physics-informed deep learning for digital materials. *Theoretical and Applied Mechanics Letters*, 11(1):100220, 2021.

[250] J Zhi, K Raju, TE Tay, and VBC Tan. Transient multi-scale analysis with micro-inertia effects using direct $FE^2$ method. *Computational Mechanics*, 67:1645–1660, 2021.

[251] TI Zohdi and P Wriggers. *Introduction to computational micromechanics (Lecture Notes in Applied and Computational Mechanics).* Springer-Verlag, Berlin, Heidelberg, 2004.

# Index

Printed in the United States
by Baker & Taylor Publisher Services